A SHORT COURSE IN
SOIL AND ROCK
SLOPE ENGINEERING

NOEL SIMONS, BRUCE MENZIES and MARCUS MATTHEWS

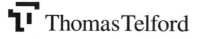

Published by Thomas Telford Publishing, Thomas Telford Ltd, 1 Heron Quay,
London E14 4JD.
URL: http://www.thomastelford.com

Distributors for Thomas Telford books are
USA: ASCE Press, 1801 Alexander Bell Drive, Reston, VA 20191-4400, USA
Japan: Maruzen Co. Ltd, Book Department, 3–10 Nihonbashi 2-chome, Chuo-ku,
Tokyo 103
Australia: DA Books and Journals, 648 Whitehorse Road, Mitcham 3132, Victoria

First published 2001

A catalogue record for this book is available from the British Library

ISBN: 0 7277 2871 7

Typeset by Academic + Technical, Bristol
Printed and bound in Great Britain by MPG Books, Bodmin

Cover photograph
Photograph of a landslide pushing over an apartment building in Kuala Lumpur, by kind
permission of Bruce Mitchell.

Preface

In our Mission Statement for this book, we determined to provide our readers with:

- a full explanation of the fundamentals of soil mechanics and rock mechanics involved in the understanding of slope engineering
- an outline of the methods used in carrying out slope stability analysis 'by hand' to enable the checking of computer outputs
- a brief introduction to software applications for slope stability analysis and where to find them
- a review of the principles of investigation and stabilization of slope failures.

A Short Course in Soil and Rock Slope Engineering is based on University of Surrey short courses that were designed not only to familiarize students with the practicalities of geotechnical engineering but also to refresh the knowledge of practising engineers and engineering geologists. In accord with the 'Short Course...' title, the book concentrates facts and figures in quick-reference charts and tables. Analytical methods are illustrated with many worked examples. Above all we have tried to get the fundamentals right and to avoid having them obscured by too much detail. We think the book will appeal to students because particular emphasis is put on explaining the fundamental soil mechanics and rock mechanics involved in understanding and designing soil and rock slopes.

The book is divided into two parts – effectively two short courses – one for soil slopes and one for rock slopes. Initially, we were attracted to the idea of integrating the common elements that exist between the analysis and design of soil and rock slopes and so take a more unified approach to soils and rocks as 'geomaterials'. We also felt, however, that we could not stray too far from the traditional expectation that soil slopes and rock slopes present different problems with different approaches to solving them. For example, for soil slopes, analysis usually proceeds by assuming the soil is a continuum within which slip surfaces may develop as well as pre-exist. Pore water pressures are dealt with implicitly in terms of effective stress. For rock slopes, however, analysis usually proceeds by assuming the rock is not a continuum but is an assembly of rigid blocks sliding along existing joints or surfaces that may or may not predispose the slope to failure. Water pressures are dealt with as force vectors. We decided finally that the traditional approach should prevail.

So that this second book in our Short Course series can stand alone, we have reproduced some material on fundamental soil mechanics from the first book, *A Short Course in Foundation Engineering*. Included are topics on effective stress, short term and long term stability, and some aspects of shear strength. We have also drawn on key classic works that are listed in the Acknowledgements. These make an excellent bibliography of recommended further reading.

Almost all analysis and design of slopes are computer-aided. The Geotechnical and Geo-environmental Software Directory (www.ggsd.com) catalogues over fifty computer programs for slope stability analysis. We list these Chapter 4, Classic methods of slope stability analysis. Nearly all of them are commercial. To be realistic, therefore, we considered it appropriate to the particular character of our book that an introduction to commercially available software was included. We canvassed the opinions of many colleagues in universities and consulting firms. We then obtained permission from GEO-SLOPE International Ltd to include with this book a CD of their software package 'SLOPE/W Student Edition' for the analysis of soil slopes. For rock slopes, we were unable to find a corresponding commercial partner. Instead, we have developed our own spreadsheet for the Hoek & Bray Comprehensive three-dimensional wedge analysis. This spreadsheet, which is called 'Surrey Wedge' and which may be used for educational purposes only, can be downloaded from the University of Surrey web site (www.surrey.ac.uk/CivEng/research/geotech/index.htm).

While classic methods of slope stability analysis have been automated by computer, and new computer methods have been devised, it is still the responsibility of the design engineer to ensure that ground properties (inputs) and design predictions (outputs) pass the sanity test! Indeed, it is the explicit duty of the engineer to check the outputs of computer programs by calculation 'by hand' i.e. in an immediately obvious and transparent way without the aid of a 'black box' (or mysterious) computer program – although a computer spreadsheet may be permitted as an aid to hand computation! These checks 'by hand' use the classic methods we describe, many of them still used to this day by commercial software packages.

Noel Simons, Bruce Menzies, Marcus Matthews
University of Surrey
Guildford 2000

Acknowledgements

Above all, we acknowledge the enormous contribution to slope engineering of Sir Alec Skempton FRS FREng.

We warmly thank our colleagues at the University of Surrey for all their support over many years: Chris Clayton, Mike Huxley, Ab Tarzi, Mike Gunn, Vicki Hope and Rick Woods. We are most grateful for the help and advice of John Hutchinson, Eddie Bromhead, Evert Hoek, Gian Paulo Giani, Tim Spink, John Krahn, David Deane, Derek Craig, John Grimes, Ray Telling, Bruce Mitchell, Gordon Wilson, Chris Pamplin, Ben Totterdell, Geoff Davis, Rob Tapply, Keith Cole, Steve Fort and Alan Clark. Of course, the comments expressed in this book are those of the authors and do not necessarily reflect the views of any of the above. We also gratefully acknowledge the important contributions to this book of past research students: Ayad Madhloom, Peter Wilkes, Malcolm Roy and the late Nick Kalteziotis. We are also most grateful to Margaret Harris who drew many of the original figures.

We acknowledge permissions from the following.

Professor J.N. Hutchinson to make verbatim extracts from the following:

- Hutchinson, J.N. (1968) Mass movement. *Encyclopaedia of Geomorphology* (*Encyclopaedia of Earth Sciences*, Series III, Ed. R.W. Fairbridge), pp. 688–695, Reinhold Publishers, New York.
- Hutchinson, J.N. (1983) Methods of locating slip surfaces in landslides. *Bulletin of the Association of Engineering Geologists*, Vol. **XX**, No. 3, pp. 235–252.
- Hutchinson, J.N. (1984) Landslides in Britain and their countermeasures. *Journal of Japan Landslide Society*, Vol. **21**, No. 1, pp. 1–23.
- Hutchinson, J.N. (1988) General Report, Morphological and geotechnical parameters of landslides in relation to geology and geohydrology. *Proc. 5th. Int. Symp. Landslides, Lausanne*, Vol. **1**, pp. 3–35, Balkema, Rotterdam.

Professor Rhodes W. Fairbridge to make verbatim extracts from Hutchinson, J.N. (1968) Mass movement. *Encyclopaedia of Geomorphology* (*Encyclopaedia of Earth Sciences*, Series III, Ed. R.W. Fairbridge), pp. 688–695, Reinhold Publishers, New York.

The Secretary for the Association of Engineering Geologists to use information from Hutchinson, J.N. (1983) Methods of locating slip surfaces in landslides. *Bulletin of the Association of Engineering Geologists*, Vol. **XX**, No. 3, pp. 235–252.

Dr Evert Hoek to use material from his notes 'Practical Rock Engineering', 1998 edition, available for download on www.rocscience.com.

Blackwell Science Ltd to publish text from pages 86–115 and figures 2.4, 2.5 and 2.7 from *Site Investigation* by C.R.I. Clayton, N.E. Simons and M.C. Matthews, ISBN 0 632 02908 0.

A.A. Balkema and from Professor Gian Paolo Giani to reprint text and figures from pages 103–129 and 191–228 of *Rock Slope Stability Analysis*, by Gian Paolo Giani. ISBN 90 541 122 9, 1992, 25cm, 374 pp., €82.50/US$87.50/£58. A.A.Balkema, P.O.Box 1675, Rotterdam, Netherlands (fax: +31.10.4135947; e-mail sales@balkema.nl).

A.A. Balkema to reprint text and figures from General Report, 'Morphological and geotechnical parameters of landslides in relation to geology and geohydrology' by J.N. Hutchinson. Reprinted from: Instability phenomena in the zone of the Alpine Arc – Extracts from the Proceedings of the 5th International Symposium on Landslides, Lausanne, 10–15 July 1988, Bonnard, Ch. (ed.) 90 6191 841 3, 1990, 30 cm, 144 pp., paper, €66.00/US$78.00/£46, A.A. Balkema, P.O. Box 1675, Rotterdam, Netherlands (fax: +31.10.4135947; e-mail: sales@balkema.nl).

E. & F.N. Spon Ltd (part of the Taylor & Francis Group) to use material from *Rock Slope Engineering* by E. Hoek and J. Bray, 1981.

The Institution of Mining & Metallurgy to reproduce text and figures from pages 230–253 of *Rock Slope Engineering*, 2nd edition, by Evert Hoek and John Bray, ISBN 0 900488 36 0.

The National Academies, Transport Research Board, to use material from: Special Report 176: *Landslides: Analysis and Control*. Transportation Research Board, National Research Council, Washington, D.C., 1978. Use of this material does not imply endorsement of a particular product, method, or practice by the Transportation Research Board.

Kluwer Academic Publishers to make extracts from text and figures from pages 261–275 of *Discontinuity Analysis for Rock Engineering* by Stephen D. Priest, ISBN 0 412 47600 2.

Taylor & Francis Books Ltd and Professor E.N. Bromhead to make use of text and figures from approximately two pages of text and two figures from *The Stability of Slopes*, 2nd edition, by E.N. Bromhead, ISBN 041925580X, published by E & F.N. Spon.

Transport Research Laboratory, UK, to make extracts from text and figures from TRRL Laboratory Report 1039: *Rock stability assessment in preliminary site investigations – graphical methods* by G.D. Matheson. Crown copyright 1983. Reproduced by permission of the Controller of HM Stationery Office.

GEO-SLOPE International Ltd to distribute a CD of their software package 'SLOPE/W Student Edition Copyright ©1991–1999 GEO-SLOPE International Ltd. All rights reserved.'

Charmouth Heritage Coast Centre, to reproduce photos of a rock fall at Charmouth, Dorset, England. Copyright: Charmouth Heritage Coast Centre. Photographer Adrian Adams.

Bruce Mitchell, to reproduce his photographs of a landslide pushing over an apartment block in Kuala Lumpur. Copyright Bruce Mitchell.

Aerofilms Ltd, to reproduce an air photo of Stag Hill, Guildford. Copyright Aerofilms Ltd.

West Dorset District Council and High-Point Rendel, to make extracts from internal documents and drawings referring to preliminary proposals for drilled drainage array trials at Lyme Regis.

Contents

PART I

Soil Slopes

On 11 December 1993 ex-US Marines pilot, Bruce Mitchell, was relaxing on the balcony of his apartment on the sixth floor of Block 3 of three apartment buildings called Highland Towers, close to a steep slope in Kuala Lumpur. He heard a rumbling noise and saw that earth was flowing out of the bottom of the slope. He snatched up his camera and took the amazing sequence of photographs shown below as the landslide breached the retaining walls and pushed over Block 1. In this tragedy, 48 people lost their lives.

a) Five to seven minutes after hearing the first rumblings, Bruce Mitchell photographed the breach of a retaining wall as earth flows across the car park.

b) Five to ten seconds later, using a camera with an auto-wind system, Bruce Mitchell took a sequence of photographs of Block 1 being pushed over by the landslide (Cover photo).

c) Block 1 crashes to the ground in a cloud of dust and debris.

d) Block 1 comes to rest in a remarkably intact state. From the moment it started to move until it hit the ground took about fifteen to twenty seconds.

© Copyright Bruce Mitchell

CHAPTER ONE

Effective stress, strength and stability

Effective stress

'*The stresses in any point of a section through a mass of soil can be computed from the* total *principal stresses* σ_1, σ_2, σ_3 *which act at this point. If the voids of the soil are filled with water under a stress u, the total principal stresses consist of two parts. One part, u, acts in the water* and *in the solid in every direction with equal intensity. It is called the* neutral stress *(or the pore water pressure). The balance* $\sigma'_1 = \sigma_1 - u$, $\sigma'_2 = \sigma_2 - u$, *and* $\sigma'_3 = \sigma_3 - u$ *represents an excess over the neutral stress u and it has its seat exclusively in the solid phase of the soil.*

This fraction of the total principal stresses will be called the effective *principal stresses . . . . A change in the neutral stress u produces practically no volume change and has practically no influence on the stress conditions for failure . . . . Porous materials (such as sand, clay and concrete) react to a change of u as if they were incompressible and as if their internal friction were equal to zero. All the measurable effects of a change of stress, such as compression, distortion and a change of shearing resistance are* exclusively *due to changes in the effective stresses* σ'_1, σ'_2, *and* σ'_3. *Hence every investigation of the stability of a saturated body of soil requires the knowledge of both the total and the neutral stresses.*'
Karl Terzaghi (1936).

Definition of effective stress

Effective stress in any direction is defined as the difference between the total stress in that direction and the pore water pressure. The term *effective stress* is, therefore, a misnomer, its meaning being a stress difference.

As pointed out by Skempton (1960), even the work of the great pioneers of soil mechanics like Coulomb, Collin, Rankine, Rasal, Bell and Forchheimer was of limited validity owing to the absence of the unifying principle of effective stress. Like all truly basic ideas the concept of effective stress is deceptively simple and yet its full significance has become apparent quite slowly. The concept was glimpsed by Lyell (1871), Boussinesq (1876) and Reynolds (1886) and observed by Fillunger (1915), Bell (1915), Westerberg (1921) and Terzaghi and Rendulic (1934). It was Terzaghi (1936) who clearly stated for the first time this basic law governing the mechanical properties of porous materials. It is remarkable therefore that even to this day the principle of effective stress is imperfectly and cursorily explained in many undergraduate textbooks and, probably as a result of this, imperfectly known and understood by many practising engineers.

The nature of effective stress

Soil is a skeletal structure of solid particles in contact, forming an interstitial system of interconnecting voids or pores. The pores are filled partially or wholly with water. The interaction between the soil structure and the pore fluid determines the unique time-dependent engineering behaviour of the soil mass.

The deformability of a soil subjected to loading or unloading is, in the main, its capacity to deform the voids, usually by displacement of water. The strength of a soil is its ultimate resistance to such loading.

Shear stresses can be carried only by the structure of solid particles, the water having no shear strength. On the other hand, the normal stress on any plane is the sum of two components: owing to both the load transmitted by the solid particles of the soil structure and to the pressure in the fluid in the void space (Bishop and Henkel, 1962).

The deformability and strength of a soil are dependent on the difference between the applied external total loading stress, σ, and the pore water pressure, u. This difference is termed the *effective stress* and is written $(\sigma - u)$.

The physical nature of this parameter may be intuitively understood by considering the saturated soil bounded by a flexible impermeable membrane as shown in Fig. 1.1 The external applied total pressure is σ normal to the boundary. The pore water pressure is u ($<\sigma$) which, being

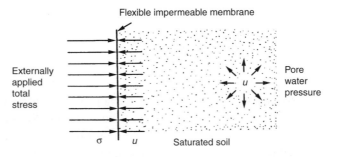

Fig. 1.1 Intuitive soil model demonstrating the nature of effective stress

a hydrostatic pressure, acts with equal intensity in all directions, giving a pressure of u normal to the boundary. By examining the stresses normal to the boundary it may be seen by inspection that the disparity in stresses $(\sigma - u)$ is transmitted across the boundary into the soil structure, assuming an equilibrium condition. Thus, *the effective stress $(\sigma - u)$ is a measure of the loading transmitted by the soil structure*.

The principle of effective stress

The principle of effective stress was stated by Bishop (1959) in terms of two simple hypotheses:

- Volume change and deformation in soils depends on the difference between the total stress and the pressure set up in the fluid in the pore space, not on the total stress applied. This leads to the expression

$$\sigma' = \sigma - u \qquad (1.1)$$

where σ denotes the total normal stress, u denotes the pore pressure, and σ' is termed the effective stress.

- Shear strength depends on the effective stress, not on the total normal stress on the plane considered. This may be expressed by the equation

$$\tau_f = c' + \sigma' \tan \phi' \qquad (1.2)$$

where τ_f denotes the shear strength, σ' the effective stress on the plane considered, c' the cohesion intercept, ϕ' the angle of shearing resistance, with respect to effective stress.

The principle of effective stress, as expressed above, has proved to be vital in the solution of practical problems in soil mechanics.

To explore more rigorously the physical nature of effective stress, consider the forces acting across a surface $X–X$ in the soil which approximates to a plane but passes always through the pore space and points of contact of the soil particles (Bishop, 1959) as shown in Fig. 1.2.

Normal stress is then equal to the average force perpendicular to this plane, per unit area, and areas are considered as projected onto the plane.

Let

σ denote the total normal stress on this plane,
σ'_i the average intergranular normal force per unit area of the plane,
u the pore water pressure,
a the effective contact area of the soil particles per unit area of the plane.

It follows that $\sigma = \sigma'_i + (1 - a)u$ whence

$$\sigma'_i = (\sigma - u) + au \qquad (1.3)$$

Thus the effective stress $(\sigma - u)$ is not exactly equal to the average intergranular force per unit area of the plane, σ'_i, and is dependent on the contact

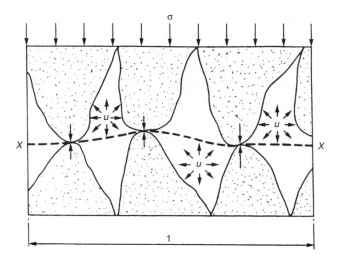

Fig. 1.2 Interparticle forces acting across a surface X–X

area between the particles. Although this area may be small it cannot be zero as this would imply infinite local contact stresses between the particles.

Consider now the deformations at the contact between two soil particles also acted on by pore water pressure (Fig. 1.3).

The force system may be regarded as being made up of two components. If P is the average force per contact and there are N contacts per unit area, then the intergranular force per unit area of the plane $X–X$ space is

$$\sigma'_1 = NP \tag{1.4}$$

Now if a homogeneous isotropic soil particle is subjected to an isotropic stress, u, over its whole surface, the deformation incurred is a small elastic reduction in particle volume without any change in shape. The soil skeleton as a whole, therefore, also reduces slightly in volume without change in shape.

The compressibility of the skeletal structure of the soil, however, is very much greater than the compressibility of the individual soil particles of which it is comprised. Hence it is only that part of the local contact stress which is in excess of the pore water pressure that actually causes

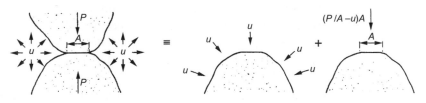

Fig. 1.3 Separation of interparticle force components

a structural deformation by volumetric strain or by shear strain or by both. This excess stress which controls structural deformation is equal to $(P/A - u)$ where A is the area of the particular contact. By summing the corresponding components of excess inter particle force an expression is obtained for σ' defined as that part of the normal stress which controls volume change due to deformation of the soil structure, whence the excess force per unit of the plane X–X is

$$\sigma' = N((P/A) - u)A$$

$$= NP - uNA$$

$$= NP - ua \quad \text{(since } NA = a)$$

$$= \sigma_i - au \tag{1.5}$$

Substituting for σ'_i from equation (1.3) gives

$$\sigma' = (\sigma - u) + au - au$$

or

$$\sigma' = (\sigma - u) \tag{1.6}$$

i.e. the effective stress is that part of the normal total stress which controls deformation of the soil structure, irrespective of the interparticle contact areas. This leads to the interesting conclusion that although the average intergranular force per unit area depends on the magnitude of 'a', volume changes due to deformation of the soil structure depend simply on the stress difference $(\sigma - u)$, whatever the nature of 'a' (Bishop, 1959).

To understand better the nature of effective stress, it is instructive to consider what it is not! It is not the intergranular stress or the inter-granular force per unit cross sectional area. To illustrate this point, let the average intergranular stress be σ_g.

For force equilibrium in the vertical direction

$$\sigma A = \sigma_g aA + u(1 - a)A$$

whence

$$\sigma_g = (\sigma - u(1 - a))/a \tag{1.7}$$

To use realistic numbers let $\sigma = 100\,\text{kPa}$, $u = 50\,\text{kPa}$, and let $a = 0.01$ (clay) and 0.3 (lead shot), respectively.

From equation (1.3) $\sigma'_i = 50.5\,\text{kPa}$ and $65\,\text{kPa}$, respectively.
From equation (1.7) $\sigma_g = 5050\,\text{kPa}$ and $216.7\,\text{kPa}$, respectively.
The effective stress is $\sigma' = (\sigma - u) = 50\,\text{kPa}$.

These values are summarized in Table 1.1. It can clearly be seen from Table 1.1 that effective stress is *not* the average intergranular stress and *not* the average intergranular force per unit area. It is simply and *exactly* equal to the total stress less the pore pressure.

Table 1.1 Intergranular force per unit area, intergranular stress and effective stresses for clay and lead shot

Soil type	Intergranular contact area per unit area	Intergranular force per unit area σ'_i: kPa	Intergranular stress σ_g: kPa	Effective stress $\sigma' = (\sigma - u)$: kPa
Clay	0.01	50.5	5050	50
Lead shot	0.3	65	216.7	50

In an elegant experiment performed on lead shot, Laughton (1955) showed clearly that, in spite of significant contact areas between the particles, volume change and shear strength were still governed by the simple expression for effective stress, namely, $\sigma' = (\sigma - u)$.

The important implication the principle of effective stress has on the strength is that a change in effective stress results in a change of strength, and the corollary follows, that if there is no change in effective stress, then there is no change in strength. While it is true that a change in volume will always be accompanied by a change in effective stress, it is not necessarily true, however, that a change in effective stress will produce a change in volume.

Consider, for example, the undrained triaxial test on a saturated soil. During the test, while there is no change in water content and therefore in volume, the pore pressures do change and alter the vertical or horizontal effective stress, or both. At failure, the effective stress throughout the sample will have changed considerably from that pertaining before the axial loading stage of the test. These changes in effective stress are accompanied by specimen deformation by the change of shape. It follows, therefore, that *the sufficient and necessary condition for a change in the state of effective stress to occur is that the soil structure deforms.* Deformation may occur by volumetric strain, by shear strain or by both. The corollary follows that deformation is induced by a change in the state of effective stress, whether or not there is a change in volume.

This implication of the principle of effective stress is of interest. Consider for example, the interrelation of stress changes in the oedometer or under uniform global loading conditions in the field, for a saturated clay.

Let $\Delta\sigma_v$ be the change in vertical total stress, $\Delta\sigma_h$ be the change in horizontal total stress, and Δu be the change in pore-water pressure. At the moment of applying the vertical stress increment there is no deformation, and it thus follows that there is no change in effective stress in any direction and therefore

$$\Delta u = \Delta\sigma_v = \Delta\sigma_h \tag{1.8}$$

This expression has been proved by Bishop (1958) for soft soils. Bishop (1973a, b) has shown that, for porous materials of very low compressibility, equation (1.8) is modified. Equation (1.8) is valid, of course, independently of the value of the pore pressure parameter, A.

As a consequence of this, during drainage the stress path followed in the oedometer is quite complex. At the start of the consolidation stage, it has been shown that the oedometer ring applies a stress increment to the sample equal to the increment in vertical stress. During consolidation, however, the horizontal stress decreases to a value, at the end of pore pressure dissipation, equal to K_0 times the effective vertical stress (Simons and Menzies, 1974).

The computation of effective stress

The computation of effective stress requires the separate determination of the total stress, σ, and of the pore water pressure, u. The effective stress is then found as

$$\sigma' = \sigma - u$$

The determination of vertical total stress

Consider the typical *at rest* ground condition shown in Fig. 1.4. This is a *global* loading condition.

Consider an element of soil at a depth D metres. The water level is at the surface. The bulk unit weight of the soil (i.e. including solids and water) is $\gamma \, \text{kN/m}^3$. The total vertical stress σ_v is computed by finding the total weight of a vertical column subtended by unit horizontal area $(1 \, \text{m}^2)$ at depth D. The weight of this column divided by its base area is $\gamma D \, \text{kPa}$ and is the vertical total stress acting on a horizontal plane at depth D.

The vertical total stress σ_v, and the horizontal total stress σ_h are principal stresses. In general, $\sigma_v \neq \sigma_h$.

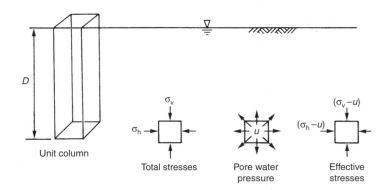

Fig. 1.4 'At rest' in situ stresses due to self-weight of soil

The determination of pore water pressure

Referring to Fig. 1.4, the pore water pressure, u, is found by considering a vertical unit column of water only. The presence of the soil structure has no effect on the pore water pressure. Thus, $u = \gamma_w D$, where γ_w, is the unit weight of water. A helpful approximation is to take $\gamma_w = 10\,\text{kN/m}^3$ (more accurately, $\gamma_w = 9.807\,\text{kN/m}^3$).

For a clay layer rapidly loaded locally, the viscous retardation of pore water flow in the fine-grained soil gives a build-up of pore pressure. Water will eventually flow out of the zone of loading influence to the ground surface and into surrounding soil unaffected by the loading. This flow or consolidation takes place under the load-induced hydraulic gradient which is itself reduced by the flow as the consolidation of the soil structure allows it to support more load. The law of diminishing returns thus applies and there is an exponential decay in the excess or load-generated pore pressure. This effect is illustrated in Fig. 1.5 where a saturated clay layer is rapidly loaded by the building-up of an embankment. The distribution of pore pressure with time (isochrones) is shown by the relative heights of piezometric head in the piezometers.

In real soils subjected to rapid local loading, the effects of deformation of the soil structure at constant volume, the compressibility of the pore fluid in practice and the dependence of the structural properties of the soil skeleton upon the mean stress, all mean that initially the loading change is shared between the soil structure and the generated pore pressure change. The generated pore pressure change is thus not only a

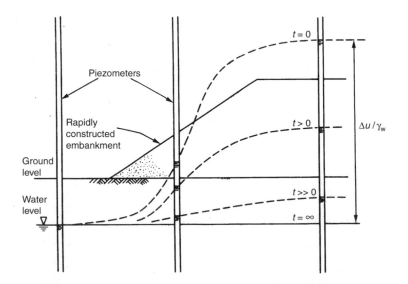

Fig. 1.5 Pore pressure response of a saturated clay to rapid local loading

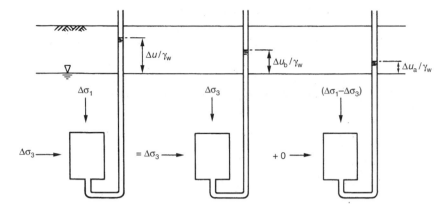

Fig. 1.6 Components of excess pore pressure generated by a loading increment $(\Delta\sigma_1 > \Delta\sigma_2 = \Delta\sigma_3)$

function of the loading change but also a function of the soil properties. These properties are experimentally determined and are called the *pore pressure parameters A and B.*

Consider the loading increment applied to a cylindrical soil element shown in Fig. 1.6. The loading change is in triaxial compression, the major principal total stress increasing by $\Delta\sigma_1$, while the minor principal (or radial) total stress increases by $\Delta\sigma_3$. An excess pore pressure, i.e. greater than the existing pore pressure, of Δu is generated by the loading increment.

The generalized loading system of Fig. 1.4 may be split into two components consisting of an isotropic change of stress $\Delta\sigma_3$ generating an excess pore pressure Δu_b and a uniaxial change of stress $(\Delta\sigma_1 - \Delta\sigma_3)$ generating an excess pore pressure Δu_a.

By the principle of superposition

$$\Delta u = \Delta u_b + \Delta u_a \tag{1.9}$$

Assuming that the excess pore pressure generated by the loading increment is a simple function of that loading increment we have,

$$\Delta u_b = B\Delta\sigma_3 \tag{1.10}$$

and

$$\Delta u_a = \bar{A}(\Delta\sigma_1 - \Delta\sigma_3) \tag{1.11}$$

where $\bar{A}$ and B are experimentally determined pore pressure parameters.

Thus, the total pore pressure change is made up of two components: one that is B times the isotropic stress change, and the other that is $\bar{A}$ times the change in principal stress difference.

Hence,

$$\Delta u = B\Delta\sigma_3 + \bar{A}(\Delta\sigma_1 - \Delta\sigma_3) \tag{1.12}$$

(Note that Skempton (1954) gives $\Delta u = B[\Delta\sigma_3 + A(\Delta\sigma_1 - \Delta\sigma_3)]$, i.e. $\bar{A} = AB$.)

The pore pressure parameters may be measured in the triaxial compression test where a cylindrical soil sample is tested in two stages. In the first stage, the sample is subjected to an increment of all-round pressure and the pore pressure increase measured. In the second stage, the sample is loaded axially and the pore pressure increase measured. For a saturated soil, $B = 1$ and $\bar{A} = A$.

WORKED EXAMPLE

Question   The strata in the flat bottom of a valley consist of 3 m of coarse gravel overlying 12 m of clay. Beneath the clay is fissured sandstone of relatively high permeability.

The water table in the gravel is 0.6 m below ground level. The water in the sandstone is under artesian pressure corresponding to a standpipe level of 6 m above ground level.

The unit weights of the soil are:

Gravel	above water table	$16 \, kN/m^3$
	below water table (saturated)	$20 \, kN/m^3$
Clay	saturated	$22 \, kN/m^3$

1   Plot total stresses, pore water pressures and effective vertical stresses against depth:
(a)   With initial ground water levels,
(b)   Assuming that the water level in the gravel is lowered 2 m by pumping, but the water pressure in the sandstone is unchanged,
(c)   Assuming that the water level in the gravel is kept as for (b), but that relief wells lower the water pressure in the sandstone by 5.5 m,
(d)   Assuming that the relief wells are then pumped to reduce the water level in the sandstone to 15 m below ground level.
Note that for (b), (c) and (d), stresses are required both for the short-term and the long-term conditions.
2   To what depth can a wide excavation be made into the clay before the bottom *blows up* (neglect side shear) assuming the initial ground water level.
(a)   With initial artesian pressure in sandstone?

      (b)   With relief wells reducing the artesian pressure to 0.6 m above ground level?

      (c)   With relief wells pumped to reduce the pressure to 15 m below ground level?

3   An excavation 9 m in depth (below ground level) is required. If a ratio of total vertical stress to uplift pressure of 1.3 is required for safety, to what depth must the piezometric head in the sandstone be lowered?

4   If the coefficient of volume change of the clay is $0.0002 \, \text{m}^2/\text{kN}$ to what extent would the clay layer in the locality eventually decrease in thickness if the artesian pressure were permanently lowered by this amount?

5   If, on the other hand, the water level in the sandstone were raised to 15 m above ground level owing to the impounding behind a dam upstream of the site, at what depth in the undisturbed clay would the vertical effective stress be least and what would this value be?

Answers 1   See Fig. 1.7. Note that rapid changes of global total stress do not cause immediate changes in effective stress within the clay layer.

      2   (a)   Let the excavation proceed to a depth D below ground level. The bottom will blow up when the total stress, diminished by excavation, equals the water pressure in the sand stone, that is, when $(15 - D)22 = 21 \times 10$ assuming the excavation is kept pumped dry.

           (b)   Thus, $D = 5.454 \, \text{m}$.

           (c)   Similarly, $(15 - D)22 = 15.6 \times 10$. Thus, $D = 7.909 \, \text{m}$.

           (d)   By inspection, $D = 15 \, \text{m}$.

      3   By inspection, $(6 \times 22)/10p = 1.3$, where p is the total piezometric head in the sandstone, whence $p = 10.153 \, \text{m}$, i.e. the piezometric head in the sandstone must be lowered to $(15 - 10.153) = 4.847 \, \text{m}$ below ground level.

      4   The change in effective stress is $(21 - 10.153)10/2 = 54.235 \, \text{kPa}$, whence the change in thickness of the clay layer is $0.0002 \times 54.235 \times 12 = 0.130 \, \text{m}$.

      5   The pore pressure at the base of the clay layer will be $(15 + 15)10 = 300 \, \text{kPa}$ giving a minimum vertical effective stress of $21.6 \, \text{kPa}$ at 15 m depth.

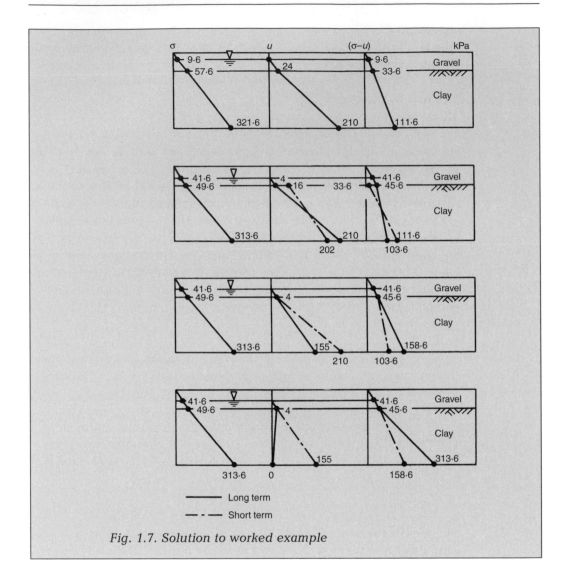

Fig. 1.7. Solution to worked example

Strength

Overview

The measurement of shear strength may be attempted by a variety of tests. These different tests produce different strengths for the one soil. The question arises therefore as to which of the various strengths may be used in a particular design problem. This chapter endeavours to set out a rational basis for answering this question by firstly discussing the essential nature of shear strength, emphasizing the common effective stress basis of both undrained and drained shear strength. The link between test model and construction prototype is then made in a statement about similitude. The stability mechanics of constructional loading changes are then briefly re-examined. Efforts to obtain similarity between the shear test models and the field prototype are considered.

The nature of shear strength

The shear strength of a soil in any direction may, for simplicity, be considered as the maximum shear stress that can be applied to the soil structure in that direction. When this maximum has been reached, the soil is regarded as having failed, the strength of the soil having been fully mobilized. The shear strength is derived from the soil's structural strength alone, the pore water having no shear strength.

The resistance of the soil structure to shear arises from the frictional resistance generated by the interparticle forces. In the soil mass the loading transmitted by the soil structure normal to the shear surface is an integrated measure of these interparticle forces. The shear strength on any plane in a soil is thus some function of the effective stress normal to that plane. Assuming a linear relationship gives:

$$\tau_f = k_1 + k_2(\sigma_n - u_f) \tag{1.13}$$

where τ_f is the shear stress at failure, i.e. shear strength,

σ_n is the total stress normal to the plane,

u_f is the pore water pressure at failure,

k_1 and k_2 are two experimentally determined constants.

Experiment has shown this expression to be substantially correct over a wide range of soils for a limited range of stresses. Common usage has $k_1 = c'$, and $k_2 = \tan \phi'$, whence

$$\tau_f = c' + (\sigma_n - u_f) \tan \phi' \tag{1.14}$$

where c' is the cohesion intercept and ϕ' is the angle of shearing resistance with respect to effective stress.

Direct and indirect measurements of shear strength

If the shear strength parameters in terms of effective stress, c' and $\tan \phi'$, are known, the shear strength on any plane may be estimated from a knowledge of the effective stress at failure normal to that plane, $(\sigma_n - u_f)$. In this way, the shear strength may be evaluated indirectly by using the experimentally determined values of c' and $\tan \phi'$ and estimating or measuring the total normal stress σ_n and the pore water pressure u_f, whence $\tau_f = c' + (\sigma_n - u_f) \tan \phi'$. As well as obtaining the strength indirectly in this way it is also possible to measure the shear strength directly. A test that does this is the direct shear box test that also provides a useful model in which to study shear strength.

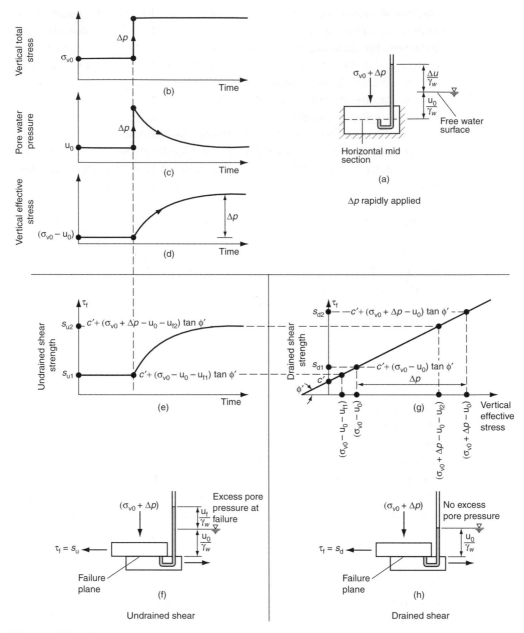

Fig. 1.8 The effect of an increase in loading on the shear strength of a saturated clay measured in the direct shear test

Consider a saturated clay specimen confined in a direct shear box and loaded vertically on a horizontal midsection with a total stress of σ_{v0} (Fig. 1.8(a)). The box is contained in an open cell that is flooded to a constant depth. The soil structure is given time to consolidate to a vertical effective stress of $(\sigma_{v0} - u_0)$, i.e. the pore pressure has the equilibrium value of u_0. The shear strength may be found directly by shearing the specimen. By shearing the specimen rapidly the undrained condition is simulated, i.e. no change in water content and therefore no change in overall volume. This is due to the inability of a fine-grained soil to deform rapidly because of the viscous resistance to displacement of water through the pores.

The undrained shear strength, s_u, in the first instance is $s_{u1} = \tau_{f1}$ where τ_{f1} is the rapidly sheared strength of the soil structure. This structure is generated by shear distortion from the parent structure that had previously been achieved by consolidation to a vertical effective stress of $(\sigma_{v0} - u_0)$. If it had been possible to measure the excess pore pressure at failure, u_{f1}, in the failure zone then it follows from equation (1.14) that

$$s_{u1} = \tau_{f1} = c' + (\sigma_{v0} - u_0 - u_{f1}) \tan \phi' \tag{1.15}$$

(see Figs. 1.8(e) and (f)) where c' and ϕ' are given.

Now consider the effect of applying an additional normal total stress of Δp (Fig. 1.8). The load increment is applied instantaneously and so the application occurs under undrained conditions. As the specimen is rigidly confined laterally, there can be no lateral strain. In addition, due to the undrained condition at the instant of loading there can be no vertical strain. The soil structure therefore does not deform and the effective stress is unaltered. The structural strength is unaffected and if the specimen is rapidly sheared, equation (1.15) would still hold.

The excess pore water pressure, Δu, generated by the application of Δp, sets up an hydraulic gradient between the inside of the specimen and the relatively lower water pressure of the flooded cell. Water accordingly flows out of the specimen (Fig. 1.8(c)). The hydraulic gradient causing the water flow is itself reduced by the flow whence the flow rate decays exponentially. The soil thus consolidates to a new structural configuration in equilibrium with the new normal effective stress of $(\sigma_{v0} - u_0 + \Delta p)$. The specimen, reduced in volume by the amount of water squeezed out, now has a denser and therefore stiffer structure. Again, if the specimen is rapidly sheared, the maximum shear stress registered is an undrained shear strength $s_{u2} > s_{u1}$. If the excess pore pressure in the shear plane at failure is u_{f2} then

$$s_{u2} = \tau_{f2} = c' + (\sigma_{v0} + \Delta p - u_0 - u_{f2}) \tan \phi' \tag{1.16}$$

It can be seen therefore that the undrained shear strength s_u simply provides a direct measure of the shear strength of a soil structure that is rapidly sheared. This strength may also be deduced from knowledge of the state of effective stress at failure and the relationship between effective stress and shear strength for the soil parameters c' and $\tan \phi'$.

Drained and undrained measurements of shear strength

If a fine-grained saturated soil is rapidly loaded (e.g. rapidly filling a large oil tank) in the short term the soil is effectively undrained due to the high viscous forces resisting pore water flow within the soil. The excess pore pressure generated by the sudden application of load dissipates by drainage or consolidation over a period of time which may, in the case of clays, extend for many tens or even hundreds of years. Hence the terms 'short' and 'sudden' are relative and a load application over several months during a construction period may be relatively rapid with the short term or end of construction condition approximating to the undrained case.

In positive loading conditions (like embankments and footings) the subsequent consolidation under the influence of the increased load gives rise to increased strength and stability. The lowest strength and therefore the critical stability condition holds at the end of construction. The critical strength is thus the undrained shear strength before post-construction consolidation.

One way of measuring this strength is to rapidly build up the load in a full-scale field test until the soil fails and this is sometimes done, particularly in earthworks, by means of trial embankments. Such full-scale testing can be costly and is appropriate only to large projects where the soil is uniform.

For conditions of variable soil and when large expenditure on soil testing is unlikely to effect economies in design, small-scale testing is more appropriate. A rapid test in the direct shear box for example represents a convenient though vastly simplified small-scale simulation or model of the likely full-scale field or prototype failure.

It was seen in the preceding section that this rapid direct measure of shear strength gave the undrained shear strength, s_u. Alternatively, if the test is carried out sufficiently slowly the distortion of the soil structure in the shear zone produces an insignificantly small excess pore water pressure. This is because any slight increase in pore pressure has time to dissipate by drainage, i.e. the slow test is drained as against undrained in the rapid test.

Hence the pore pressure virtually remains at u_0 throughout the test (Figs. 1.8(g) and (h)). The soil structure in the shear zone is able to change its overall volume by drainage and hence the shear-distorted structure in the drained test will be different to that of the undrained test, giving a different strength. The two distinct structures achieved by consolidating the soil specimen under vertical total stresses of σ_{v0} and $(\sigma_{v0} + \Delta p)$ give rise to two distinct drained shear strengths:

$$s_{d1} = c' + (\sigma_{v0} - u_0) \tan \phi' \tag{1.17}$$

and

$$s_{d2} = c' + (\sigma_{v0} + \Delta p - u_0)\tan\phi' \qquad (1.18)$$

where $s_{d2} > s_{d1}$ (Fig. 1.8(g)). The essential feature is that generally $s_{u1} \neq s_{u2} \neq s_{d1} \neq s_{d2}$ for although it is the one soil which is being tested, the soil structures at failure in the failure zones are different in each of the four cases depending on whether the test is drained or undrained and whether the specimen was tested before or after consolidation.

In the drained direct shear test the pore pressure u_0 is known and therefore the normal effective stress is known whence the effective stress shear strength parameters c' and $\tan\phi'$ may be deduced from two or more such tests. Hence this test not only directly measures the drained shear strength for the particular consolidation pressures, σ_n, of each test but, in furnishing c' and $\tan\phi'$, allows the shear strength to be estimated for any loading condition. Thus the drained shear strength may be predicted for any $(\sigma_n - u_0)$ while the undrained shear strength may be predicted for any $(\sigma_n - u_f)$, assuming that the pore pressure at failure u_f could be estimated or measured.

If the load increment is now rapidly removed a reverse process occurs to that described above (Fig. 1.9). The rapid load reduction gives rise to a fall in pore water pressure and, as before, the instantaneous effective stress and therefore strength is unaltered and if the specimen is rapidly sheared equation (1.16) still holds. The negative excess pore pressure generated by the reduction of Δp sets up an hydraulic gradient between the inside of the specimen and the relatively higher water pressure of the flooded cell. Water is therefore sucked into the specimen dissipating the negative excess pore pressure (Fig. 1.9(c)). The soil structure thus swells to a configuration in equilibrium with the new normal effective stress of $(\sigma_{v0} - u_0)$. The specimen, increased in volume by the amount of water sucked in, now has a less densely packed and therefore softer structure. Again, if the specimen is rapidly sheared the maximum shear stress registered is an undrained shear strength $s_{u3} < s_{u2}$ (Fig. 1.9(e)). If the excess pore pressure at failure is u_{f3} then, giving due regard to the sign of u_{f3},

$$s_{u3} = c' + (\sigma_{v0} - u_0 - u_{f3})\tan\phi' \qquad (1.19)$$

If drained tests are carried out the shear strengths will again be different. The two distinct structures achieved by swelling the soil specimen from a vertical total stress of $(\sigma_{v0} + \Delta p)$ to one of σ_{v0} give rise to two distinct drained shear strengths

$$s_{d2} = c' + (\sigma_{v0} + \Delta p - u_0)\tan\phi' \qquad (1.20)$$

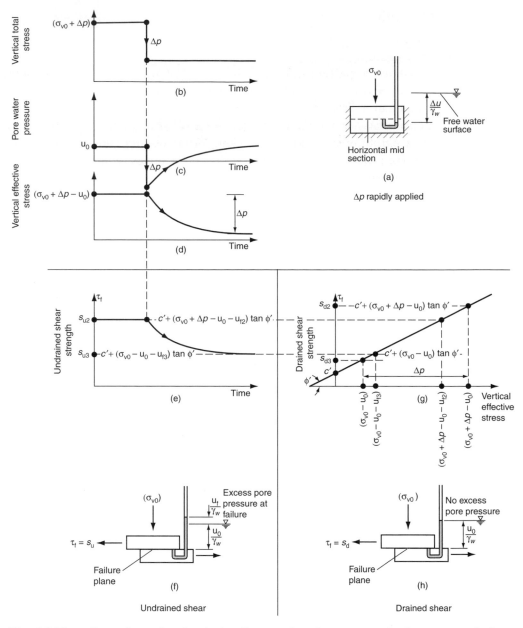

Fig. 1.9 The effect of a reduction in loading on the shear strength of a saturated clay measured in the direct shear test

and

$$s_{d3} = c' + (\sigma_{v0} - u_0) \tan \phi' \qquad (1.21)$$

respectively, where $s_{d2} > s_{d3}$.

The applicability of shear test data to field stability problems

Considerations of similitude

From the foregoing section it was seen that a given soil shear test specimen may have many strengths depending on whether the test specimen is undrained or drained, and whether the shearing takes place before, during or after either consolidation or swelling. In addition, the measurement of strength in shear tests is influenced significantly by test type, e.g. triaxial, in situ shear vane, plate bearing, penetrometer, etc.; sampling disturbance; size of test specimen or test zone; orientation of the test specimen or test zone, i.e. the effects of anisotropy; time to undrained failure; time between sampling and testing (Simons and Menzies, 2000).

The selection of appropriate shear strength data for the prediction of field stability therefore requires some guiding principle unless empirical corrections based on experience are to be made. Apart from such corrections *the use of shear test data in stability analyses is appropriate only provided there is similitude between the shear test model, the analytical model and the field prototype*. The first steps, therefore, in ensuring similitude between test model and field prototype require that:

- prior to testing the effective stresses and structural configuration of the test specimen or zone are identical to those in situ before testing disturbance, i.e. the soil specimen or zone is undisturbed
- the size of the test specimen or test zone is representative of the soil in the mass
- during the test the structural distortions and rates of distortion of the soil mass are similar to those which would arise in the field
- at failure in the test, the distortions and rates of distortion of shear surfaces are similar to those which would arise in a full-scale failure.

Finally, of course, the principle of similitude requires a realistic analytical model in which the shear test data may be used to give an estimate of the stability of the field prototype. Of considerable importance in this respect is the phenomenon of progressive failure.

Short term and long term stability

The stability of foundations and earthworks in saturated fine-grained soil is time dependent. This is because the average sizes of the interconnecting pores are so small that the displacement of pore water is retarded by viscous forces. The resistance that a soil offers to water flow may be measured in terms of the soil permeability, which is the velocity of flow through the soil under a unit hydraulic gradient.

Permeability is the largest quantitative difference between soils of different time dependent stability (Bishop and Bjerrum, 1960). A sand and a normally-consolidated clay, for example, may have similar effective stress shear strength parameters c' and $\tan \phi'$ but the permeability of the clay is several orders of magnitude lower. The stability of the clay is thus time dependent whereas the more permeable sand reacts to loading changes almost immediately.

If saturated clay is loaded, as may occur in soils supporting building foundations and earth embankments, an overall increase in mean total stress occurs. In a fine-grained soil like clay, the viscous resistance to pore water expulsion prevents the soil structure from rapidly contracting. In the short term loading condition therefore there is a change in effective stress due to shear strain only together with an increase in pore pressure. With time, this excess pore pressure is dissipated by drainage away from the area of increased pore pressure into the surrounding area of lower pore pressure unaffected by the construction. This flow of pore water causes a time dependent reduction in volume in the zone of influence, the soil consolidating and the soil structure stiffening, giving rise to decreasing settlement and increasing strength. The minimum factor of safety thus occurs in the short term undrained condition when the strength is lowest.

If saturated clay is unloaded, as may occur in an excavation or cutting, an overall reduction in mean total stress occurs. In a fine-grained soil like clay, the viscous resistance to pore water flow prevents the soil structure, relieved of some of its external loading, from rapidly expanding by sucking in pore water from the surrounding soil. With time, this suction is dissipated by drainage into the area of lowered pore pressure from the surrounding area of relatively higher pore pressure unaffected by the excavation. This flow of pore water causes an increase in soil volume in the zone of influence, soil swelling and soil structure softening. The minimum factor of safety occurs at the equilibrium long term condition when the strength is lowest.

Whether the soil is loaded or unloaded, the stability is generally represented in terms of a factor of safety that is the integrated amount by which the available soil strength may be reduced around a hypothetical shear surface before limiting equilibrium occurs. If a field failure occurs the factor of safety is unity and the average in situ shear strength may be estimated from a back-analysis of the slipped mass. At the design stage, however, it is rare that such conveniently apt and accurate shear strength data are available. It is more usual to rely on small-scale measurements of shear strength. It is necessary to determine which of the various methods of in situ and laboratory shear strength determinations are appropriate to the stability problem considered.

Shear test data applicable to the loading condition

The loading case gave a critical stability condition in the short term, the minimum strength and factor of safety occurring at the end of loading. In this undrained condition the stressed zone does not immediately change its water content or its volume. The load increment does, however, distort the stressed zone. The effective stresses change along with the change in shape of the soil structure. Eventually the changes in structural configuration may no longer produce a stable condition and the consequent instability gives rise to a plastic mechanism or plastic flow and failure occurs.

The strength is determined by the local effective stresses at failure normal to the failure surfaces. These are conditioned by and generated from the structural configuration of the parent material (which is itself conditioned by the preloading in situ stresses) and its undrained reaction to deformation. A first step in meeting the complex similitude requirements is to ensure that the shear test is effectively undrained.

The undrained shear test may be used to give a direct measure of shear strength, namely the undrained shear strength s_u, or it may be used to give an indirect measure of shear strength if the pore water pressures are measured by providing c' and $\tan \phi'$. It is therefore possible to analyse the stability of the loaded soil by:

- using the undrained shear strength s_u in a total stress analysis, or by
- using the effective stress shear strength parameters c' and $\tan \phi'$ in an effective stress analysis.

The effective stress analysis requires an estimation of the end of construction pore pressures in the failure zone at failure, whereas the total stress analysis requires no knowledge of the pore pressures whatsoever.

In general civil engineering works the soil loading change is applied gradually during the construction period. The excess pore pressures generated by the loading are thus partially dissipated at the end of construction. The end of construction pore pressures and the increased in situ shear strength can be measured on site if the resulting increased economy of design warrants the field instrumentation and testing. On all but large projects this is rarely the case. In addition, the loading is localized, allowing the soil structure to strain laterally, the soil stresses dissipating and the principal stresses rotating within the zone of influence.

In the absence of sound field data on the end of construction shear strengths and pore pressures and in the face of analytical difficulties under local loading, an idealized soil model possessing none of these difficulties is usually invoked for design purposes. This consists of proposing that the end of construction condition corresponds to the idealized case of the perfectly undrained condition. Here the soil is considered to be fully saturated with incompressible water and is sufficiently rapidly loaded so that in the short term it is completely undrained. Pre-failure and failure distortions of the soil mass in the field are, *by implication if not in fact*, simulated by the test measuring the undrained shear strength. It follows that if the shear strength of the soil structure is determined under rapid loading conditions prior to construction this undrained shear strength may be used for short term design considerations. No knowledge of the pore pressures is required, the undrained shear strength being used in a total stress analysis.

Shear test data applicable to the unloading condition
The unloading case gives a critical stability condition in the long term, the minimum strength and lowest factor of safety occurring some consider-able time after the end of unloading or construction. It may be readily seen that the long term case approximates to a drained condition in that the pore pressure reduction generated by the unloading is gradually redistributed to the equilibrium pore pressures determined by the steady state ground water levels. Of course the redistribution of pore pressure (under no further change in loading) is accompanied by a change in effective stress as the soil structure swells towards its long term configuration.

The hydraulic gradient causing the flow of pore water into the expand-ing soil structure is itself reduced by the flow, thus giving an exponential decay in the flow rate. The fully softened structural configuration is thus approached very slowly indeed. If the long term equilibrium structure is just weak enough for a failure to occur then, in this special case, the long term factor of safety is unity. Here the effective stresses and hence strength are determined by the equilibrium pore pressures (Fig. 1.10, curve A).

If a model shear test is to be used to provide appropriate shear strength data with which to analyse the stability of this condition, then the simili-tude requirements could indicate the use of a slow or drained test. The drained shear test may be used to give a direct measure of shear strength or it may be used to give an indirect measure of shear strength by provid-ing c' and $\tan \phi'$. Using the effective stress shear strength parameters c' and $\tan \phi'$ in an effective stress analysis requires an estimation of the pore pressures. In the special case of limiting equilibrium in the long term, the pore pressures are fixed by the steady state ground water levels. The analysis may therefore simply proceed on that basis.

For the more general case where the real long term factor of safety is less than unity, i.e. a failure occurs before the pore pressures have reached their equilibrium value (Fig. 1.10, curve B), the instability is still predicted

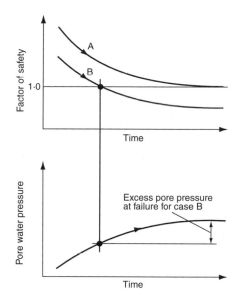

Fig. 1.10 The variation with time of overall stability and pore water pressure at a characteristic location for a cutting. Curve A indicates limiting equilibrium in the long term. Curve B indicates limiting equilibrium before the long term condition is reached

by an effective stress analysis using the long term equilibrium pore pressure distribution. If the long term factor of safety is less than unity then it becomes unity and instability arises some time prior to the long term condition. The effective stress analysis based on the equilibrium pore pressure thus provides a real check on stability even when the actual failure may occur before pore pressure equilibrium is reached and before the soil behaviour is fully modelled by the drained shear test.

Factors affecting the evaluation of field stability

Shear test similitude

Soil can only fail under conditions of local loading where the loading in the zone of influence distorts the soil mass as a whole. Beneath a rapidly constructed embankment, for example, the previously horizontal ground surface deflects, the zone of influence distorting without at the instant of loading changing its volume. The major principal stress direction is orientated vertically under the embankment. Towards the toe of the embankment the principal stress directions rotate until the major principal stress is horizontal. If model shear tests are to be used to predict the undrained shear strength in these varying stress and deformation zones the tests must, ideally, simulate the actual field stress and deformation paths.

25

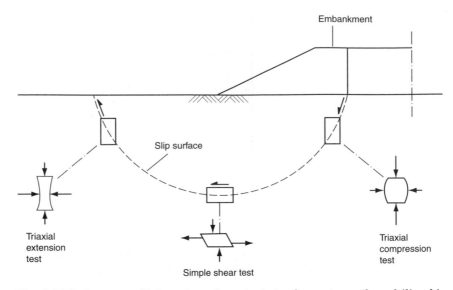

Fig. 1.11 Relevance of laboratory shear tests to shear strength mobilized in the field (after Bjerrum, 1972)

To approximate to similitude, Bjerrum (1972) suggests using different modes of shear test to evaluate the undrained shear strength for different areas of the distorted soil. Thus the triaxial compression test simulates distortion directly under the embankment where the shear surface is inclined near the major principal stress direction, the simple shear test simulates distortion where the shear surface is nearly horizontal, and the triaxial extension test simulates distortion near the toe of the embankment where the shear surface is inclined near the minor principal stress direction (Fig. 1.11).

The alternative to matching the shear test to the field failure zone on the basis of like soil distortions is to adopt a purely empirical approach. Bjerrum (1972) compared the stability of embankments that had failed with the predicted stabilities based on in situ shear vane measurements the results of which were used in limit analyses. Depending on the plasticity of the clay the shear vane overestimated the field strength by up to 100%. There is a clear lack of geometrical similitude between the vane which shears the soil on an upright circumscribing cylinder, and the field prototype which may fail along a circular arc in the vertical plane. The disparity between the strength measured by the shear vane and the field strength is ascribed by Bjerrum to the combined effects of anisotropy, progressive failure and testing rate.

Analytical similitude
In the foregoing section the term 'field strength' refers simply to a number that gives a realistic estimate of field stability when used in a traditional

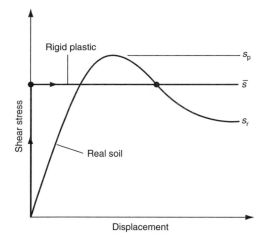

Fig. 1.12 Shear stress–displacement relationship as measured in the direct shear test for an ideal rigid–plastic material and for a real soil. The Residual Factor (Skempton, 1964) is $R = (s_p - s)/(s_p - s_r)$

limit analysis, assuming a rigid–plastic shear stress–displacement relationship which does not vary with direction (Fig. 1.12). The limit analytical model does not therefore allow for the effect of progressive failure. Most naturally occurring soils are strain softening. Irrespective of whether the shear test is drained or undrained, they possess a shear stress–deformation relationship that is characterized by a peak followed by a reduction in strength to an ultimate or residual strength. In normally-consolidated clays, the reduction in strength from peak to residual may be slight while in over-consolidated clays the reduction is marked giving a brittle behaviour.

In a homogeneous soil which is not loaded, direct shear tests on specimens sampled along a circular arc through the soil will have different stress–deformation relationships due to the variation of strength with depth and with orientation. Distorting the soil in the area of the circular arc by a local surface loading will modify the stress–deformation relationships. During loading the stresses vary in the zone of influence and hence the shear strength around the circular arc varies according to the effective stresses normal to the circular arc. If the circular arc becomes a slip surface some local regions will fail. Here the shear stresses tangential to the slip surface have exceeded the local soil strength generated by the local normal effective stresses. The strength in these failed regions reduces according to the strain softening behaviour. In the pre-failure regions, on the other hand, the shear stresses, increased by the load shedding of the strain softening zones will not have exceeded the soil strength available.

A state of limiting equilibrium is attained when the reduction in strength of the elements of the slip surface in the failure zone just begins to exceed the increase in stress taken by those elements in the pre-failure zone (Bishop, 1971a, b). Some of the failed elements may have reduced in strength to the residual value but this is not strictly necessary for limiting equilibrium in a strain softening material. Indeed, the large displacements necessary to achieve ultimate residual strength in heavily over-consolidated clays (Bishop *et al.*, 1971) would suggest that failure conditions in such a material, in the sense of a factor of safety of unity, might obtain without any element of the slip surface reaching the residual strength (Bishop, 1971b).

Conventional stability analyses do not model progressive failure. On the one hand, to take the peak strength as acting on all elements simultaneously around the slip surface overestimates the factor of safety. On the other hand, to take the residual shear strength as acting simultaneously around the slip surface underestimates the factor of safety and is clearly inappropriate in any event as the residual shear strength holds on an established shear surface, i.e. after failure has occurred.

Summary

The applicability of shear test data to field stability problems rests on the principle that similitude exists between the shear test model, the analytical model and the field prototype. Testing and analytical methods must be directed towards ensuring such similitude and assessing the effects of any lack of similitude that may arise. These points will be taken up in more detail in the following chapters.

CHAPTER TWO

Short term stability, undrained strength of clays

Short term stability

> 'One of the main reasons for the late development of soil mechanics as a systematic branch of civil engineering has been the difficulty in recognizing that the difference between the shear characteristics of sand and clay lies not so much in the difference between the frictional properties of the component particles, as in the very wide difference – about one million times – in permeability. The all-round component of a stress change applied to a saturated clay is thus not effective in producing any change in the frictional component of strength until a sufficient time has elapsed for water to leave (or enter), so that the appropriate volume change can take place.'
> **Bishop and Bjerrum (1960)**

The interaction of soil structure and pore water

The uniquely time dependent engineering behaviour of fine-grained saturated soils is derived from the interaction of the compressible structural soil skeleton and the relatively incompressible pore water. Rapid changes in external loading do not immediately bring about a volume change due to the viscous resistance to pore water displacement. Therefore, the soil structural configuration does not immediately change and thus, by Hooke's Law, the structural loading does not change.

However, while the compressible soil structure requires a volume change to change its loading, the relatively incompressible pore water may change its pressure without much volume change. The external loading change is therefore reflected by a change in pore pressure. With time, this 'excess' pore pressure will dissipate, volume change occurring by pore water flow until the consequent change in structural configuration brings the structural loading into equilibrium with the changed external loading.

This process may be examined by the spring–dashpot analogy demonstrated in Fig. 2.1.

The soil structure is modelled by a spring, the soil voids modelled by the chamber under the piston and the soil permeability modelled by the lack

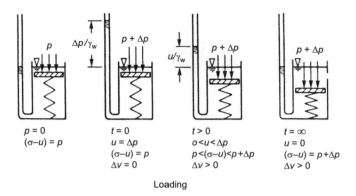

$p = 0$
$(\sigma-u) = p$

$t = 0$
$u = \Delta p$
$(\sigma-u) = p$
$\Delta v = 0$

$t > 0$
$o < u < \Delta p$
$p < (\sigma-u) < p + \Delta p$
$\Delta v > 0$

$t = \infty$
$u = 0$
$(\sigma-u) = p + \Delta p$
$\Delta v > 0$

Loading

Fig. 2.1 Spring–dashpot analogy for soil consolidation

of fit of the piston in the cylinder – thus a soil of high permeability is modelled by a piston which allows a lot of leakage whereas a soil of low permeability is modelled by a piston which allows very little leakage. It is assumed the piston is frictionless. Pore pressure is indicated by the water level in a standpipe whose bore is very much less than that of the piston. Initially the piston is uniformly loaded by a loading intensity p including the weight of the piston.

The instant immediately after $(t = 0)$ rapidly increasing the loading by Δp, the spring (soil structure) is unaffected because insufficient time has elapsed for viscous flow past the piston to reduce the volume of the chamber (pores) under the piston and thus allow the spring to compress further and carry some more load. The loading increment, Δp, is thus initially carried by an equal increase in pore pressure. As time elapses, flow takes place, the piston displacing downwards, the applied loading increment Δp being shared between the spring and the pore pressure. The hydraulic gradient causing flow from the area of high pore pressure under the piston to the area of zero pore pressure over the piston, is itself reduced by the flow as the spring is allowed to compress further and take more load. The law of diminishing returns thus applies and there is an exponential decay in pore pressure and change in length of the spring. Ultimately, the pore pressure dissipates to the equilibrium value and the total load is fully supported by the spring.

The loading condition

If a saturated clay is loaded, such as may occur in soils supporting building foundations and earth embankments (Fig. 2.2), an overall increase in mean total stress occurs (Fig. 2.3(a)). In a fine-grained soil like clay, the viscous resistance to pore water expulsion prevents the soil structure from rapidly contracting. In the short term loading condition, therefore, there is a change in effective stress due to shear strain only together with an increase

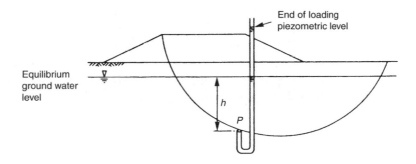

Fig. 2.2 Pore pressure generated on a potential slip surface by embankment loading

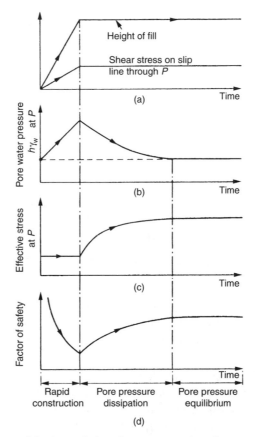

Fig. 2.3 Variation with time of the shear stress, local pore pressure, local effective stress, and factor of safety for a saturated clay foundation beneath an embankment fill (after Bishop and Bjerrum, 1960)

in pore pressure (Fig. 2.3(b) and (c)). With time, this excess pore pressure is dissipated by drainage away from the area of increased pore pressure into the surrounding area of lower pore pressure unaffected by the construction. This flow of pore water causes a time dependent reduction in volume in the zone of influence, the soil consolidating and the soil structure stiffening, giving rise to decreasing settlement and increasing strength. The minimum factor of safety occurs in the short term undrained condition when the strength is lowest (Fig. 2.3(d)).

In this undrained condition the stressed zone does not immediately change its water content or its volume. The load increment does, however, distort the stressed zone. The effective stresses change along with the change in the shape of the soil structure. Eventually the changes in the structural configuration may no longer produce a stable condition and the consequent instability gives rise to a plastic mechanism or plastic flow and failure occurs.

The strength is determined by the local effective stresses at failure normal to the failure surfaces. These are conditioned by and generated from the structural configuration of the parent material (which is itself conditioned by the preloading in situ stresses) and its undrained reaction to deformation. A first step, therefore, in fulfilling the complex similitude requirements that accordingly arise is to ensure that the shear test used to find the strength is effectively undrained.

The undrained shear test may be used to give a direct measure of shear strength, namely the undrained shear strengths s_u, or it may be used to give an indirect measure of shear strength, if the pore water pressures are known together with c' and $\tan \phi'$. It is therefore possible to analyse the stability of the loaded soil by:

(i)   using the undrained shear strength, s_u, in a total stress analysis, or

(ii) using the effective stress shear strength parameters c' and $\tan \phi'$ in an effective stress analysis.

The use of (ii) requires an estimation of the end of construction pore pressures in the failure zone at failure, whereas the use of (i) requires no knowledge of the pore pressure whatsoever.

In general civil engineering works, the soil loading change is applied gradually during the construction period. The excess pore pressures generated by the loading are thus partially dissipated at the end of construction. The end of construction pore pressures and the increased in situ shear strength can be measured on site if the resulting increased economy of design warrants the field instrumentation and testing. On all but large projects this is rarely the case. In addition, the loading is localized, allowing the soil structure to strain laterally, the soil stresses dissipating and the principal stresses rotating within the zone of influence.

In the absence of sound field data on the end of construction shear strengths and pore pressures and in the face of analytical difficulties under local loading, an idealized soil model possessing none of these difficulties is usually invoked for design purposes. This consists of proposing that the end of the construction condition corresponds to the idealized case of the perfectly undrained condition. Here the soil is considered to be fully saturated with incompressible water and is sufficiently rapidly loaded that, in the short term, it is completely undrained. Prefailure and failure distortions of the soil mass in the field are, by implication if not by fact, simulated by the test measuring the undrained shear strength. It follows that if the shear strength of the soil structure is determined under rapid loading conditions prior to construction, this undrained shear strength may be used for short term design considerations. No knowledge of the pore pressures is required, the undrained shear strength being used in a total stress analysis (the so-called $\phi = 0$ analysis).

A note on the $\phi = 0$ analysis

'It is a simple task to place a clay specimen in a shearing apparatus and cause a shear failure. A numerical value of shearing strength, which has acceptable precision, may readily be obtained if proper technique, a representative sample, and a satisfactory apparatus are used. The point which has too often been insufficiently appreciated by testing engineers is that the shearing strength, both in the laboratory specimen and in the clay in nature, is dependent on a number of variables. Before meaning can be attached to shearing strengths determined in the laboratory, the engineer who is to interpret the test results must have at his command an understanding of the factors or variables on which the strength is dependent, and he must make adjustments for every factor which occurs differently in the test than in nature.'
D. W. Taylor (1948)

The concept of the $\phi = 0$ analysis

The $\phi = 0$ analysis is a stability analysis which derives its name from the total stress interpretation of the unconsolidated–undrained triaxial test. In this test, three test specimens of saturated soil taken from the same core barrel (usually clay, but it may be any soil) are tested in unconsolidated–undrained triaxial compression at three different cell pressures. The resulting compression strengths (deviator stresses at failure) are the same because the initial isotropic state of effective stress of each test specimen is unchanged by applying cell pressure. Plotting Mohr's circles in terms of total stress (the effective stresses are not known) give circles of the same diameter (Fig. 2.4). An envelope to the circles is, of course,

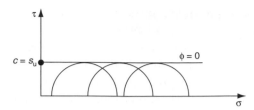

Fig. 2.4 Unconsolidated–undrained triaxial test Mohr's circles showing $\phi = 0$ failure envelope

a horizontal straight line. Interpreting these results using Coulomb's (1773) equation,

$$\tau_f = c + \sigma \tan \phi \tag{2.1}$$

gives $\tau_f = c = c_u \; (= s_u \text{ the undrained shear strength})$ (2.2)

and $\phi = \phi_u = 0$ (2.3)

This apparent insensitivity of shear strength to loading changes led Skempton (1948) to conclude that shear strength was uniquely dependent on water content and if water content did not change then strength did not change. Accordingly, it was believed at that time that as soil strength was unaffected by loading changes in the short term, the strength measured before construction could be used to predict the stability after construction. Thus, the $\phi = 0$ analysis emerged and appeared to work well for both foundations and cuts. Analysis of failed cuts using the $\phi = 0$ method (Lodalen, Norway; Frankton, New Zealand; Eau Brink, England) gave factors of safety of unity thereby apparently confirming the validity of the method. Amazingly, these early case histories gave fortuitously atypical results and it was eventually realized that the critical long term stability of cuts could not be predicted using the $\phi = 0$ analysis because the strength diminishes with long term swelling.

Thus for cuts, long term or 'drained' strengths must be used and this can be predicted in terms of effective stress using the Terzaghi–Coulomb equation:

$$s_d = c' + (\sigma - u) \tan \phi' \tag{2.4}$$

provided the long term pore water pressures are known (they normally are because the ground water regime, initially disturbed by the construction activity, has regained a steady state).

For foundations, however, critical stability is in the short term because the strength increases with consolidation in the long term. In the short term the pore water pressures are unknown and thus an effective stress analysis cannot be made unless pore pressures can be estimated. Accordingly, a $\phi = 0$ analysis must be carried out using the undrained shear strength s_u.

The applicability of the $\phi = 0$ analysis

From the principle of effective stress it is known that effective stress (and hence strength) will not change provided the soil structure does not deform. This is the case in the unconsolidated–undrained triaxial test during the cell pressure increment stages which are isotropic. The combination of the undrained condition (no volumetric strain) and the isotropic stress change (no shear strain) ensures no deformation and hence no change in the initial state of effective stress. Thus, the $\phi = 0$ condition corresponds to a 'no change in the initial state of effective stress' condition. Of course, the strength of the soil skeleton is mobilized in the unconsolidated–undrained triaxial test by axial deformation changing the effective stresses – it is just that this strength is dependent on the initial state of effective stress.

From the principle of similitude it follows that the concept inherent in the $\phi = 0$ condition (i.e. no change in strength if initial state of effective stress is unchanged), and demonstrated in the unconsolidated–undrained triaxial test, must be matched to a corresponding field condition if the s_u measured in this way is to be used in a $\phi = 0$ analysis to predict the field stability. This means that the analysis is only wholly appropriate to a field condition where the construction activity is both undrained and causes an isotropic increase in total stress. This occurs in the laboratory, for example, in the oedometer at the instant of loading change. In the field this condition may be approximated by rapid global loading such as may occur in the short term at shallow depth beneath the centre of a wide embankment. Towards the toe of the embankment shear strain will occur with consequent change in effective stress, thereby progressively invalidating the concept inherent in the $\phi = 0$ analysis. The extent to which this occurs may or may not be significant. Certainly, where narrow foundations such as strip footings are concerned, shear strains will occur throughout the zone of influence and so make the $\phi = 0$ analysis not strictly appropriate. Accordingly, high factors of safety on shear strength (typically $F = 3$) are taken to deal with this discrepancy and with any other factors causing a lack of similitude.

Undrained strength of clays

Field measurements of undrained shear strength

The triaxial and unconfined compression tests and the laboratory direct shear box test rely on obtaining samples of soil from the ground by sampling from boreholes or trial pits and sealing and transporting these samples to the laboratory. The degree of disturbance affecting the samples will vary according to the type of soil, sampling method and skill of the operator. At best, there will be some structural disturbance

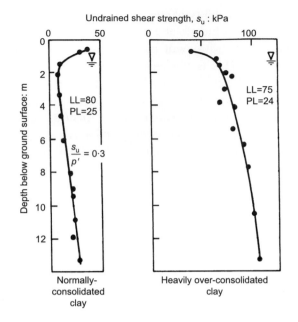

Fig. 2.5 Typical variations of undrained shear strength with depth, after Bishop and Henkel (1962)

simply from the removal of the in situ stresses during sampling and laboratory preparation even if these in situ stresses are subsequently reapplied as a first stage of the test. There is, therefore, considerable attraction in measuring shear strength in the field, in situ.

The determination of the undrained shear strength is only appropriate in the case of clays which, in short term loading, may approximate in the field to the undrained condition. A typical variation of undrained shear strength with depth is shown in Fig. 2.5, for both normally consolidated and heavily over-consolidated clay.

An indication of undrained shear strength may be obtained from plasticity tests. For example, Fig. 2.6 shows the relationship between the ratio of the undrained shear strength to the effective overburden pressure, s_u/p' and the plasticity index, PI, for several normally consolidated marine clays.

As shown in Table 2.1, the relationship between consistency and strength may be generalized to give a rough guide of strength from field inspection.

The in situ shear vane
The field shear vane is a means of determining the in situ undrained shear strength. This consists of a cruciform vane on a shaft (Fig. 2.7). The vane is inserted into the clay soil and a measured increasing torque is applied to

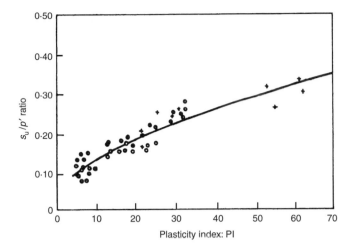

Fig. 2.6 Relationship between s_u/p' *and plasticity index, after Bjerrum and Simons (1960)*

the shaft until the soil fails as indicated by a constant or dropping torque by shearing on a circumscribing cylindrical surface. The test is carried out rapidly. Now, if s_{u_v} is the undrained shear strength in the vertical direction, and s_{u_h} is the undrained shear strength in the horizontal direction, then the maximum torque is

$$T = \frac{\pi D^2}{2}\left(Hs_{u_v} + \frac{D}{3}s_{u_h}\right) \tag{2.5}$$

where H is the vane height and D is the vane diameter, and assuming peak strengths are mobilized simultaneously along all vane edges.

This equation in two unknowns, s_{u_v} and s_{u_h}, can only be solved if the torque is found for two vanes with different height to diameter ratios.

It is often incorrectly assumed that the soil is isotropic and

$$s_{u_v} = s_{u_h} = s_u$$

Table 2.1 Consistency–strength relationship from field inspection (after BS 8004: 1986)

Consistency	Field indications	Undrained shear strength: kPa
Very stiff	Brittle or very tough	>150
Stiff	Cannot be moulded in the fingers	75–150
Firm	Can be moulded in the fingers by strong pressure	40–75
Soft	Easily moulded in the fingers	20–40
Very soft	Exudes between the fingers when squeezed in the fist	<20

Fig. 2.7 Representation of the shear vane

whence,

$$T = \frac{\pi D^2}{2}\left(H + \frac{D}{3}\right)s_u = ks_u$$

where k is a geometrical constant of the vane.

The in situ shear vane may be used in inspection pits and down bore-holes for the extensive determination of in situ strength profiles as part of a site investigation programme.

Some types of shear vane equipment have the extension rods in an outer casing with the vane fitting inside a driving shoe. This type of vane may be driven or pushed to the desired depth and the vane extended from the shoe and the test carried out. The vane may then be retracted and driving continued to a lower depth.

It must be emphasized that the in situ vane provides a direct measure of shear strength and because the torque application is usually hand-operated, it is a relatively rapid strength measure, therefore giving the undrained shear strength.

The large diameter plate loading test

In some over-consolidated clays such as London clay, the release of previous overburden pressure allows the soil mass to expand vertically causing cracks or fissures to form. In stiff fissured clay of this kind, there-fore, the difficulty in shear testing is to ensure that the specimen or zone tested is representative of the fissured soil mass as a whole and not reflecting the behaviour of intact lumps. Small-scale testing can, in these circumstances, be dangerously misleading, as demonstrated in Fig. 2.8. In the penetration test, a penetrometer with a conical point is loaded in a standard way. The penetration is measured and this provides indications

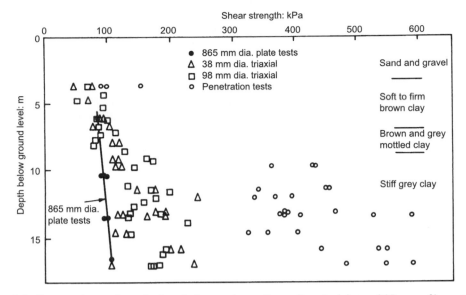

Fig. 2.8 Comparison of undrained shear strengths estimated from 865 mm diameter plate tests and triaxial tests on 38 mm and 98 mm diameter specimens in London clay at Chelsea, after Marsland (1971a)

of strength of intact lumps. The triaxial compression tests on 38 mm and 98 mm diameter specimens fail to include sufficiently the effect of the fissures. Only the large diameter (865 mm) in situ plate loading tests, carried out in the bottom of a borehole, tests a truly representative zone of soil. In this test, the plate is loaded and its deflection measured to give bearing pressure settlement relationships. At failure, bearing capacity theory is invoked to provide the field undrained shear strength. The line representing this undrained shear strength in Fig. 2.8 is the *lower bound* of the scatter of the triaxial test results and is the most appropriate value to use in a bearing capacity calculation.

Factors affecting the measurement of shear strength
The factor which most significantly affects the measurement of shear strength is the mode of test, that is, whether the test is drained or undrained and this has been considered previously. Further significant factors which affect the measurement of shear strength, particularly undrained shear strength, are the type of test, that is, whether by direct shear box, triaxial compression, in situ shear vane, etc; the effects of the orientation of the test specimen or test zone, that is, the effect of aniso-tropy; time to failure; sampling disturbance and size of test specimen or test zone; time between sampling and testing. These factors are now discussed.

Strength anisotropy

Soil strength anisotropy arises from the two interacting anisotropies of geometrical anisotropy, that is, preferred particle packing and stress anisotropy. The geometrical anisotropy arises at deposition when the sedimented particles tend to orientate with their long axes horizontal seeking packing positions of minimum potential energy. The horizontal layering or bedding which results is further established by subsequent deposition which increases the overburden pressure. The stress anisotropy arises because of a combination of stress history and the geometrical anisotropies of both the particles themselves and of the packed structure they form. The net effect is a clear strength and stress–strain anisotropy.

Consider, for example, the drained triaxial compression tests carried out on a dense dry rounded sand in a new cubical triaxial cell described by Arthur and Menzies (1972). The cubical specimen was stressed on all six faces with flat, water-filled, pressurized rubber bags. The test specimen was prepared by pouring sand through air into a tilted former. In this way, it was possible to vary the direction of the bedding of the particles which was normal to the direction of deposition. A clear stress–strain anisotropy was measured, the strain required to mobilize a given strength being greater for the bedding aligned in the vertical major principal stress direction than for the conventional case of the bedding aligned horizontally in the test specimen.

Parallel tests were carried out by Arthur and Phillips (1972) in a conventional triaxial cell testing a prismatic specimen with lubricated ends. It can be seen from Fig. 2.9(a) that remarkably similar anisotropic strengths have been measured in different apparatus. The major differences in type of apparatus include the control of deformation (one stress control, the other, displacement control), the geometric proportions of the specimens, and the application of the boundary stresses (one with uniform pressures on each face and one with rigid ends). It would appear, therefore, that drained triaxial compression tests on a dry, rounded sand give similar measures of the strength anisotropy despite fundamental differences in the type of triaxial apparatus used. It is of interest to note that fitting an ellipse to the strength distribution of Fig. 2.9(a) adequately represents the variation. An elliptical variation of strength with direction was first proposed by Casagrande and Carillo (1944) in a theoretical treatment. Lo (1965) found an elliptical-like variation in the undrained shear strength of a lightly over-consolidated Welland clay (Fig. 2.9(b)). Simons (1967) on the other hand, found a non-elliptical variation in the undrained shear strength of a heavily over-consolidated London clay (Fig. 2.9(c)). Results summarized by Bishop (1967) are similar (Fig. 2.9(d)).

Geometrical anisotropy, or *fabric* as it is sometimes called, not only gives rise to strength variations with orientation of the test axes but is also

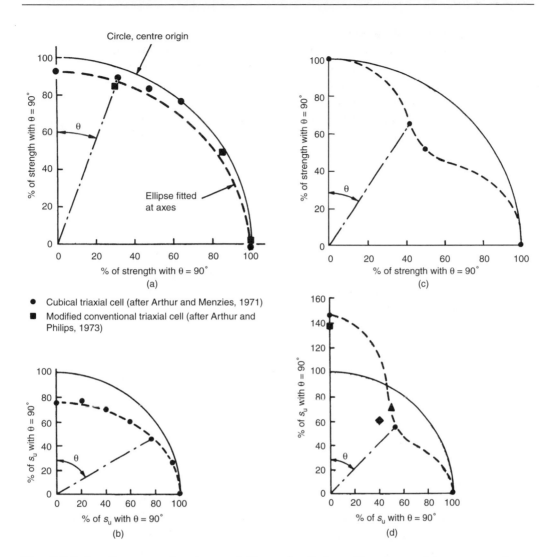

Fig. 2.9 Polar diagrams showing variations of soil strength measured in compression tests; θ denotes inclination of bedding with respect to vertical axis of test specimen. (a) drained tests on dense rounded Leighton Buzzard sand, after Arthur and Menzies (1972); (b) undrained tests on lightly over-consolidated Welland clay, after Lo (1965); (c) undrained tests on heavily over-consolidated blue London clay, after Simons (1967); (d) undrained tests on heavily over-consolidated London clay, after Bishop (1967)

probably partly the cause of undrained strength variations between test type. Madhloom (1973) carried out a series of undrained triaxial compression tests, triaxial extension tests and direct shear box tests using specimens of a soft, silty clay from King's Lynn, Norfolk. The soil was obtained by using a Geonor piston sampler. Samples were extruded in

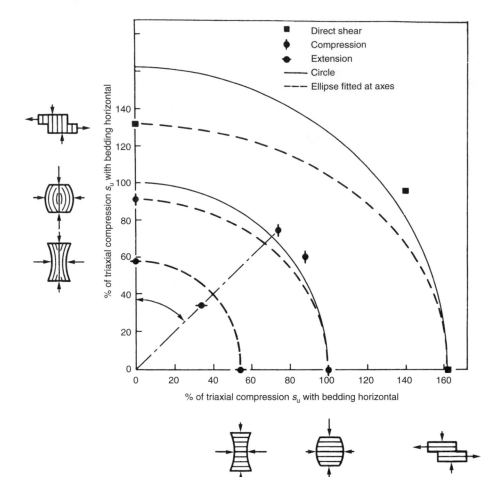

Fig. 2.10 Polar diagram showing the variation in undrained shear strength with test type and specimen orientation for a soft clay from King's Lynn, Norfolk, after Madhloom (1973)

the laboratory and hand-trimmed to give test specimens in which the bedding was orientated at different angles to the specimen axes. A polar diagram showing the variation of undrained shear strength with test type and specimen orientation is given in Fig. 2.10.

It is clear that the magnitude of undrained strength anisotropy in clays is much greater than drained strength anisotropy in sands. It can be seen that generally for this type of clay the triaxial compression test indicates a strength intermediate between that indicated by the triaxial extension test and the direct shear box test. This was not the case, however, in tests on soft marine clays reported by Bjerrum (1972) and given in Table 2.2. Here the direct shear box (and the corrected shear vane) indicated

Table 2.2 Comparison between the results of compression and extension tests, direct simple shear tests, and in situ vane tests on soft clay, after Bjerrum (1972(a))

Type of soil	Index properties: %			
	Water content w	Liquid limit w_l	Plastic limit w_p	Plasticity index PI
Bangkok clay	140	150	65	85
Matagami clay	90	85	38	47
Drammen plastic clay	52	61	32	29
Vaterland clay	35	42	26	16
Studentertunden	31	43	25	18
Drammen lean clay	30	33	22	11

Table 2.2 Continued

Type of soil	Triaxial test τ/p_0'		Simple shear test τ_f/p_0'	Vane tests s_u/p_0'	
	Compression	Extension		Observed	Corrected for rate
Bangkok clay	0.70	0.40	0.41	0.59	0.47
Matagami clay	0.61	0.45	0.39	0.46	0.40
Drammen plastic clay	0.40	0.15	0.30	0.36	0.30
Vaterland clay	0.32	0.09	0.26	0.22	0.20
Studentertunden	0.31	0.10	0.19	0.18	0.16
Drammen lean clay	0.34	0.09	0.22	0.24	0.21

strengths intermediate between the triaxial extension and compression tests.

Bjerrum (1972) found that using the undrained shear strength measured by the vane in a conventional limit analysis gave varying estimates of the actual stability depending on the plasticity of the clay. The disparity between $s_{u_{(field)}}$ and $s_{u_{(vane)}}$ may be partly accounted for by the combined effects of anisotropy and testing rate. The correlation

$$s_{u_{(field)}} = s_{u_{(vane)}} k_v$$

for soft clays is given in Fig. 2.11.

Time to undrained failure

As demonstrated by Bjerrum, Simons and Torblaa (1958), the greater the time to undrained failure, the lower will be undrained strength (Fig. 2.12(a)). It is therefore necessary to take this factor into account when using the results of in situ vane tests or undrained triaxial compression tests, with a failure time of the order of 10 min, to predict the short term stability of cuttings and embankments, where the shear stresses leading

43

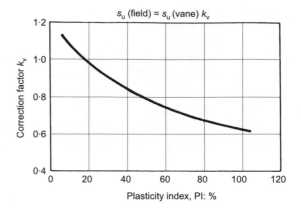

Fig. 2.11 Vane strength correction factor for soft clays, after Bjerrum (1973)

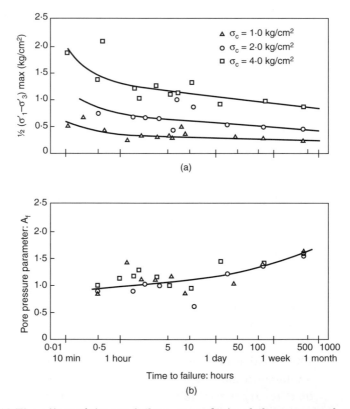

Fig. 2.12 The effect of time to failure on undrained shear strength and pore pressure generation, after Bjerrum, Simons and Torblaa (1958); (a) undrained shear strength plotted against time to failure; (b) pore pressure parameter A_f plotted against time to failure

44

to failure may be gradually applied over a period of many weeks of construction.

The greater the plasticity index of the clay, the greater is the reduction factor which should be applied to the results of the tests with small times to failure. As shown in Fig. 2.12(b) most of the reduction in undrained shear strength is because of an increase in the pore water pressure as the time to failure increases.

A further factor to be considered is the elapsed time between taking up a sample, or opening a test pit, and performing strength tests. No thorough study of this aspect has been made to date but it is apparent that the greater the elapsed time, for a stiff, fissured clay, the smaller is the measured strength. Marsland (1971b) noted that from loading tests made on 152 mm diameter plates at Ashford Common, strengths measured 4–8 hours and 2.5 days after excavation were approximately 85% and 75%, respectively, of those measured 0.5 hour after excavation. Other evidence was provided by laboratory tests on 38 mm dia. specimens cut from block samples of fissured clay from Wraysbury, which were stored for different periods before testing. Strengths of specimens cut from blocks stored for about 150 days before testing were only about 75% of the strengths of specimens prepared from blocks within 5 days of excavation from the shaft. This could be attributed to a gradual extension of fissures within the specimens.

Sampling disturbance and test specimen size

If the test specimen is disturbed by the sampling process, the measured undrained shear strength will generally be lower than the in situ value for a given test apparatus and procedure. Thin-walled piston samples jacked into the ground cause very little disturbance and this technique, together with careful handling in the field, during transit, and in the laboratory, are believed to give reasonably reliable measurements of the undrained shear strength of clay. Equally well, hand-cut block samples of clay taken from open excavation may be used. In Table 2.3, which shows results reported by Simons (1967), the undrained shear strength of various sized test specimens of London clay is compared with that obtained from the customary 38 mm dia. × 76 mm triaxial specimens extracted from a U-4 sampling tube. The strength of 38 mm dia. × 76 mm triaxial specimens taken from hand-cut blocks is 143% of the strength obtained from 38 mm dia. × 76 mm specimens taken from U-4 sampling tubes. This difference results from the greater disturbance caused by using the U-4 sampling tubes.

From Table 2.3 it may also be seen that the size of specimen is of crucial importance. With stiff, fissured clays, the size of the test specimen must be large enough to ensure that the specimen is fully representative of

Table 2.3 Effect of test specimen size and time to failure on undrained shear strength of triaxial compression test specimens cut from intact blocks of blue London clay, after Simons (1967)

Size of triaxial test specimen: mm	Number of tests	Water content w: %	Time to failure t_f: min.	s_u: kPa	s_u ($w = 28$%): kPa	Strength ratio
305×610	5	28.2	63	48.8	50.7	0.62
152×305	9	27.1	110	51.5	46.0	0.56
102×203	11	27.7	175	47.9	46.4	0.57
38×76 (U-4)	36	26.9	8	93.4	81.9	1.00
38×76 (blocks)	12	28.1	7	116.3	117.3	1.43
13×25 (intact)	19	26.6	10	262.4	219.3	2.68

the structure of the clay in the mass. If specimens are too small, the measured strength will be greater than that which can be relied on in the field.

In addition to the laboratory tests given in Table 2.3, Simons (1967) compared the results obtained from an analysis of a slip on the same site with 610 mm × 610 mm square in situ shear box tests, the 305 mm dia. × 610 mm vertical triaxial specimens and the 38 mm dia. × 76 mm vertical specimens from U-4 samples, the latter representing standard practice. These results are given in Table 2.4. Corrections have been made for the different water contents and times to failure, assuming purely undrained shear. The main points to note are as follows:

- The standard 38 mm dia. × 76 mm specimens give a strength 185% times that indicated by an analysis of the slip. If no corrections are made for water content and time to failure, this ratio would be 310%.
- The 305 mm dia. × 610 mm triaxial specimens show a strength 21% higher than that indicated by the slip analysis. Of course, part of this

Table 2.4 Comparison of undrained shear tests on London clay, with strength estimated from a field failure, after Simons (1967)

Test	Water content w: %	Time to failure t_f: min.	s_u: kPa	s_u ($w = 28$%): kPa	s_u ($w = 28$%) ($t_f = 4000$ min): kPa	Strength ratio
Slip	29.3	4000	30.1	35.4	35.4	1.00
610 mm × 610 mm shear box	28.1	71	47.9	48.3	41.2	1.16
305 mm × 610 mm triaxial	28.2	63	48.8	50.8	43.1	1.21
38 mm × 76 mm triaxial	26.9	8	93.4	81.8	65.6	1.85

difference is due to the different inclinations of the failure surfaces as discussed later.

The 610 mm × 610 mm in situ shear box tests give a strength 16% higher than that from slip analysis.

Bearing in mind the approximate nature of the corrections made for water content and time to failure, and the possibility that slight restraint imposed by the shear box may have resulted in a higher measured strength, reasonable agreement between the in situ shear box tests and the slip analysis is indicated.

To summarize, the standard undrained triaxial tests carried out on 38 mm dia. × 76 mm specimens taken from U-4 samples greatly over-estimate the in situ strength of the London clay as indicated by an analysis of the end of construction slip. Much better agreement is obtained from the results of triaxial specimens 100 mm dia. × 200 mm high and larger, and 610 mm × 610 mm in situ shear tests.

Table 2.5 End of construction failures of footings and fills on a saturated clay foundation, after Bishop and Bjerrum (1960)

Locality	Data of clay					Safety factor S_u-analysis	Reference
	W	LL	PL	PI	$\dfrac{W - PL}{PI}$		
1. Footings, loading tests							
Loading test, Marmorerá	10	35	15	20	−0.25	0.92	Haefeli, Bjerrum
Kensal Green	–	–	–	–	–	1.02	Skempton 1959
Silo, Transcona	50	110	30	80	0.25	1.09	Peck, Bryant 1953
Kippen	50	70	28	42	0.52	0.95	Skempton 1942
Screw pile, Lock Ryan	–	–	–	–	–	1.05	Morgan 1944, Skempton 1950
Screw pile, Newport	–	–	–	–	–	1.07	Wilson 1950
Oil tank, Fredrikstad	45	55	25	30	0.67	1.08	Bjerrum, Øverland 1957
Oil tank A, Shellhaven	70	87	25	62	0.73	1.03	Nixon 1949
Oil tank B, Shellhaven	–	–	–	–	–	1.05	Nixon (Skempton 1951)
Silo, USA.	40	–	–	–	–	0.98	Tschebotarioff 1951
Loading test, Moss	9	–	–	–	–	1.10	NGI
Loading test, Hagalund	68	55	20	35	1.37	0.93	Odenstad 1949
Loading test, Torp	27	24	16	8	1.39	0.96	Bjerrum 1954b
Loading test, Rygge	45	37	19	18	1.44	0.95	Bjerrum 1954b
2. Fillings							
Chingford	90	145	36	109	0.50	1.05	Skempton, Golder 1948
Gosport	56	80	30	50	0.48	0.93	Skempton 1948a
Panama 2	80	111	45	66	0.53	0.93	Berger 1951
Panama 3	110	125	75	50	0.70	0.98	Berger 1951
Newport	50	60	26	34	0.71	1.08	Skempton, Golder 1948
Bromma II	100	–	–	–	1.00	1.03	Cadling, Odenstad 1950
Bocksjön	100	90	30	60	1.17	1.10	Cadling, Odenstad 1950
Huntington	400	–	–	–	–	0.98	Berger 1951

Case records (Bishop and Bjerrum, 1960)

Bishop and Bjerrum (1960) concluded that, with few exceptions, the end of construction condition is the most critical for the stability of foundations and that for saturated clays this may may be examined more simply by the $\phi = 0$ analysis. From the field tests and full-scale failures tabulated in Table 2.5, they also concluded that an accuracy of $\pm 15\%$ can be expected in the estimate of factor of safety where the undrained shear strength is measured by undrained triaxial tests (or unconfined compression tests) on undisturbed samples, or from vane tests in the field.

Table 2.6 Long-term failures in cuts and natural slopes analysed by the $\phi = 0$ analysis, after Bishop and Bjerrum (1960)

Locality	Type of slope	Data of clay					Safety factor, $\phi = 0$ analysis	Reference
		W	LL	PL	PI	Liquidity index $\dfrac{W - PL}{PI}$		
1. Over-consolidated, fissured clays								
Toddington	Cutting	14	65	27	38	−0.34	20	Cassel, 1948
Hook Norton	Cutting	22	63	33	30	−0.36	8	Cassel, 1948
Folkestone	Nat. slope	20	65	28	37	−0.22	14	Toms, 1953
Hullavington	Cutting	19	57	24	33	−0.18	21	Cassel, 1948
Salem, Virginia	Cutting	24	57	27	30	−0.10	3.2	Larew, 1952
Walthamstow	Cutting	–	–	–	–	–	3.8	Skempton, 1942
Sevenoaks	Cutting	–	–	–	–	–	5	Toms, 1948
Jackfield	Nat. slope	20	45	20	25	0.00	4	Henkel/Skempton, 1955
Park Village	Cutting	30	86	30	56	0.00	4	Skempton, 1948b
Kensal Green	Cutting	28	81	28	53	0.00	3.8	Skempton, 1948b
Mill Lane	Cutting	–	–	–	–	–	3.1	Skempton, 1948b
Bearpaw, Canada	Nat. slope	28	110	20	90	0.09	6.3	Peterson, 1952
English Indiana	Cutting	24	50	20	30	0.13	5.0	Larew, 1952
SH 62, Indiana	Cutting	37	91	25	66	0.19	1.9	Larew, 1952
2. Over-consolidated, intact clays								
Tynemouth	Nat. slope	–	–	–	–	–	1.6	Imperial College
Frankton, N.Z.	Cutting	43	62	35	27	0.20	1.0	Murphy, 1951
Lodalen	Cutting	31	36	18	18	0.72	1.01	N.G.I.
3. Normally-consolidated clays								
Munkedal	Nat. slope	55	60	25	35	0.85	0.85	Cadling/Odenstad, 1950
Säve	Nat. slope	–	–	–	–	–	0.80	Cadling/Odenstad, 1950
Eau Brink cut	Cutting	63	55	29	26	1.02	1.02	Skempton, 1945
Drammen	Nat. slope	31	30	19	11	1.09	0.60	N.G.I.

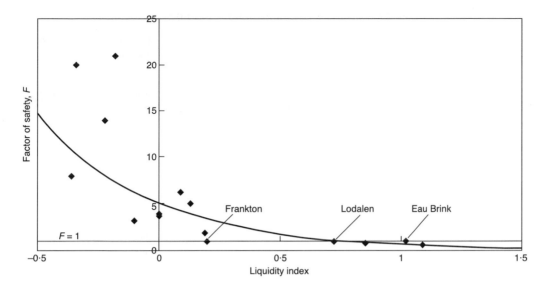

Fig. 2.13 Variation of factor of safety with Liquidity Index for long-term failures in cuts and natural slopes analysed by $\phi = 0$ analysis. Note how untypical are the cases of Frankton, Lodalen, Eau Brink and Drammen!

Bishop and Bjerrum also considered long-term failures in cuts and natural slopes. As shown in Table 2.6, the $\phi = 0$ analysis is unable to predict a sensible factor of safety near unity except for the anomalous cases of Frankton NZ, Lodalen, Eau Brink cut and Drammen. (A second investigation of the slip at Drammen by Kjaernsli and Simons (1962) gave a minimum factor of safety of 0.47.) As shown in Fig. 2.13, there is a weak correlation of factor of safety with soil plasticity showing how inappropriate the short-term $\phi = 0$ analysis is to what is essentially a long-term stability problem. This is, of course, to be expected from the soil mechanics considerations of short-term stability given at the beginning of this chapter.

Summary of main points

(a) The undrained strength is appropriate to short term, 'end of construction', conditions only and may be used in a total stress analysis (or $\phi = 0$ analysis) to assess the stability of footings and fills, i.e. for positive loading conditions and for the short-term undrained condition for temporary cuttings and excavations.

(b) The value of undrained shear strength measured in a test will normally be different from test type to test type (e.g. see Fig. 2.10) and from the idealized rigid–perfectly plastic value back-analysed from a field failure. Empirical correlations between a particular

49

test type (e.g. vane) and construction type (e.g. embankment foundation) failure can help here (e.g. see Fig. 2.11).

(c)   The undrained shear strength is not appropriate to long term stability conditions and should not be used in assessing the long-term stability of cuts and excavations (i.e. unloading conditions) and natural slopes where critical stability is in the long term (see Chapter 3).

CHAPTER THREE

Long term stability, drained strength of clays

Long term stability

The interaction of soil structure and pore water

As pointed out in Chapter 2, the uniquely time dependent engineering behaviour of fine-grained saturated soils is derived from the interaction of the compressible structural soil skeleton and the relatively incompressible pore water. Rapid changes in external loading do not immediately bring about a volume change due to the viscous resistance to pore water displacement. Therefore, the soil structural configuration does not immediately change and thus, by Hooke's Law, the structural loading does not change.

However, while the compressible soil structure requires a volume change to change its loading, the relatively incompressible pore water may change its pressure without much volume change. The external loading change is therefore reflected by a change in pore pressure. With time, this 'excess' pore pressure will dissipate, volume change occurring by pore-water flow until the consequent change in structural configuration brings the structural loading into equilibrium with the changed external loading. This process may be examined using the spring–dashpot analogy demonstrated in Fig. 3.1.

The soil structure is modelled by a spring, the soil voids modelled by the chamber under the piston and the soil permeability modelled by the lack of fit of the piston in the cylinder – thus a soil of high permeability is modelled by a piston which allows much leakage whereas a soil of low permeability is modelled by a piston which allows very little leakage. It is assumed the piston is frictionless. Pore pressure is indicated by the water level in a standpipe whose bore is very much smaller than that of the piston. Initially the piston is uniformly loaded by a loading intensity $p + \Delta p$, including the weight of the piston.

The instant immediately after rapidly decreasing the loading by Δp, as shown in Fig. 3.1, the spring is again unaffected because insufficient time has elapsed for viscous flow past the piston to increase the volume of the chamber under the piston and thus allow the spring to expand and shed some load.

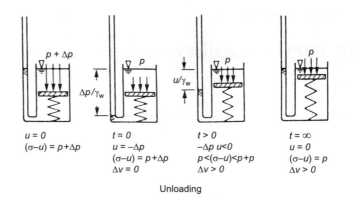

$$u = 0$$
$$(\sigma - u) = p + \Delta p$$

$$t = 0$$
$$u = -\Delta p$$
$$(\sigma - u) = p + \Delta p$$
$$\Delta v = 0$$

$$t > 0$$
$$-\Delta p \; u < 0$$
$$p < (\sigma - u) < p + p$$
$$\Delta v > 0$$

$$t = \infty$$
$$u = 0$$
$$(\sigma - u) = p$$
$$\Delta v > 0$$

Unloading

Fig. 3.1 Spring–dashpot analogy for soil swelling

The loading reduction Δp is thus initially reflected by a numerically equal decrease in pore pressure. As before, as time elapses, flow takes place, the piston displacing upwards, the loading reduction Δp being shared between the spring and the pore pressure. Ultimately, the negative pore pressure increases to the equilibrium value and the loading in the spring reduces to p.

The generation of pore pressure in the loading of real soils

The stability considerations of foundations and earthworks in saturated fine grained soils are highly time dependent. This is because the average size of the interconnecting pores are so small that the displacement of pore water is retarded by viscous forces. The resistance that a soil offers to water flow is its 'permeability' which is the velocity of flow under a unit hydraulic gradient. It can be seen from Table 3.1 that permeability is the largest quantitative difference between soils of different time dependent stability (as pointed out by Bishop and Bjerrum (1960)).

Note in Table 3.1 that the sand and normally-consolidated clay marked with an asterisk have similar shear strength parameters but the permeability of the clay is several orders of magnitude lower, thereby accounting for its unique time dependency whereas the more permeable sand reacts to loading changes almost immediately.

The unloading condition

If a saturated clay is unloaded, such as may occur in an excavation or cutting, an overall reduction in mean total stress occurs. In a fine-grained soil like clay, the viscous resistance to pore water flow prevents the soil structure, partially relieved of its external loading, from rapidly expanding and sucking in pore water from the surrounding soil. With time, this suction is dissipated by drainage into the area of lowered pore pressure

Table 3.1 Effective stress strength parameters and permeabilities for soils of widely varying particle size, after Bishop and Bjerrum (1960)

Soil	Permeability: m/s	c': kPa	ϕ': degrees
Rockfill	5	0	45
Gravel	5×10^{-4}	0	43
Medium sand	–	0	33
Fine sand	1×10^{-6}	0	20–35*
Silt	3×10^{-7}	0	32
Normally-consolidated clay of low plasticity	1.5×10^{-10}	0	32*
Normally-consolidated clay of high plasticity	1×10^{-10}	0	23
Over-consolidated clay of low plasticity	1×10^{-10}	8	32
Over-consolidated clay of high plasticity	5×10^{-11}	12	20

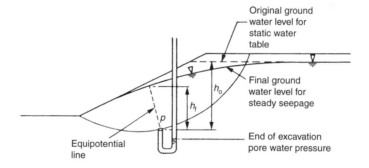

Fig. 3.2 Short term and long term pore pressures in a cutting

from the surrounding area of higher pore pressure unaffected by the excavation. This migration of pore water causes an increase in soil volume in the zone of influence, the soil swelling and the soil structure softening, giving rise to a reduction in strength. The minimum factor of safety occurs at the equilibrium long-term condition.

The time dependent behaviour of fine-grained soils whose in situ total stresses are subject to change may be usefully considered under conditions of unloading and loading. For example, consider the time-dependent stability of a cutting as represented in Fig. 3.2.

The reduction in the in situ total stresses (Fig. 3.3(a)) causes a reduction in pore water pressure dependent on the actual change in principal stress difference and the appropriate value of A (Fig. 3.3(b)). The consequent migration of pore water causes the soil structure to swell reducing the strength and hence stability (Fig. 3.3(d)).

53

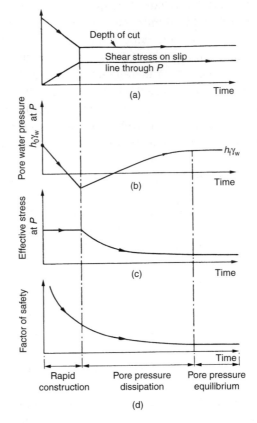

Fig. 3.3 Variation with time of the shear stress, local pore pressure, local effective stress, and factor of safety for a saturated clay excavation, after Bishop and Bjerrum (1960)

Drained strength of clays

Overview

It has been pointed out that the long term stability of clay slopes must be analysed in terms of effective stress using the relevant drained shear strength of the clay. The nature and determination of this strength are now considered in terms of the two major influences on operational drained strength. These are residual or ultimate strength, and progressive failure.

A number of investigations of slides in natural or man-made slopes have shown that the average shear stress along the failure surface on over-consolidated plastic clays and clay shales is considerably smaller than the peak shear strength measured in relevant shear tests in the laboratory. The failure of all such slopes, unless they have already failed in the past, is probably progressive.

As early as 1936, Terzaghi postulated a mechanism by which stiff fissured clays might grow progressively softer as a result of fissures and cracks associated with their structure. Although the concept of the residual strength has appeared in soil mechanics literature since 1937, the significance and relevance of it to the analysis of the stability of slopes in over-consolidated clays and clay shales was fully appreciated first by Skempton. In his Rankine Lecture in 1964 Skempton presented strong evidence of correlations between residual shear strength obtained in the laboratory and average shear strength along slip surfaces in natural slopes in stiff fissured clays.

A number of papers have since been published which confirm the validity of the residual shear strength concept. Although the concept is relevant to all clays, it is of particular practical importance in over-consolidated clays and clay shales where the decrease in strength from peak to residual is large.

Residual strength

If a specimen of clay is placed in a shearing apparatus and subjected to displacements at a very slow rate (drained conditions) it will initially show increasing resistance with increasing displacement. However, under a given effective pressure, there is a limit to the resistance the clay can offer, and this is termed the 'peak strength', s_f. With further displacement the resistance or strength of clay decreases. This process, which Skempton (1964) refers to as 'strain softening', is not without limit because ultimately a constant resistance persists, regardless of the magnitude of displacement. This value of ultimate resistance is termed 'residual strength', s_r.

If several similar tests are conducted under different effective pressures, the peak and residual strengths when plotted against the effective normal pressure (as shown in Fig. 3.4) will show a straight line relationship, at least within a limited range of normal stress. Peak strength can therefore be expressed by

$$s_f = c' + \sigma' \tan \phi' \tag{3.1}$$

and the residual shear strength by

$$s_r = c'_r + \sigma' \tan \phi'_r \tag{3.2}$$

The value of c'_r is generally very small, but even so may exert a significant influence on the calculated factor of safety and depth of the corresponding slip surface. Thus, in moving from peak to residual, the cohesion intercept approaches zero. During the same process the angle of shearing resistance can also decrease. During the shearing process, over-consolidated clays tend to expand, particularly after passing the peak. Thus, the loss

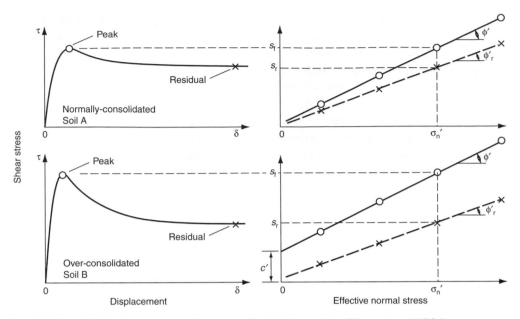

Fig. 3.4 Simplified shear strength properties of clay, after Skempton (1964)

of strength in passing from peak to residual is partly due to an increase in water content. A second factor that equally contributes in the post-peak reduction of the strength is the development of thin bands or domains in which the clay particles are orientated in the direction of shear, as noted by Skempton (1964).

In general, the difference between peak and residual strength depends on soil type and stress history and is most marked for heavily over-consolidated fissured clays. For normally-consolidated clays this difference is generally small. Thus, the concept of residual strength is of particular importance in the case of the long-term stability of slopes of over-consolidated fissured clay.

Factors influencing residual strength

The difference in strength between peak and residual for over-consolidated clays increases with clay content and the degree of over-consolidation. Skempton (1964) has shown that the residual strength decreases with increasing clay fraction and that at a given effective stress the residual strength is practically independent of the past stress history. In fact, Kenney (1967), Bishop *et al.* (1971) and Townsend and Gilbert (1973) have shown that fully remoulded samples gave essentially the same residual strength as undisturbed samples of the same soil at the same normal stress. Furthermore, the residual shear strength has been found to be independent of the loading sequence (stress history) in multi-stage tests, because the same value ϕ'_r has been shown to exist, no matter if σ'_n is increased or decreased (see equation (3.2)) within a limited stress range.

Although it is generally believed that the amount of clay in the material controls the magnitude of the residual strength, Kenney (1967) investigated the influence of mineralogy on the residual strength and showed that it is the type of clay mineral present that is the governing factor. With the same ion concentration in the pore fluid, the residual friction angles reported by Kenney were about 4° for sodium montmorillonite, 10° for calcium montmorillonite, about 15° for kaolinite and from 16° to 24° for hydrous mica or illite.

The value of ϕ'_r was also found by Kenney to depend on ion concentration in the pore fluid and to increase as salt concentration increases. In the case of sodium montmorillonite the residual friction angle increased from about 4° for negligible salt dissolved in the pore fluid to 10° with 30 g/litre sodium chlorite in the pore fluid. From his investigations on several natural soils, pure minerals and mineral mixtures, Kenney concludes that the residual shear strength is primarily dependent on mineral composition and, to a lesser degree, on the system chemistry and the effective normal stress, and that it is not directly related to plasticity or grain size of the soil. In general, according to Kenney, massive minerals such as quartz, feldspar and calcite exhibit high values of $\phi'_r > 30°$. For micaceous minerals (e.g. hydrous mica, illite) $15° < \phi'_r < 26°$. Soils containing montmorillonite exhibit low values of $\phi'_r < 10°$.

Although ϕ'_r and plasticity index (PI or I_p) may not be related directly, subsequent investigations by Voight (1973) have indicated that there is a definite statistical relationship between ϕ'_r and I_p, the general trend being that ϕ'_r decreases with increasing I_p. Correlations of residual shear friction angle with liquid limit have also been observed by Mitchell (1976).

The value of ϕ'_r may also exhibit a variation with the effective normal stress. For the brown London clay tested by Bishop *et al.* (1971), ϕ'_r varied from 14° at $\sigma_n = 7$ kPa down to 8° at $\sigma'_n = 250$ kPa, the increase in ϕ'_r below about 70 kPa being very marked. The general trend appears to be of ϕ'_r decreasing with increasing σ'_n. This may be attributed to the increased pressure at the interparticle contact points and the increased number of interparticle contacts per unit area on the slip surface as σ'_n increases. Mitchell (1976) has also discussed the stress dependency of ϕ'_r exhibited by some clays. Chandler (1966) showed that the field values of ϕ'_r of Upper Lias clay are strongly stress dependent, decreasing with increasing normal effective stress.

Non-linear Mohr–Coulomb failure envelopes have been observed for some clays, the curvature being more marked at low pressures. The cohesion intercept obtained by some investigators could be due to this curvature. Skempton and Petley (1967) have shown that, in some cases, above a certain value of the normal stress, ϕ_r' can be considered independent of normal stress and a linear envelope can be fitted. Similar observations have been made by Townsend and Gilbert (1973) for some clay shales.

The residual shear strength has been found to decrease very slightly with decreasing rates of shear. For most practical purposes it can be considered independent of the rate of shearing (Kenney, 1967; Skempton, 1965; La Gatta, 1970; Garga, 1970).

Determination of residual strength

The residual shear strength is not only of practical importance in relation to the analysis of long-term stability of slopes, natural or man-made, but it may also be considered to be a fundamental property of the particular soil. Therefore, it is important in the laboratory to measure residual strength accurately. It is commonly assessed by:

- reversing shear box tests
- triaxial tests
- ring shear tests
- back analysis of a field failure on a pre-existing failure surface where post-slip movements occur and where post-slip piezometric levels are known.

Figure 3.5 compares multiple reversing direct shear box tests with test results from triaxial and ring shear apparatus.

Reversing shear box tests

In the case where a pre-existing surface is to be tested, large displacements have already reduced the strength to the residual value, and testing can be conveniently accomplished by employing either the shear box or the triaxial apparatus. Over recent years the direct shear box has been widely used and numerous sets of values of residual strength have been reported (Hermann and Wolfskill, 1966; Bishop and Little, 1967; Cullen and Donald, 1971).

Skempton (1964) determined the residual shear strength of soils by repeatedly shearing a specimen in a direct shear machine. After completing the first traverse, with a displacement of about 7.5 mm, the upper half of the shear box was pushed back to its original position and then pulled forward again, this process being repeated until the strength of the clay had dropped to a steady (residual) value.

Kenney (1967), who performed reversed direct shear tests on remoulded soil, followed a somewhat different procedure. A specimen with a moisture content exceeding the liquid limit was placed within a confining ring and between two circular Carborundum plates to consolidate. The specimen had an initial thickness of about 2.5 mm and a diameter of 8 cm. When consolidation was complete, the confining ring was removed and the sample, with a thickness of about 1 mm, was sheared forwards and backwards with a travel of 2 to 2.5 mm each side of the centre. Modified shear box devices have also subsequently been developed by many investigators.

The most serious drawback of the direct shear test in a measurement of residual strength is that laboratory conditions do not simulate the field conditions of a large relative displacement uninterrupted by changes in direction. Successive back and forth displacements may not be equivalent to a total displacement of the same amount in one direction. Although the effect of reversals is not exactly known in the direct shear test, it is believed to be accompanied by some degree of lack of perfect reorientation or disturbance of the previously orientated particles. Bishop *et al.* (1971) noticed from direct shear tests on slip surfaces in blue London clay that ϕ'_r at the second forward travel is greater than that during and at the end of the first forward travel. Similarly, Cullen and Donald (1971) found that the residual strength for some of the soils tested is about 10% lower in the first forward travel compared with subsequent reversals. In such cases the lower value of residual strength at the end of the first travel has been accepted. Area correction problems may also arise, especially if the shear box travel is large.

The values of residual strength obtained in the direct shear box may be high compared with the corresponding values in the ring shear apparatus. It does not follow, however, that the estimation of the residual strength in the direct shear box always leads to significant errors.

Skempton (1964) and Skempton and Petley (1967) have shown that the measured residual strength in the reversing shear box correlates closely with the average mobilized strength calculated for a number of field failures in over-consolidated clays where movement has occurred along existing slip surfaces. Noble (1973) has also measured residual strengths in the reversing shear box which were compatible with the observed behaviours of three landslides in the United States. In addition, Skempton and Petley (1967) demonstrated that in reversing shear tests performed on initially unsheared clays the residual strength was in good agreement with both shear box and triaxial tests on natural slip surfaces.

The residual strength obtained for Curaracha Shale by Bishop *et al.* (1971) in the direct shear box and the ring shear apparatus was practically identical and they suggested that pre-cut samples in the direct shear box on hard materials, such as shales, give better estimates of ϕ'_r than for softer materials, such as clays, since the two halves of the box can be well separated and 20 or 30 reversals can be imposed with less squeezing.

A series of tests on clay shales were conducted by Townsend and Gilbert (1973) in the ring shear apparatus, the rotation shear apparatus and direct shear box (pre-cut samples). The results showed close agreement and Townsend and Gilbert concluded that the direct shear test can be conveniently used for hard over-consolidated shales.

The preparation of undisturbed samples for the direct shear test is easier than for any other type of test. Testing along discontinuities such as principal slip surfaces and joints may not lead to any errors due to reversal effects because the residual strength may be reached before the end of the first traverse. In such cases, due to the overall test simplicity, the direct shear test is preferable.

Triaxial tests

The conventional triaxial cell has also been used for the measurement of the residual strength by a number of investigators. Chandler (1966) measured the residual strength of Keuper Marl by cutting a shear plane in the sample at an angle of approximately $(45° + \phi'_r/2)$ to the horizontal and testing in the triaxial apparatus as suggested by Skempton (1964). Leussink and Muller-Kirchenbauer (1967), Skempton and Petley (1967) and Webb (1969), among others, have published values of the residual strength obtained in the triaxial apparatus.

When employing the triaxial apparatus for the determination of the residual strength, special techniques and analyses must be used to take account of the following factors:

- horizontal thrust on the loading ram
- restraint of the rubber membrane
- change of the cross-sectional area.

After the peak strength has been reached in a triaxial test specimen the post-peak deformation is generally localized to a thin zone between the sliding blocks. The shear and normal stresses in this zone are functions of the vertical and horizontal loads on the end of the ram and the cell pressure. At the end faces of the test specimen, horizontal frictional forces can be mobilized due to end restraints. This could lead to erroneous results, especially for the deformation beyond the peak.

In order to maintain an even pressure along the failure plane, Chandler (1966) employed a modified form of a triaxial cell with a loading cap freed to move laterally without tilting by the use of a number of ball bearings between the top cap and a special plate on the loading ram. Leussink and Muller-Kirchenbauer (1967) minimized the horizontal load on the loading ram using a triaxial apparatus with a free moving pedestal. Bishop *et al.* (1965) and Webb (1969) considered the effects of the horizontal components of load on the measured test parameters.

The apparent increase in strength due to membrane restraint in triaxial tests where failure occurs on a single plane, has also been taken into account. The restraint provided by the membrane has been examined by using dummy specimens of plasticine (Chandler 1967) and Perspex (Blight 1967). The change in cross-sectional area resulting from movement along the shear plane has also been determined and incorporated in the analysis.

Polishing of the inclined cut plane in triaxial tests by a flat spatula or glass plate produces quite a strong orientation of particles and the residual strength obtained by testing polished cut plane triaxial test specimens has always been found to be lower than the values of residual strength measured in direct shear tests (Garga 1970).

Ring shear tests

Among the difficulties of obtaining the residual strength in the triaxial apparatus is that sufficient movement may not be obtained to achieve the residual stage on other than existing discontinuities or pre-cut planes. Herrmann and Wolfskill (1966), who also used triaxial tests which were continued to large strains and triaxial tests on specimens with a pre-cut inclined plane, concluded from their results that neither type of triaxial test was able even to begin to approach the residual state.

Skempton and Hutchinson (1969) noted that, for some clays, a true residual stage is reached only after large displacements (of the order of 1 m) and the residual strength obtained in reversal or cut plane tests is considerably higher than this 'ultimate' residual strength obtained in the ring shear apparatus. La Gatta (1970), using data of previous investigators obtained from repeated reverse direct shear tests, has re-plotted the stress ratio (τ/σ'_n) versus the logarithm of the displacement and showed clearly that in many cases a constant residual strength was not reached and further displacement was necessary to establish the residual strength. Shear strength data for triaxial, shear box and ring shear tests are compared in Fig. 3.5.

The ring and rotational shear tests are the only tests in which very large and uniform deformations can be obtained in the laboratory and have been used in soil mechanics for many years to investigate the shear strength of clays at large displacements (Tiedemann, 1937; Haefeli, 1938). Several designs of the apparatus and results have also been reported by De Beer (1967), Sembelli and Ramirez (1969), La Gatta (1970), Bishop *et al.* (1971) and Bromhead (1979).

The ring shear apparatus described by Bishop *et al.* (1971) may be used to determine the full shear strength displacement of an annular soil specimen subjected to a constant normal stress, confined laterally and

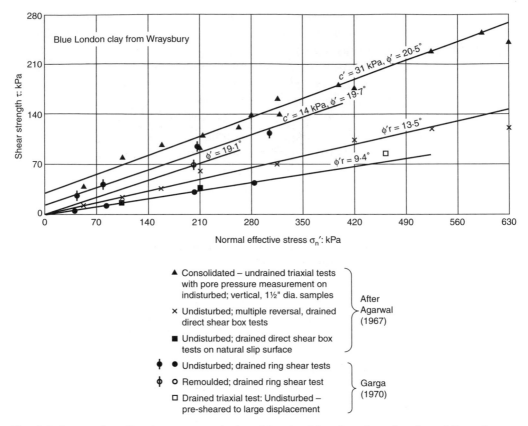

Fig. 3.5 Strength–effective stress relationships for blue London clay from Wraysbury, after Bishop et al. *(1971)*

ultimately caused to rupture on a horizontal plane of relative motion. The apparatus may be considered as a conventional shear box extended round into a ring (Fig. 3.6). Consequently large displacements (e.g. 1 m) may be obtained in one direction so that the residual strength may be accurately determined (e.g. see Fig. 3.7).

Tests in the ring shear apparatus give values of ϕ'_r generally lower than those obtained in the direct shear test or the triaxial test. The slip surface obtained after completing a ring shear test is generally more smooth and polished due to more complete orientation of particles.

Independent tests on blue London clay in the rotational Harvard apparatus by La Gatta (1970) and in the ring shear apparatus described by Bishop *et al.* (1971) gave essentially the same value of $\phi'_r = 9.3°$ $(c'_r = 0)$, whereas the average value of ϕ'_r obtained in the direct shear test was 3–4° higher, as shown in Fig. 3.5.

The two advantages in determining the residual shear strength in a torsion or ring shear apparatus are that the cross-sectional area of the

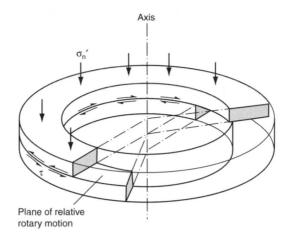

Fig. 3.6 Diagrammatic view through a test specimen of the ring shear apparatus, after Bishop et al. (1971)

sample remains constant during testing and that the sample can be subjected to any uninterrupted displacement in one direction.

The reversal direct shear test may considerably overestimate the residual shear strength and generally gives values of ϕ_r' higher than those determined in the ring shear tests. This is due to the disturbance of particle orientation and the change in direction of principal stresses in each reversal, so that attainment of the residual state is not achieved. Polished cut plane triaxial tests give values of residual strength lower than the values measured in the direct shear box and may approach the values obtained in the ring shear tests.

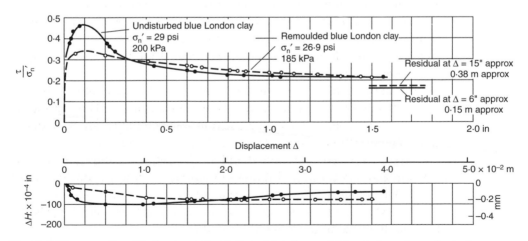

Fig. 3.7 Stress ratio–displacement relationships from drained ring shear tests for undisturbed and remoulded blue London clay from Wraysbury, after Bishop et al. (1971)

It has been suggested that an accurate determination of residual shear strength can be made only by plotting τ/σ'_n against the logarithm of the displacement and taking as residual strength that value of τ/σ'_n which corresponds to zero slope of the curve (La Gatta, 1970; Garga, 1970).

The simplicity of design of the Bromhead ring shear apparatus (Bromhead (1979) – see Figs. 3.8(a) and (b)) together with its ease of use compared to the complex ring shear apparatus of Bishop *et al.* (1971) has ensured that this is the most widely used ring shear apparatus in both research and commercial testing laboratories.

Back-analysis

As pointed out by Skempton (1977), residual shear strength can be calculated with some accuracy where post-slip movements occur provided piezometric levels are known. After the slip in 1949 at Sudbury Hill, no remedial works were undertaken; the toe of the slip was merely trimmed back. Further small movements occurred in succeeding winters, and were similarly treated. Piezometer levels during these post-slip movements were known and it was therefore possible to calculate with some accuracy the residual shear strength and average normal effective pressure. The results are compared in Fig. 3.9 with those obtained from back-analysis of the first time slide. It can be seen from Fig. 3.9 that the strength of the first time slide is significantly greater than the residual strength.

The influence of clay fraction on residual strength

Lupini *et al.* (1981) carried out tests on different soil mixtures where the gradings of the soils could be varied. The proportions of platy particles to rotund particles were confirmed as controlling the type of residual shearing mechanism. They demonstrated three modes of residual shear: a turbulent mode in soils with a high proportion of rotund particles or with platy particles of high interparticle friction; a sliding mode in which a low shear strength shear surface of strongly orientated low-friction platy particles forms; and a transitional mode involving both turbulent and sliding shear (see Fig. 3.10). The practical importance of this work is that it indicates that a comparatively minor change in clay fraction may significantly affect the residual shear strength. In any one deposit, therefore, e.g. London clay, if the clay fraction changes across the London Basin, this may indicate a change in residual strength from assumed values.

Comment (Bromhead, 1979)

In his book *The Stability of Slopes*, Bromhead (1979) remarks that peak strength and its measurement were far better understood in

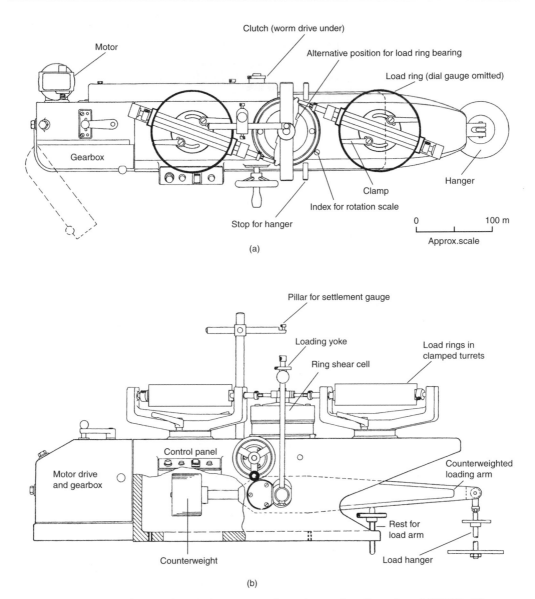

Fig. 3.8(a) Bromhead ring shear apparatus: plan view, after Bromhead (1979). The torque is measured by the proving rings. These are mounted in turrets so that they can be swung out of the way for sample preparation (the test cell is removable). Two stops are provided on the torque arm to allow easy setting of the radius at which the proving rings act. (b) Bromhead ring shear apparatus: elevation and general layout. Ordinarily, this apparatus is mounted on a combined stand and small table, but is equally at home on a workbench, after Bromhead (1979)

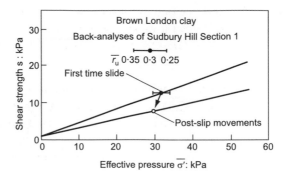

Fig. 3.9 Back-analysis of Sudbury Hill Section I, after Skempton (1977)

geotechnical circles than residual strength. This is due to a number of factors, including the following.

- The apparatus for measuring peak strength is commonplace in industrial laboratories as well as research establishments, and is in daily use.
- Many peak strength tests are *undrained* and hence are quick and cheap, leading to their routine use.
- Residual strength as a concept is relatively recent, and is viewed as being applicable to 'landslides' and not as an aspect of behaviour with far wider ramifications.

Furthermore, the view is prevalent that simple peak strength measurements in the laboratory genuinely represent field strengths, whereas residual strength measurement is in some way still 'experimental'. In fact the opposite is true!

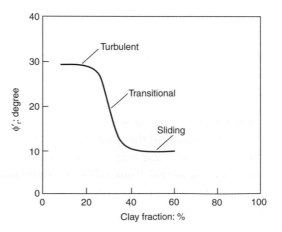

Fig. 3.10 Sand–bentonite mixtures; peak and residual friction coefficients against clay fraction, after Lupini et al. (1981)

The essence of peak strength measurement in the laboratory is to test a representative soil sample in such a way that it preserves its fabric. Peak strength testing, however, is beset with major problems, including

- disturbance, which usually *decreases* the measured strength
- failure to follow the correct stress path in the test, which can either decrease or increase the measured strength relative to the field
- unrepresentative sampling, which *increases* the strength relative to that which is likely to be operative in the field
- poor testing technique, leading to partial or inadequate drainage in 'undrained' or 'drained' tests respectively, so affecting the measured strengths
- failure to take into account progressive failure in brittle soils (e.g. by doing multistage tests on them) so that their behaviour is modified by systematic changes in the soil fabric as the test progresses
- errors in choosing strain rates, which cannot later be rectified.

In comparison, residual strength tests are measuring a soil property that is largely independent of the stress path followed, and in the ring shear apparatus it is possible to allow for, or rectify, many of the factors listed above because of the unlimited strain capacity of the machine.

Definitions

In summary, the following definitions should be noted.

- *Residual strength* is the *drained* strength after sufficient movement has occurred on a failure surface to orientate the clay particles into a parallel position.
- The *residual factor* is given by

$$R = \frac{s_f - \bar{s}}{s_f - s_r}$$

where $\bar{s}$ is the average or operational strength back-analysed from a field failure, s_f is the peak failure strength obtained from a drained peak strength test (e.g. a drained triaxial test) and s_r is the residual strength obtained from a drained ring shear test.

- The *brittleness index* is given by

$$I_B = \frac{s_f - s_r}{s_f}$$

with s_f and s_r as defined above.

Progressive failure

Overview

In conventional analysis of slope stability problems, it is assumed that the peak shear strength of the soil is fully mobilized simultaneously along the whole length of the failure surface. Thus, the soil is treated as a rigid–plastic material and the actual soil stress–strain relationship does not enter into the method of analysis. The true stress–strain curve of a soil considerably deviates from that of a rigid–plastic material and the ratio of strength to shear stress is not uniform along the entire length of a potential slip surface. In such cases the state of limiting equilibrium is associated with non-uniform mobilization of shearing resistance and thus with progressive failure. As progressive failure invalidates conventional stability limit analysis, it is important to specify the necessary conditions for a progressive failure to take place.

Necessary conditions for progressive failure

Terzaghi and Peck (1948) and Taylor (1948) have associated progressive failure with non-uniform stress and strain conditions and redistribution of shear stress along a potential sliding surface.

If an element within a soil which possesses a strain softening stress–strain curve is sheared beyond the peak failure strain it will lose part of its sustained stress. This part of stress must be shed to the neighbouring elements, which in turn may be brought past the peak by this additional stress and thus the process of progressive failure can initiate. Thus, local redistribution of stress can occur if the soil exhibits a brittle behaviour.

On the other hand, even if the soil stress–strain relationship exhibits a strong brittle behaviour, progressive failure cannot initiate if the stress and strain distribution within the soil mass is uniform.

Therefore, the development of a sliding surface by progressive failure is possible if the following three conditions are satisfied:

- the soil exhibits a brittle behaviour with a marked decrease in strength after failure strain
- stress concentrations take place
- the boundary conditions are such that differential strain may take place.

When all three conditions are satisfied, the likelihood of progressive failure is large and a potential slip surface may develop along which the average shear strength lies between peak and residual strength values.

The influence of normal stress on the post-peak stress–strain behaviour

Drained triaxial test results on over-consolidated clays performed under different values of σ_3' have shown that the stress–strain curves can show plastic or brittle behaviour according to the magnitude of the confining pressure. Although variations due to clay type or the stress range used in the tests may exist, the general trends are, with increasing confining pressure (Bishop *et al.*, 1965; Lo, 1972):

- the magnitude of the post-peak strength reduction decreases
- the rate of decrease in strength after passing the peak decreases
- the strain to reach the peak increases.

The reduction of strength in passing from peak to residual may be expressed by the brittleness index I_B (Bishop, 1967):

$$I_B = \frac{s_f - s_r}{s_f} \tag{3.3}$$

This index depends on the normal pressure and generally decreases with increasing normal pressure. Furthermore, the shearing displacement required to reach the residual strength strongly depends on the normal pressure, soil type and conditions. Mitchell (1976) has shown that shearing displacements of only 1 or 2 mm are necessary to reach the residual state for clay materials in contact with smooth steel or other polished hard surfaces. The required displacements for clay against clay were measured to be several centimetres. Herrmann and Wolfskill (1966) have found that the displacement required to reach the residual condition decreased considerably as the normal stress increased. Subsequent results in a rotary shear apparatus carried out by La Gatta (1970) have not shown this trend. With blue London clay the displacement required to reach the residual state was found by La Gatta (1970) to be about 400 mm at a normal pressure of $\sigma'_n = 100$ kPa and 200 kPa, about 300 mm at $\sigma'_n = 400$ kPa, and 100 mm for $\sigma'_n = 800$ kPa. Tests in the ring shear apparatus conducted by Garga (1970) showed clearly that greater displacements are required to reach the residual state when the sample is sheared under low effective stresses than those required under higher effective normal stresses. For blue London clay the displacements varied from about 500 mm at a normal stress of 42 kPa to approximately 125 mm at a normal stress of 280 kPa.

Bishop *et al.* (1971) found that once the residual strength has been established under a given normal stress, subsequent rebound at lower normal stress requires further displacement to re-establish the residual strength. They point out that this effect should be allowed for when carrying out stability analyses of slides on pre-existing surfaces where re-initiation of slides takes place under a normal effective stress lower than that obtaining when the original surface was formed. The field value of ϕ'_r for clays such as brown London clay which show marked stress dependency in the re-initiated slide tends to be greater than that measured in the laboratory.

69

Apparently, the magnitude of the normal effective stress considerably influences the post-peak behaviour. This also emphasizes the necessity for an accurate determination of the strength envelopes at the in situ stress level and the danger of extrapolating from test results at higher stresses. Determination of values of peak and residual strengths to be incorporated in the analysis of slope stability problems should be carried out in the laboratory under conditions simulating those in the field.

The influence of testing rate

Results on the effect of rate of shearing on the peak drained strength are rather limited. Bishop and Henkel (1962) reported drained triaxial tests remoulded Weald clay specimens in which the time to failure was varying between 1 day to 2 weeks. The tests showed a decrease in strength of about 5% per tenfold increase in testing time. Tests on a normally-consolidated, undisturbed marine clay carried out by Bjerrum *et al.* (1958) showed that with times to failure of up to a month the drained shear strength was independent of the test duration when the latter was greater than 1 day. The authors suggested that the expected reduction of rheological component in this case was offset by an increase in true cohesion as a result of secondary consolidation as the time to failure increased. Constant stress level creep tests under drained conditions with duration up to $3\frac{1}{2}$ years were performed by Bishop and Lovenbury (1969) on undisturbed brown London clay and a normally-consolidated Pancone clay from Italy. Their results indicated that on an engineering time scale little decrease in strength from peak to residual can be accounted for by the time dependent component of strength. Samples at stress levels below the residual strength were found to creep and there was no threshold value of stress below which time dependent axial deformation did not take place.

There are some clays which may exhibit considerable drained strength reduction with time to failure. Bjerrum (1969) suggested that the effect of time to failure on the peak drained strength must be considered in the case of plastic clays. Drained triaxial tests on St Vallier clay, Canada, reported by Lo (1972) showed that the drained shear strength decreased logarithmically with time to failure, amounting to a decrease in strength of about 12% per log cycle of time. The drained strength decrease with the logarithm of time was expressed by Lo using the equation suggested by Hvorslev (1960):

$$s_t = s_f - k \log_{10} t/t_0 \tag{3.4}$$

where s_t = drained strength measured at time to failure t
s_f = drained strength measured in time to failure t_0 in rapid conventional tests
k = rate of decrease of strength per logarithmic cycle of time.

In addition, Lo noted that the rate of post-peak reduction was changed with time to failure. Consequently, for some clays, the time effect may constitute an important mechanism of progressive failure.

The influence of discontinuities

The majority of over-consolidated clays contain numerous discontinuities such as fissures, bedding planes, joints and faults. If, in addition, they have been sheared by landsliding or tectonic forces, shear zones will be formed containing minor shears and, usually, one or more principal slip surfaces (Skempton 1966). Non-fissured, intact, over-consolidated clays such as boulder clays and clay tills are relatively rare. Since discontinuities represent local zones or surfaces of reduced shear strength which reduce the strength of the clay mass, it is expected that the stability of slopes in over-consolidated clays will be largely controlled by the strength along these discontinuities.

Terzaghi in 1936 gave the first explanation of the softening action of fissures in stiff clays with time. He pointed out the dangers of progressive failure if fissures and joints open up as a result of small movements consequent upon removal of lateral support when the excavation was made.

Skempton (1964) suggested that, in addition to allowing the clay to soften, the joints and fissures cause concentrations of shear stress which locally exceed the peak strength of the clay and lead to progressive failure. According to Skempton and La Rochelle (1965) fissures can adversely influence the strength of over-consolidated clays as follows.

- Open fissures may form a portion of a failure surface across which no shear resistance can be mobilized.
- Closed fissures may form a portion of a failure surface on which only the residual strength can be mobilized.
- Fissures, whether open or closed, may adversely influence the stress within a slope, increasing the likelihood of progressive failure.

The shear strength along the different types of discontinuities generally depends on the amount of relative displacement which these planes of weakness have undergone. A tentative classification of discontinuities according to their occurrence and relative shear movement has been presented by Skempton and Petley (1967). According to this classification, principal displacement shears such as those found in landslides, faults and bedding-plane slips have undergone large displacements (more than 100 mm) and their surfaces appear polished. Minor shears such as Riedel, thrust and displacement shears of limited extent are described as non-planar and slickensided along which small displacements (less than 10 mm) have occurred. Joint surfaces, including systematic joints, displayed 'brittle fracture' texture with little or no relative shear movement.

According to investigations by Skempton and Petley (1967) and Skempton *et al.* (1969), the strength along principal slip surfaces is at or near the residual. Along minor shears the strength may be appreciably higher than residual. On joint surfaces c' is small and ϕ' is approximately the same as at peak for intact clay indicating that the fracture which produced the joint virtually eliminated the cohesion but reduced the friction angle ϕ' by only a very slight amount. Movements of not more than 5 mm, however, are sufficient to bring the strength along the joint to the residual and to polish the joint.

The influence of the initial stress state

It is well established from both field and laboratory studies that in over-consolidated clays and clay shales the in situ horizontal stress may exceed the overburden pressure and in shallow depths the ratio of horizontal effective stress to vertical effective stress (K_0) may become large enough so that the soil approaches a state of passive failure. Skempton (1961) used an indirect method to estimate the in situ stresses in the London clay at Bradwell and found that K_0 varied considerably with depth, increasing from a value of about 1.5 at a depth of 30 m to a value of 2.5 at 3 m.

Using finite element methods, Duncan and Dunlop (1969) examined the effect of initial lateral stresses in excavated slopes. The soil was treated as a homogeneous linear elastic material. The process of excavation was simulated analytically in one step and the distribution of shear stresses was calculated. Two different soils were examined: a soil with $K_0 = 0.81$, representative of a normally-consolidated clay, and a soil with $K_0 = 1.60$ (over-consolidated clay). The value of K_0 was found to influence greatly the magnitudes of the post-excavation shear stresses, which were much greater in the over-consolidated soil. The maximum shear stress in the region of the toe of the slope from which progressive failure was most likely to be initiated, was about ten times greater for the over-consolidated soil. The higher stresses were large enough so that failure could be expected and Duncan and Dunlop concluded that the high initial horizontal stresses in heavily over-consolidated clays and shales increase the likelihood of progressive failure in these materials. Lo and Lee (1972) also showed the crucial dependence on the in situ effective stresses of analytical models incorporating strain-softening soil behaviour.

Thickness of shear zone

The displacement required to reach failure in slopes mainly depends on the strain which corresponds to the peak strength and on the thickness of failure zone. For a relatively thin failure zone the total movement before failure will be small whereas the required movement before failure occurs will be appreciably larger if the failure zone is relatively thick.

By analysing the measurements of horizontal movements obtained by a slope indicator, Gould (1960) found that failures in landslides in over-consolidated clays in the California coast region occurred within a narrow zone of 6 mm to 20 mm in thickness. In the landslide of Jackfield, England, described by Henkel and Skempton (1955), the failure zone was approximately 50 mm thick. The water content of the clay in this zone was 10% greater than in the adjacent material outside the failure zone. Skempton and Petley (1967) have observed, in a large landslide in stiff fissured clay at Guildford, England, that the shear zone had a width of about 6 mm which contained numerous minor shears. The actual slip surface consisted of a band about 50 μm wide in which the particles were strongly orientated. At Walton's Wood, England, they observed that the shear zone had a width of about 20 mm and the particles were strongly orientated within a band about 20–30 μm wide. The increase in water content in the shear zone was about 3%.

Mechanism of progressive failure

In a field failure, the average shear strength is the average value of the strengths of all the elements around the slip surface. This strength will lie between the peak and residual strengths. Skempton (1964) has compared the average (or operational or back-analysed) shear strength ($\bar{s}$) occurring at failure of several natural slopes and cuttings to the peak (s_f) and residual (s_r) shear strengths of specimens from the failure zone (surface) corresponding to the average effective normal stress, and defined the residual factor R by the equation

$$R = \frac{s_f - \bar{s}}{s_f - s_r} \tag{3.5}$$

If the average field shear strength equals the residual strength, $R = 1.0$; R is zero if the average field shear strength equals the peak shear strength. Skempton found a residual factor $R = 0.08$ for a landslide in a natural slope at Selset in a uniform non-fissured and unweathered clay, indicating that the average strength mobilized along the total failure surface was very close to the peak strength; and $R = 1$ for two landslides in natural slopes consisting of fissured, jointed and weathered clay (London clay and Coalport Beds respectively) indicating that the average strength mobilized at failure was very close to the residual value.

It is suggested that fissures and joints, apart from their weakening effect in the soil mass, act as stress concentrators which can overstress locally the soil beyond the peak strength and hence a progressive failure may be initiated. Although a slip may occur before the residual value is reached

73

everywhere within the mass, continued sliding will cause the average strength to decrease toward that limiting value.

In clays without fissures or joints, however, the post-peak reduction is very small, or even negligible. Compacted clay fills as used in embankments and earth dams may belong in this category. If a failure has already taken place any subsequent movement on the existing slip surface will be controlled by the residual strength, no matter what type of clay is involved.

Bjerrum (1967) has suggested a mechanism of progressive failure which is not associated with the presence of fissures in the clay. During the process of drained loading and after long periods of time, strain energy is stored in the soil mass. Depending upon the nature of the soil the strain energy may be stored or released upon unloading. If weak diagenetic bonds have been developed, the stored energy is soon released after unloading. If the bonds are strong the strain energy can only be released if the bonds are destroyed as a result of weathering during long periods of time. Bjerrum classified diagenetic bonds as weak, strong or permanent.

The rate of release of strain energy upon unloading is slower in soils with strong bonds, and if the bonds are permanent the energy may never be released. As the strain energy is released by the disintegration of diagenetic bonds, due to the process of weathering, progressive failure of a soil mass is initiated and continues retrogressively from the face of the slope. Bjerrum (1967) proposed a classification, shown in Table 3.2, of over-consolidated clays and shales on the basis of the likelihood of progressive failure.

It can be seen from the classification in Table 3.2 that the danger of progressive failure is greater in the case of over-consolidated plastic clays possessing strong diagenetic bonds (e.g. shales) which have been subjected to weathering and the gradual release of the strain energy in their bonds.

Consequently, heavily consolidated plastic clays and shales have initially large lateral stresses and show a high tendency for lateral expansion.

Table 3.2 Relative danger of progressive failure, after Bjerrum (1967)

Soil	Relative danger of progressive failure
Over-consolidated plastic clay with weak bonds	
Unweathered	High
Weathered	High
Over-consolidated plastic clay with strong bonds	
Unweathered	Low
Weathered	Very high
Over-consolidated clay with low plasticity	Very low

This could result in stress concentrations at the toe of an excavation or cut, local shear failure and gradual development of a continuous sliding surface.

Based on studies of a great number of case histories on over-consolidated clays, James (1971) emphasized the necessity of relatively large deformations in the field to produce progressive failure and the reduction of strength to near residual conditions. As a result, James suggested that a cutting in over-consolidated clay or clay shale designed with strength parameters $\phi' = \phi'_{peak}$ and $c' = 0$ can ensure long term stability against first time failures and will not be subject to progressive failure in the majority of cases encountered in practice. This suggestion is in accordance with observations made by Henkel and Skempton (1955), Skempton and Delory (1957) and Skempton (1970).

As emphasized by Peck (1967) and Bishop (1971), a complete understanding of the problem of progressive failure would require a finite element solution for a strain-softening material. Lo and Lee (1973a) have presented a finite element solution for the determination of stresses and displacements in slopes of strain-softening soils. The various factors that influence the extent and propagation of the overstressed zone were investigated and typical results obtained with slope geometry and soil properties commonly encountered in practice indicated that the extent of the overstressed zone defined by the residual factor increases with the inclination and height of the slope and with the magnitude of the in situ stress as defined by the coefficient of earth pressure at rest.

Role of case records in assessing field strength

Reference should be made to Chapter 5, Classic case records of slope failures, where the assessment of field strength is dealt with in considerable detail.

Summary of main points

(a)   The drained strength (as embodied in the effective stress parameters c' and ϕ') is appropriate to any stability condition where the pore pressures and hence effective stresses are known. For cut slopes and natural slopes, pore pressures are normally only known in the long term when pore pressures have come into a steady state determined by the boundary conditions. Critical stability of slopes is in the long term where the factor of safety is a minimum.

(b)   The value of drained strength (as embodied in the effective stress parameters c' and ϕ') will normally be different from test type to test type and from values back-analysed from a field failure (e.g. see Fig. 5.20).

75

(c)   The drained strength is not appropriate to short term stability conditions where the pore pressures are not usually known, and should not be used in assessing the stability of temporary works such as excavations. An exception to this rule is the case of rapid drawdown of partially submerged slopes where the pore pressures are known in the short term (see Chapter 4, Classic methods of slope stability analysis).

CHAPTER FOUR
Classic methods of slope stability analysis

Overview

Historically, the stability of slopes has been assessed by methods of calculation 'by hand' – before the advent of computers, spreadsheets and bespoke software packages. The most widely accepted of these classic methods are described below. Principally, these are by Bishop (1955) and Janbu (1957). The chore of calculation 'by hand' was speeded up by the introduction of stability coefficients and design charts by Bishop and Morgenstern (1960) and Hoek and Bray (1977). Computer programs have automated the methods of Bishop and Janbu and others, *but not replaced them*. On the contrary, these classic methods continue, except now they are in computer-automated form so that numerous trial slip and slope geometries can be rapidly evaluated.

Wider accessibility to ('main frame') computing in the late 1950s to the early 1960s gave rise to methods of analysis that could only be solved by computer-automated numerical methods, principally by Little and Price (1958) – at that time, this was thought to be the most complicated program ever written for a computer of British origin (Bromhead, 1985) – and Morgenstern and Price (1965, 1967). Now commercial computer programs for slope stability analysis are widely available and some can link with other programs for stress distribution and for seepage (e.g. see SLOPE/ W Student Edition CD supplied with this book).

While classic methods have been automated by computer and new computer methods have been devised, it is still the responsibility of the design engineer to ensure that ground properties (inputs) and design predictions (outputs) pass the sanity test! Indeed, it is the explicit duty of the engineer to check the outputs of computer programs by 'hand' calculation, i.e. in an immediately obvious and transparent way without the aid of a 'black box' (or mysterious) computer program – although perhaps a computer spreadsheet may be permitted! These checks 'by hand' will use the same classic methods described below, many of which are still used to this day by commercial software packages.

Chronology

> 'Following the success of Bishop's theory [of slope stability analysis], and with good results obtainable from the routine method [Bishop-simplified], efforts were made to develop a similar theory applicable to slides with any shape of slip surface. The chronology of the early developments is confused, but appears to have been as follows. Bishop's paper was presented at a conference a year before it appeared in print. Separate researches led Janbu and Kenney to the same result (Kenney working under Bishop at Imperial College). Janbu published first, in 1955, but in an incorrect form, and Kenney's thesis appeared the following year. Subsequently, Janbu republished a corrected form of the equations, but in a form relatively inaccessible to English readers. In the meantime, Bishop had persuaded Price, who had been one of the authors of the first stability analysis computer program (see Little and Price, 1958), to try to program Kenney's equations for non-circular slips. It was then found that the basic equations could give rise to numerical problems when evaluated to high precision that would not appear at slide-rule accuracy [i.e. to within about 5% at best]: he and Morgenstern (Morgenstern and Price, 1965, 1967) then developed a more sophisticated method (again at Imperial College). This time, they were secure in the knowledge that complexity in computation was no longer a bar to widespread use of a method because of the growing availability of computers. Janbu developed his method further, and published his generalized procedure of slices in 1973, and a number of other methods also appeared in print throughout the late 1960s and 1970s.'
> **Bromhead (1985)**

There follows a summary in lecture-note form of the methods of slope stability analysis common throughout the geotechnical literature.

Simple cases

Dry slope in sand

Referring to Fig. 4.1, assume that the stresses on the vertical sides of the slice are equal and opposite. Resolving forces parallel to the slope give the disturbing force $W \sin \beta$. Resolving forces normal to the slope give the resultant normal force $W \cos \beta$.

The factor of safety against sliding is then

$$F = \frac{\text{Resisting force}}{\text{Disturbing force}} = \frac{W \cos \beta \tan \phi'}{W \sin \beta} = \frac{\tan \phi'}{\tan \beta}$$

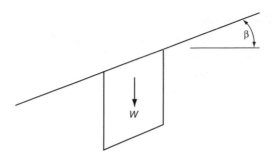

Fig. 4.1 Definition sketch for a dry slope in sand

whence, for limiting equilibrium,

$$\beta_{\max} = \phi'$$

Fully submerged slope in sand

Referring to Fig. 4.2, consider a slice of width a and depth d. The buoyant weight of the slice is:

$$W' = ad\gamma' \qquad (\gamma' = \gamma - \gamma_w)$$

The effective stress normal to the base of the slice is:

$$\sigma'_n = ad\gamma' \cos \beta \frac{\cos \beta}{a}$$

The shear stress along the sloping base of the slice is:

$$\tau = ad\gamma' \sin \beta \frac{\cos \beta}{a}$$

The factor of safety against sliding may be derived as

$$F = \frac{d\gamma' \cos^2 \beta \tan \phi'}{d\gamma \sin \beta \cos \beta}, \quad \text{whence}$$

$$F = \frac{\tan \phi'}{\tan \beta}$$

Note that the expression for factor of safety for the fully submerged slope is identical to that for a dry slope.

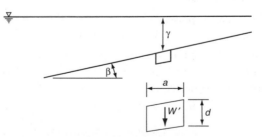

Fig. 4.2 Definition sketch for a fully submerged slope in sand

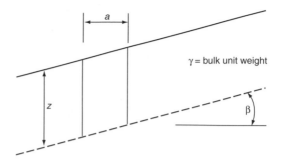

Fig. 4.3 Definition sketch for a semi-infinite slope, $\phi = 0$ analysis

Semi-infinite slope: $\phi = 0$ analysis

Referring to Fig. 4.3, consider a slice of width b and depth z. The disturbing force is $\gamma z b \sin \beta$. The restoring force is

$$s_u \frac{a}{\cos \beta}$$

The factor of safety against sliding is therefore

$$F = \frac{s_u}{\gamma z \cos \beta \sin \beta}$$

WORKED EXAMPLE

Consider a natural slope in over-consolidated London clay near limiting equilibrium.

Take $z = 7\,\text{m}$, $s_u = 100\,\text{kPa}$, $\gamma = 20\,\text{kN/m}^3$, $\beta = 10°$ which are typical values. Then

$$F = \frac{s_u}{\gamma z \cos \beta \sin \beta}$$

$$= \frac{100}{20 \times 7 \times \cos 10° \sin 10°}$$

$$= 4.2$$

This is a nonsense for London clay slopes which can fail at this inclination! This calculation serves to illustrate how inappropriate a $\phi = 0$ analysis is when applied to natural slopes.

Now consider a natural slope in soft normally consolidated Norwegian clay.

Take $z = 3\,\text{m}$, $s_u = 10\,\text{kPa}$, $\gamma = 18\,\text{kN/m}^3$, $\beta = 15°$, as typical values for a slope about to fail. Now $F = 0.74$.

This calculation again confirms that a $\phi = 0$ analysis should not be applied to assess the stability of a natural slope whether in an over-consolidated fissured clay or a normally-consolidated intact clay.

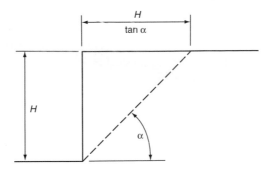

Fig. 4.4 Definition sketch for a vertical bank, $\phi = 0$ analysis, no tension crack

$\phi = 0$ analysis for vertical cut: no tension crack

In Fig. 4.4, the disturbing force is $1/2\gamma H^2 \sin \alpha / \tan \alpha$. The restoring force is $s_u H / \sin \alpha$.

Taking the critical angle of sliding as $\alpha_{crit} = 45°$, then $H_c = 4s_u/\gamma$. Alternatively, assuming a slip circle, the corresponding result would be $H_c = 3.85 s_u/\gamma$. These predictions would be unsafe in practice because real clay soils are weak in tension.

$\phi = 0$ analysis for vertical cut: with tension crack

Refer to Fig. 4.5. Assume the depth of tension crack $D = H_s/2$, following Terzaghi (1943).

The disturbing force is

$$\gamma \frac{3}{4} H_c \times \frac{H_c}{2} \times \sin 45°$$

The restoring force is

$$s_u \frac{H_c}{2} \times \frac{1}{\sin 45°}$$

At failure when $F = 1$, the critical height is $H_c = 2.67 s_u/\gamma$. This gives a more realistic prediction for the height of a temporary vertical cut in clay.

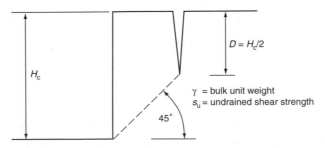

Fig. 4.5 Definition sketch for a vertical bank, $\phi = 0$ analysis, with tension crack

WORKED EXAMPLE

Consider the different approaches to analysing the short term stability of a vertical bank illustrated in Fig. 4.6.

Case (a) may be summarized as follows.

60°: Disturbing $= \frac{1}{2} \times 10 \times 5.774 \times 20 \times \sin 60° = 500 \, kN$
Restoring $= 60 \times 11.55$ $= 693 \, kN$
F $= 693/500$ $= 1.30$

45°: Disturbing $= \frac{1}{2} \times 10 \times 10 \times 20 \times \sin 45°$ $= 707 \, kN$
Restoring $= 14.14 \times 60$ $= 848 \, kN$
F $= 848/707$ $= 1.2$

30°: Disturbing $= \frac{1}{2} \times 10 \times 17.3 \times 20 \times \sin 30°$ $= 865 \, kN$
Restoring $= 60 \times 20$ $= 1200 \, kN$
F $= 1200/865$ $= 1.39$

For case (b) we have:
For a dry slope:
Disturbing $= 5 \times 7.5 \times 20 \times \sin 45°$ $= 530 \, kN$
Restoring $= 7.07 \times 60$ $= 4.24 \, kN$
F_{dry} $= 424/530$ $= 0.80$

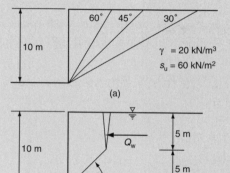

(a)

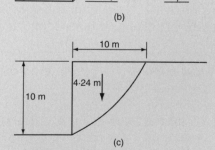

(b)

(c)

Fig. 4.6 Definition sketches for worked example (a) assuming different angles of slip planes, (b) taking a 45° slip plane and a water-filled tension crack, (c) assuming slip plane is the quadrant of a circle

For a wet slope (tension crack filled with water)

Water force	$= \frac{1}{2} \times 5 \times 5 \times 10$	$= 125\,\text{kN}$
Extra disturbing	$= 125 \times \sin 45°$	$= 88.4\,\text{kN}$
F_{wet}	$= 424/530 + 88.4$	$= 0.69$

It can be seen that the filling of the tension crack with water has a significant effect on stability.

For case (c) we have (noting the centroid of a quadrant of a circle radius R is located $4R/3\pi$ from the centre of the circle):

Area	$= \pi \times 10^2/4$	$= 78.5\,\text{m}^2$
Perimeter	$= \pi \times 10/2$	$= 15.71\,\text{m}$

Taking moments gives:

Disturbing	$= 78.5 \times 20 \times 4.24$	$= 6656.8\,\text{kN/m}$
Restoring	$= 15.71 \times 60 \times 10$	$= 9424\,\text{kN/m}$
F	$= 9424/6656.8$	$= 1.42$

In summary we have:

case (a) $F = 1.39$ (60°), 1.20 (45°), 1.39 (30°).
case (b) $F = 0.80$ (dry), 0.69 (wet)
case (c) $F = 1.42$

Semi-infinite slope: effective stress analysis

Refer to Fig. 4.7. For flow parallel to slope, an effective stress analysis gives the following.

Soil strength is $s = c' + \sigma' \tan \phi'$.

Shear stress $= \tau = \gamma z \sin \beta (1/\cos \beta)$

$$= \gamma z \sin \beta \cos \beta$$

Normal stress $= \sigma = \gamma z \cos \beta (1/\cos \beta)$

$$= \gamma z \cos^2 \beta$$

Pore water pressure $= u = m z \gamma_w \cos \beta \cos \beta$

$$= m z \gamma_w \cos^2 \beta$$

The normal effective stress on the base of the slice is:

$$\sigma' = (\gamma - m\gamma_w)z \cos^2 \beta$$

Whence the factor of safety against sliding is

$$F = \frac{c' + (\gamma - m\gamma_w)z \cos^2 \beta \tan \phi'}{\gamma z \sin \beta \cos \beta}$$

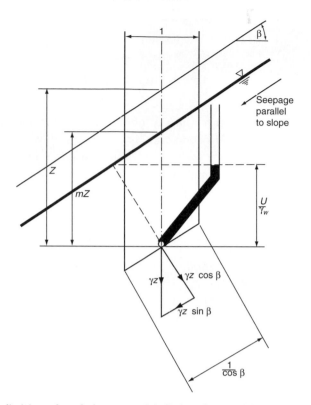

Fig. 4.7 Definition sketch for a semi-infinite slope with seepage parallel to the slope, effective stress analysis

If $c' = 0$, the critical slope is given by

$$\tan \beta_c = \frac{(\gamma - m\gamma_w) \tan \phi'}{\gamma}$$

If in addition $m = 1$, i.e. the water table is at the ground surface, then

$$\tan \beta_c = \frac{\gamma'}{\gamma} \tan \phi'$$

If flow is horizontal, however, then

$$u = \gamma_w z \quad \text{and}$$

$$F = \left(1 - \frac{\gamma_w}{\gamma \cos^2 \beta}\right) \frac{\tan \phi'}{\tan \beta}$$

compared with

$$F = \left(1 - \frac{\gamma_w}{\gamma}\right) \frac{\tan \phi'}{\tan \beta}$$

for parallel flow and $c' = 0$.

WORKED EXAMPLE

Referring to Fig. 4.8, take $\gamma = 19\,\mathrm{kN/m^3}$, $c' = 2\,\mathrm{kPa}$. Assume the slope is on the point of sliding, i.e. $F = 1$. What is the value of the angle of shearing resistance with respect to effective stress, ϕ'?

Now the factor of safety against sliding is:

$$F = \frac{c' + (\gamma - m\gamma_w)z\cos^2\beta\tan\phi'}{\gamma z\sin\beta\cos\beta}, \quad \text{whence}$$

$$\tan\phi' = \frac{27.045 - 2}{69.838}$$

$$= 0.3586$$

$$\phi' = 19.7°$$

If $c' = 0$, $\phi' = 21.2°$ and if z is infinite, $\phi' = 24.2°$.

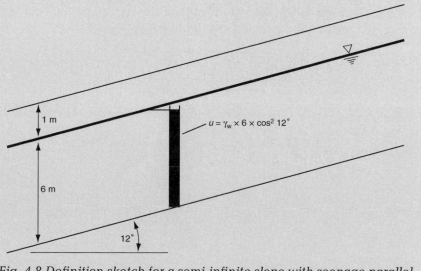

$u = \gamma_w \times 6 \times \cos^2 12°$

1 m

6 m

12°

Fig. 4.8 Definition sketch for a semi-infinite slope with seepage parallel to the slope, worked example

Anisotropy of permeability

The effect of anisotropy with respect to permeability on the stability of a long slope in a frictional soil (i.e. $c' = 0$), for a condition of seepage parallel to the slope surface and the water table at the ground surface may be assessed using the equation of Telling (1988):

$$F = [1 - (\gamma_w/\gamma)\beta\sec^2 i]\tan\phi'/\tan i$$

wherein $\beta = 1/(1 + k^2 \tan^2 i)$; $k = \sqrt{k_H/k_V}$, where k_H is the horizontal permeability, and k_V is the vertical permeability; i is the slope angle and γ_w, γ and ϕ' are as usually defined.

By way of illustration, the variation of F with k is as follows for $i = 14°$ and $\phi' = 27°$:

$k = 1$	$F = 1.02$
$k = 10$	$F = 1.56$
$k = 20$	$F = 1.73$

Slip circle analysis: $\phi = 0$ method

Note that the $\phi = 0$ analysis is appropriate only to temporary short term excavations where the undrained or short term strength is assumed to be controlling stability. The slip circle $\phi = 0$ method is outlined in Fig. 4.9 where the slope is divided into convenient slices and the factor of safety is the ratio of the sum of the restoring moments to the sum of the disturbing moments. So, referring to Fig. 4.9,

$$F = \frac{R \sum s_u l}{\sum Wx}$$

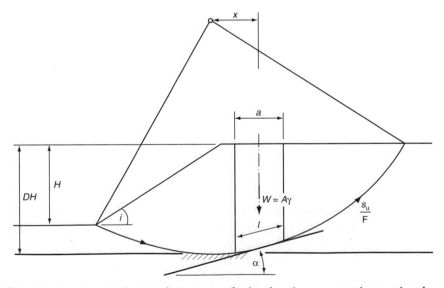

Fig. 4.9 Key sketch for total stress analysis of a slope assuming a circular slip surface, method of slices

Noting that $x = R\sin\alpha$, $l = a\sec\alpha$ and $W = pa$ gives

$$F = \frac{\sum s_u \sec\alpha}{\sum p \sin\alpha}$$

for equal slice widths, which is a convenient form for hand calculation. Solutions are available for simple geometry and for:

- constant undrained shear strength (Taylor, 1937)

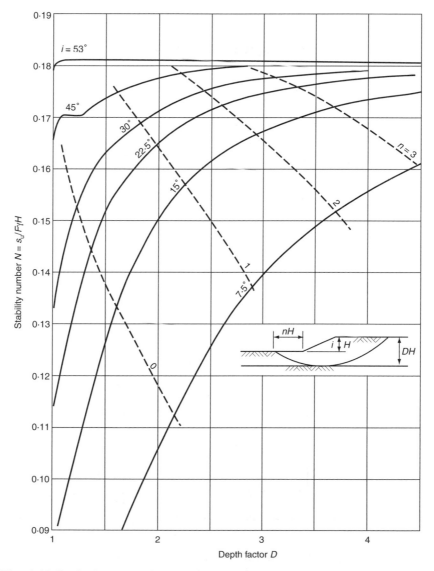

Fig. 4.10 Taylor's curves for stability numbers

- shear strength increasing linearly with depth from zero at ground surface (Gibson and Morgenstern, 1962).

Taylor's solution

This is for simple slopes with no seepage and no tension cracks. It is derived from slip circle analysis with s_u constant with depth. Referring to Fig. 4.10, solutions are expressed in terms of a 'stability number' N where:

$$N = \frac{s_u}{\gamma H} \quad \text{for } F = 1 \text{ whence in practice}$$

$$F = \frac{s_u}{N \gamma H} \quad \text{for slope angles less than } 54°$$

It is advisable to use this method as a first approximation for total stress ($\phi = 0$) case. Stability numbers for slope angles greater than 54° are given in Table 4.2.

Notes on Taylor's charts for $\phi = 0$ case

(1) for $\beta > 54°$, critical circles are toe circles and $H_c = f(\beta)$
(2) for $\beta < 54°$, critical circles are of great depth and H_c is independent of β
(3) for $\beta < 54°$ and $D \neq \infty$, $H_c = f(\beta)$.

WORKED EXAMPLES

(1) Take $h = 10\,\text{m}$, $s_u = 50\,\text{kPa}$, $d = \infty$, $\gamma = 20\,\text{kN/m}^3$.
For $\beta < 54°$, $s_u/\gamma H = 0.18$

$$F = \frac{50}{0.18 \times 20 \times 10} = 1.38$$

For $\beta = 90°$, $s_u/\gamma H = 0.26$

$$F = \frac{50}{0.26 \times 20 \times 10} = 1.96$$

(2) Take $s_u = 30\,\text{kPa}$, $\gamma = 18\,\text{kN/m}^3$, $F = 1.3$.
Determine H_{max} for $\beta = 45°$ and 90° and for $D = 1$ and ∞. The solution is set out in Table 4.1.

$$H_{max} = \frac{s_u}{\gamma N F}$$

Table 4.1 Values of $s_u/\gamma H = N$

	$D = 1$	$D = \infty$
$\beta = 45°$	0.165	0.181
H_{max}	7.77 m	7.08 m
$\beta = 90°$	0.26	0.26
H_{max}	4.93 m	4.93 m

Note: using a plane failure surface for $\beta = 90°$ gives

$$H_{max} = 4\frac{s_u}{\gamma} = \frac{s_u}{0.25\gamma}$$

i.e. $H_{max} = \dfrac{30}{0.25 \times 18 \times 1.3} = 5.12\,\text{m}$

Stability numbers for a $\phi = 0$ analysis are presented in Table 4.2.
Note: $N = s_u/\gamma H$

β = slope angle

Assumptions are:

- constant s_u
- no tension cracks
- slip circle analysis.

Table 4.2 Stability numbers for $\phi = 0$ analysis

β	Plane	Circle	
90°	0.250	0.261	⎫
75°	0.192	0.219	⎬ toe circles are critical circles
60°	0.144	0.191	⎭
45°	0.104	0.181 (0.170)* (0.166)**	
30°	0.067	0.181 (0.156)* (0.134)**	
15°	0.033	0.181 (0.145)* (0.088)**	

*toe circles, **toe circles and $D = 1$

Gibson and Morgenstern (1962)

The assumptions in Fig. 4.11 are:

- total stress analysis
- slip circle
- no tension crack

89

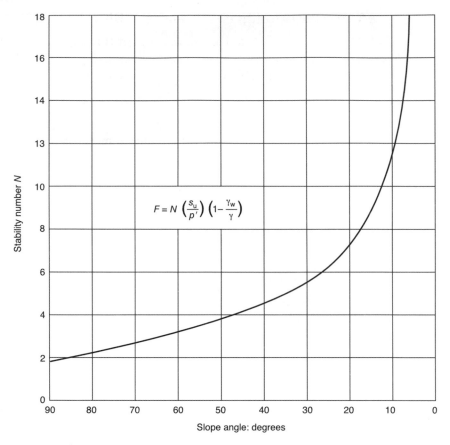

Within the figure:
$$F = N \left(\frac{s_u}{p'}\right)\left(1 - \frac{\gamma_w}{\gamma}\right)$$

Axis labels: Stability number N (vertical), Slope angle: degrees (horizontal)

Fig. 4.11 Gibson and Morgenstern's curves for stability numbers

- s_u increasing linearly with depth from zero at ground surface approximating to a normally consolidated clay
- F depends *not* on H, but on slope angle, s_u/p', and on γ, the unit weight. F is given by

$$F = N\frac{s_u}{\gamma z} = N\frac{s_u}{p'}\left(1 - \frac{\gamma_w}{\gamma}\right)$$

WORKED EXAMPLE

If $s_u/p' = 0.3$, $\gamma = 19\,\text{kN/m}^3$, what slope would be required to give $F = 1.3$ for an excavated cutting?

Substituting in the equation above gives $N = 9.1485$ and hence the required slope is $12°$.

Effective stress slip circle analysis: Bishop (1955)

The slip circle failure mechanism is shown in Fig. 4.12(a) and the forces acting on an individual slice are shown in Fig. 4.12(b). The polygon of forces is given in Fig. 4.12(c). Key points to note are as follows.

- The factor of safety

$$F = \frac{\text{available strength}}{\text{strength required for equilibrium}}$$

and thus the mobilized strength is given by

$$s = \frac{c'}{F} + (\sigma_n - u)\frac{\tan \phi'}{F}.$$

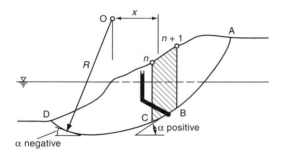

(a)

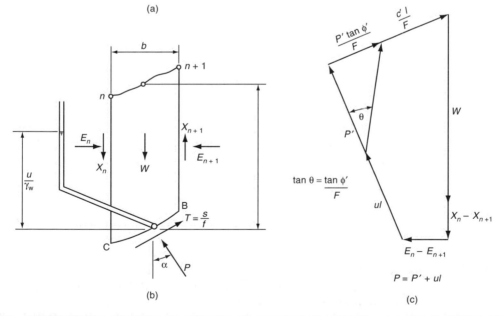

(b)

(c)

Fig. 4.12 Definition sketches for effective stress analysis of a slope, method of slices, (a) slip circle failure mechanism, (b) forces acting on an individual slice, (c) force polygon

- A generalized analytical solution is not possible and a numerical solution is required (Bishop and Morgenstern, 1960).
- To estimate values of σ_n at each point on failure surface we use the slices method of Fellenius (1936), first published at the Large Dams Congress.

Referring to Fig. 4.12, the analysis then proceeds as follows. The mobilized strength is

$$s = \frac{c'}{F} + (\sigma_n - u)\frac{\tan \phi'}{F} \tag{4.1}$$

The total normal stress at the base of a slice is

$$\sigma_n = P/l \tag{4.2}$$

whence $s = \dfrac{c'}{F} + [(P/l) - u]\dfrac{\tan \phi'}{F}$ \hfill (4.3)

The shear force, s, acting on base of slice is sl. Taking moments about O gives, for limiting equilibrium

$$\sum Wx = \sum SR = \sum slR \tag{4.4}$$

from equation (4.3)

$$\sum Wx = R \sum \frac{c'l}{F} + (P - ul)\frac{\tan \phi'}{F}$$

i.e. $F = \dfrac{R}{\sum Wx} \sum [c'l + (P - ul)\tan \phi']$ \hfill (4.5)

to obtain P resolve normal to slip surface

$$P = (W + X_n - X_{n+1})\cos \alpha - (E_n - E_{n+1})\sin \alpha \tag{4.6}$$

Inserting equation (4.6) into equation (4.5) gives

$$F = \frac{R}{\sum Wx} \sum \{c'l + \tan \phi'(W \cos \alpha - ul)$$

$$+ \tan \phi'[(X_n = X_{n+1})\cos \alpha - (E_n - E_{n+1})\sin \alpha]\} \tag{4.7}$$

Since there are no external forces on the slope, it follows that the sum of the internal forces must be zero, i.e.

$$\sum (X_n - X_{n+1}) = 0 \tag{4.8a}$$

$$\sum (E_n - E_{n+1}) = 0 \tag{4.8b}$$

but the terms in $X_n \ldots$ and $E_n \ldots$ do not disappear; they only do so if $\phi' = \text{constant}$ and $\alpha = \text{constant}$, i.e. we have sliding on a plane.

One proposal, the United States Bureau of Reclamation (USBR) solution, is to assume that

$$\sum \tan \phi'[(X_n - X_{n+1})\cos \alpha - (E_n - E_{n+1})\sin \alpha] = 0$$

Then, putting $x = R \sin \alpha$ gives

$$F = \frac{1}{\sum W \sin \alpha} \sum [c'l + \tan \phi'(W \cos \alpha - ul)] \tag{4.9}$$

This form of solution gives results on the conservative side, particularly for deep circles.

In simple terms, we have resolved the total weight W normal to the slice and then subtracted the force due to pore water pressure ul, see equation (4.9).

Following Taylor, we could consider the buoyant weight of the slice to be $W - ub$ and resolve this, i.e.

$$(W - ub) \cos \alpha = W \cos \alpha - u \cos \alpha \frac{l}{\sec \alpha}$$

$$= W \cos \alpha - ul \cos^2 \alpha$$

and obtain

$$F = \frac{1}{\sum W \sin \alpha} \sum [c'l + \tan \phi'(W \cos \alpha - ul \cos^2 \alpha)] \tag{4.10}$$

To avoid these errors, we return to equation (4.5) above. Put $(P - ul) = P'$, and resolve vertically to eliminate the E_n terms as follows:

$$W + X_n - X_{n+1} = ul \cos \alpha + P' \cos \alpha + P' \frac{\tan \phi'}{F} \sin \alpha + \frac{c'}{F} l \sin \alpha$$

then

$$P' = \frac{W + X_n - X_{n+1} - l[u \cos \alpha + (c' \sin \alpha/F)]}{[\cos \alpha + (\tan \phi' \sin \alpha)/F]} \tag{4.11}$$

Substituting this in equation (4.5) and putting $l = b \sec \alpha$, $x = R \sin \alpha$ yields:

$$F = \frac{1}{\sum W \sin \alpha} \sum \{c'b + \tan \phi'[W - ub + (X_n - X_{n+1})]\}$$

$$\times \frac{\sec \alpha}{1 + [(\tan \phi' \tan \alpha)/F]} \tag{4.12}$$

This converges rapidly.

With sufficient accuracy for most practical purposes, we can take $X_n - X_{n+1} = 0$ and obtain the Bishop-simplified solution (Bishop 1955).

For the special case of equal slice widths, equation (4.12) can be written

$$F = \frac{[\sum c' + (p - u) \tan \phi']/m_\alpha}{\sum p \sin \alpha} \tag{4.13}$$

where

$$m_\alpha = \cos \alpha \left(1 + \frac{\tan \alpha \tan \phi'}{F}\right) \tag{4.14}$$

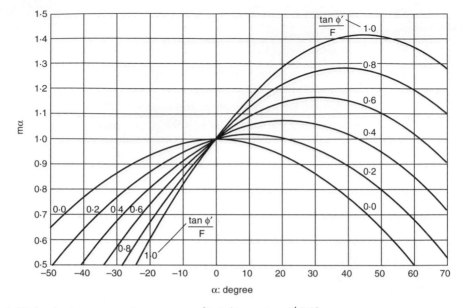

Fig. 4.13 Janbu's curves of $m_\alpha = \cos\alpha[1 + (\tan\alpha\tan\phi'/F)]$

and p is the total vertical stress at base of slice. Values of m_α are given in Fig. 4.13. This form of solution is well suited to hand calculations.

There are therefore three forms of solution which have been in common use. These are:

- $F = \dfrac{1}{\sum W \sin\alpha} \sum [c'l + \tan\phi'(W\cos\alpha - ul)]$ – the USBR solution

- $F = \dfrac{1}{\sum W \sin\alpha} \sum [c'l + \tan\phi'(W\cos\alpha - ul\cos^2\alpha)]$ – this is known as Taylor's solution (resolving buoyant weight) and gives values of F greater than those given by equation (4.9)

- $F = \dfrac{[\sum c' + (p - u)\tan\phi']/m_\alpha}{\sum p \sin\alpha}$ – this is the Bishop-simplified form given by Janbu, Bjerrum and Kjaernsli (1971).

A worked example using the Bishop-simplified form is given in Fig. 4.14. Two cases quoted by Bishop (1955) gave values of F as shown in Table 4.3.

According to Bishop (1955) the differences in F between Bishop-rigorous and Bishop-simplified methods are generally less than 1%.

Circular analyses compared

Whitman and Bailey (1967) considered four cases. The results are given in Table 4.4.

WORKED EXAMPLE

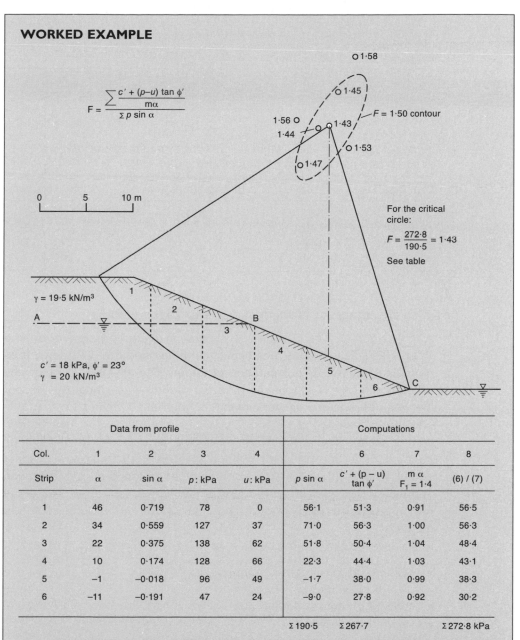

$$F = \frac{\sum \dfrac{c' + (p-u)\tan\phi'}{m\alpha}}{\sum p \sin\alpha}$$

○ 1·58

○ 1·45

1·56 ○

1·44

○ 1·43

○ 1·53

○ 1·47

F = 1·50 contour

For the critical circle:

$$F = \frac{272 \cdot 8}{190 \cdot 5} = 1 \cdot 43$$

See table

0   5   10 m

$\gamma = 19 \cdot 5$ kN/m³

A

1
2
3
B
4
5
6
C

$c' = 18$ kPa, $\phi' = 23°$
$\gamma = 20$ kN/m³

	Data from profile				Computations		
Col.	1	2	3	4	6	7	8
Strip	α	$\sin\alpha$	p : kPa	u : kPa	$p\sin\alpha$	$c' + (p-u)\tan\phi'$	$\dfrac{m\alpha}{F_1 = 1 \cdot 4}$	(6) / (7)
1	46	0·719	78	0	56·1	51·3	0·91	56·5
2	34	0·559	127	37	71·0	56·3	1·00	56·3
3	22	0·375	138	62	51·8	50·4	1·04	48·4
4	10	0·174	128	66	22·3	44·4	1·03	43·1
5	−1	−0·018	96	49	−1·7	38·0	0·99	38·3
6	−11	−0·191	47	24	−9·0	27·8	0·92	30·2
					Σ 190·5	Σ 267·7		Σ 272·8 kPa

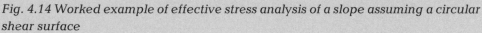

Fig. 4.14 Worked example of effective stress analysis of a slope assuming a circular shear surface

Table 4.3 Factors of safety found by the USBR, simplified and rigorous methods for two cases, after Bishop (1955)

Example	USBR	Bishop-simplified	Bishop-rigorous
1	1.38	1.53	1.60–1.61
2	1.5	1.84	1.92

Table 4.4 Factors of safety found by the USBR, simplified and rigorous methods for four cases, after Whitman and Bailey (1967)

Example	USBR	Bishop-simplified	Bishop-rigorous
1	1.49	1.61	1.58–1.62
2	1.10	1.33	1.24–1.26
3	0.66	0.70–0.82	0.73–0.78
4	1.14	2.00	2.01–2.03

Slip circle analysis: Morgenstern and Price (1965)

A more rigorous form of slip circle analysis is given by Morgenstern and Price (1965). This is not a 'hand calculation' method but is carried out by numerical means in a computer program. The method involves the solution of a pair of simultaneous partial differential equations – one for force equilibrium and one for moment equilibrium – using a two-variable Newton approximation method. It is widely regarded as the 'benchmark' against which other methods are measured, particularly calculations 'by hand'. This is because their method satisfies both force and moment equilibrium at the same time whereas other methods do not. Whitman and Bailey (1967) used the Morgenstern and Price method as a datum of measurement in their review and comparison of various computer methods.

The key elements of the method are:

- failure surface of general shape
- two-dimensionality
- force and moment equilibrium
- computer analysis
- based on assumptions interrelating the slice side forces X and E.

It can be seen that large errors can arise with the USBR method and we recommend therefore that use of the USBR method should be abandoned. A difficulty with the Bishop-simplified method can arise if

$$\left(1 + \tan \alpha \frac{\tan \phi'}{F}\right)$$

becomes negative or 0. This can occur if α is negative and $\tan\phi'/F$ is large. If

$$\left(1 + \tan\alpha\frac{\tan\phi'}{F}\right)$$

is less than 0.2 the use of other methods is recommended.

Stability coefficients for earth slopes

Bishop and Morgenstern (1960), using the Bishop-simplified method, have produced charts of stability coefficients from which a factor of safety for an effective stress analysis can be rapidly obtained. These charts are given in Appendix 1. The general solutions are based on the assumption that the pore pressure u at any point is a simple proportion r_u of the overburden pressure γh. This proportion is the pore pressure ratio $r_u = u/\gamma h$ and is regarded as being constant throughout the cross-section. This is called a homogeneous pore pressure distribution.

For a simple soil profile and specified shear strength parameters, the factor safety, F, varies linearly with the magnitude of the pore pressure expressed by the ratio r_u such that

$F = m - nr_u$

(see Fig. 4.15) where m and n are termed 'stability coefficients' for the particular slope and soil properties. The use of pore pressure ratio, r_u,

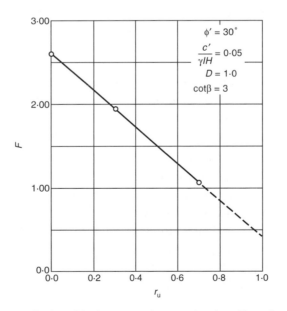

Fig. 4.15 Linear relationship between factor of safety F and pore pressure ratio r_u, after Bishop and Morgenstern (1960)

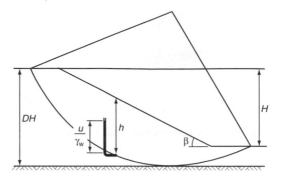

Fig. 4.16 Specification of parameters, after Bishop and Morgenstern (1960)

permits the results of stability analysis to be presented in dimensionless form.

There are seven variables (see Fig. 4.16):

- the height of the slope H
- the depth to a hard stratum DH
- the pore pressure ratio $r_u = u/\gamma h$
- the slope angle β
- the cohesion intercept with respect to effective stress c'
- the angle of shearing resistance with respect to effective stress ϕ'
- the bulk unit weight γ.
- for a given value of $c'/\gamma H$, F depends on $\cot \beta$, D, r_u and ϕ'. Three values of $c'/\gamma H$ have been used: 0, 0.025, 0.05
- slope inclination varies from 2:1 to 5:1
- ϕ' ranges from $10°$ to $40°$
- $D = 1.00, 1.25, 1.50$
- If $r_u > r_{ue}$, where r_{ue} is indicated by the dotted line on the charts, the critical circle is at greater depth
- use linear extrapolation in range $c'/\gamma H$ 0.05 to 0.10 (Fig. 4.17).

The values of the stability coefficients have been plotted against the cotangent of the slope angle (see Appendix 1, Figs. A1.1 to A1.8) for ϕ' varying between $10°$ and $40°$ with values of the dimensionless parameters $c'/\gamma H$ and D specified for each figure. The bold lines show values of m and n at intervals of $2.5°$. The broken lines are those of equal r_u (denoted by r_{ue}). To calculate the factor of safety of a section whose $c'/\gamma H$ lies within the range covered by these figures, it is necessary only to apply the equation $F = m - nr_u$, to determine the factor of safety of the two nearest values of $c'/\gamma H$ and then perform a linear interpolation between these values, for the specified value of $c'/\gamma H$.

For a given set of parameters (β, ϕ', $c'/\gamma H$), there is a value of the pore pressure ratio for which the factor of safety, when D equals 1.00, is the

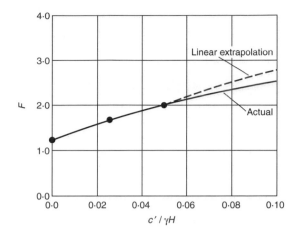

Fig. 4.17 Factor of safety – stability number relationship showing extrapolation in the range of c′/γH *0.05 to 0.10*

same as the factor of safety when D equals 1.25. If the design value of the pore pressure ratio is higher than r_{ue} for the given section and strength parameters, then the factor of safety with a depth factor $D = 1.25$ has a lower value than with D equal to 1.00. This argument can be extended to discern whether the factor of safety with D equal to 1.50 is more critical than with D equal to 1.25.

WORKED EXAMPLES

(1) Use stability coefficients to investigate the slope shown in Fig. 4.14. The data are $\cot \beta = 2.5{:}1$, $c' = 18\,\text{kPa}$, $\phi' = 23°$, $H = 12\,\text{m}$, $\gamma = 19.5{-}20.0\,\text{kN/m}^3$, say $19.8\,\text{kN/m}^3$. Hence, referring to Fig. 4.14:

$$r_u = 0, \frac{37}{127}, \frac{62}{138}, \frac{66}{128}, \frac{49}{96}, \frac{24}{47}$$

$$= 0, 0.29, 0.45, 0.52, 0.51, 0.51$$

for each slice, respectively, whence the average r_u (linear interpolation) $= 0.38$, or alternatively, the average r_u (using areas) $= 0.39$.

$$\frac{c'}{\gamma H} = \frac{1.8}{19.8 \times 12} = 0.076$$

Therefore do calculations for $c'/\gamma H = 0.025$ and 0.050 and use linear extrapolation (see Fig. 4.17).

For $d = 1.00$, $c'/\gamma H = 0.05$

$m = 1.79$, $n = 1.42$, $r_{ue} = 0.45$

Therefore deeper circles are not critical since $r_u = 0.39 < 0.45$

$F = 1.79 - 0.39 \times 1.42 = 1.24$

For $D = 1.00$, $c'/\gamma H = 0.025$

$m = 1.52$, $n = 1.35$, $r_{ue} = 0.8$

Therefore deeper circles are not critical

$F = 1.52 - 0.39 \times 1.35 = 0.99$

for $c'/\gamma H = 0.076$

$$F = 1.24 + \frac{0.076 - 0.050}{0.050 - 0.025} \times (1.24 - 0.99)$$

$$= 1.24 + 0.26$$

$$= 1.50$$

This can be compared with $F = 1.43$ by slip circle analysis (see Fig. 4.14).

(2) Consider the stability of an earth dam where $\gamma = 20 \, \text{kN/m}^3$, $c' = 23 \, \text{kPa}$, $\phi' = 25°$, $H = 35 \, \text{m}$, $r_u = 0.5$, side slopes 4 horizontal to 1 vertical on a rock foundation.

Now

$c'/\gamma H = 23/20 \times 35 = 0.033$

$c'/\gamma H$	m	n
0.025	2.40	2.15
0.050	2.75	2.20

For $\phi' = 25°$, slope $4:1$

for $c'/\gamma H = 0.025$, $F = 2.40 - 0.5 \times 2.15 = 1.325$
for $c'/\gamma H = 0.050$, $F = 2.75 - 0.5 \times 2.20 = 1.65$
for $c'/\gamma H = 0.033$, $F = 1.325 + (1.65 - 1.325) \times 8/25 = 1.43$ using linear interpolation.

(3) Check Lodalen (see Fig. 5.10) using Bishop and Morgenstern's stability coefficients.

$H = 18\,\text{m, say.}$

$c'/\gamma H = 9.8/18.7 \times 18 = 0.029$

$r_u = 0.4$

for $c'/\gamma H = 0.025, D = 1.0$

$\qquad m = 1.45, n = 1.40$

$\qquad F = 1.45 - 0.4 \times 1.40 = 0.89$

for $c'/\gamma H = 0.05, D = 1.0$

$\qquad m = 1.75, n = 1.50$

$\qquad F = 1.75 - 0.4 \times 1.50 = 1.15$

for $c'/\gamma H = 0.029$

$\qquad F = 1.89 + 4/25(1.15 - 0.89)$

$\qquad = 0.93$

Sevaldson (1956) found $F = 1.05$.

Non-homogeneous pore pressure ratio distribution

If the r_u value varies from point to point in a slope, it is necessary to use some average value of r_u when using stability coefficients. One way of doing this is illustrated in Fig. 4.18.

For section area 1:

$$\text{average: } r_u = \frac{h_1 r_{u1} + h_2 r_{u2} + h_2 r_{u2}}{h_1 + h_2 + h_3}$$

$$\text{overall average: } r_u = \frac{\sum_1^n A_n r_{un}}{\sum_1^n A_n}$$

where A_n is the area of any slice.

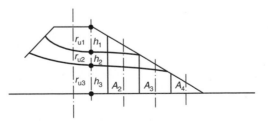

Fig. 4.18 Definition sketch for averaging non-homogeneous pore pressure ratio distribution, after Bishop and Morgenstern (1960)

101

Hoek and Bray design charts

Hoek and Bray (1982) presented a series of design charts. They considered five different ground water flow conditions as shown in Appendix 2, Fig. A2.1. The results of stability calculations based on Bishop's semi-rigorous slip circle method are presented in five design charts corresponding to the five ground water flow conditions. These charts are given in Appendix 2, Fig. A2.2 to Fig. A2.6.

Use of Hoek and Bray charts

The method uses the following steps:

- decide on groundwater conditions
- calculate $c'/\gamma H \tan \phi'$
- find this value on outer circular scale and follow radial line to inter-section with curve corresponding to slope angle
- read off $\tan \phi'/F$ or $c'/\gamma HF$ and calculate F (see also Figs 10.22 and 10.23).

WORKED EXAMPLE

(1) Check the example given in Fig. 4.14 using Hoek and Bray. Using chart number 3,

$$\frac{c'}{\gamma H \tan \phi'} = \frac{18}{19.8 \times 12 \times \tan 23°} = 0.1785$$

Now slope angle $= \arctan 1/2.5 = 21.8°$ whence $\tan \phi'/F = 0.3$ (from chart number 3) whence $F = \tan 23°/0.3 = 1.41$.

Compare this with $F = 1.43$ by Bishop-simplified slip circle analysis, and $F = 1.50$ by Bishop and Morgenstern's stability coefficients.

(2) Check Lodalen (see Fig. 5.10) using Hoek and Bray data. Using chart number 3,

$$c'/\gamma H \tan \phi' = 9.8/18.7 \times 18 \times \tan 27.1$$

$$= 0.0569$$

slope angle $= 26.6°$

$$\frac{\tan \phi'}{F} = 0.5 \quad \therefore \quad F = 1.02$$

$$\frac{c'}{\gamma HF} = 0.028 \quad \therefore \quad F = 1.04$$

say $F = 1.03$, which can compared with $F = 1.05$ as obtained for Lodalen by Sevaldson (1956).

Partially submerged slopes

Refer to Fig. 4.19. The net disturbing moment is that due to the soil less the moment about O of the mass of water acting on BAF. Imagine a section of water bounded by a free surface at EBG and outlined by EFGB as shown in Fig. 4.20. Now the moment of the mass of water FABE must equal the moment of the mass of water BAFG. But the weight of a mass of saturated soil less the weight of a mass of water occupying the same volume is the submerged unit weight. Therefore the resultant disturbing moment due to the mass of soil less the moment due to the water acting on BAF is given by using bulk unit weight above the level of free water surface and submerged unit weight below as depicted in Fig. 4.21.

Using the boundary EBG in this way implies nothing about magnitudes of pore water pressures inside slope – it is only used to give the statically equivalent disturbing moment, i.e. the statics of Fig. 4.21(a) of our partially submerged slope are equivalent to the representation in Fig. 4.21(b).

If we are doing an effective stress analysis, however, not only must the moments be correct, but also the effective stresses and hence the shear strength must be correct. To achieve this, we must operate with water pressures referred to the external water level as summarized in Fig. 4.22,

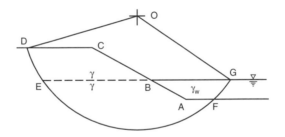

Fig. 4.19 Definition sketch for a partially submerged slope. The net disturbing moment is the moment due to the mass of soil (using saturated unit weight γ) less the restoring moment due to the water (unit weight γ_w)

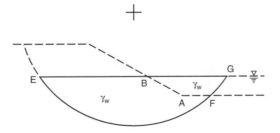

Fig. 4.20 Definition sketch showing a cylindrical surface beneath a free water level corresponding to the geometry of the partially submerged slope shown in Fig. 4.19

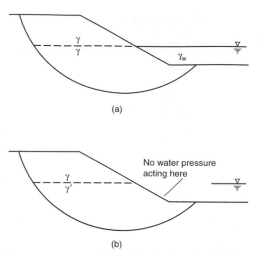

(a)

(b)

Fig. 4.21 The partially submerged slope shown in (a) is statically equivalent to the slope shown in (b) using submerged unit weight, γ', below the level line of submergence, and also measuring pore water pressures from this level line of submergence as datum

i.e. the pore pressure is measured from the external level water surface as datum. The bulk unit weight is used above the external water level and the submerged unit weight is used below the external water level. *Operating in this way, we automatically and correctly take into account the effect on stability of the external water pressure.*

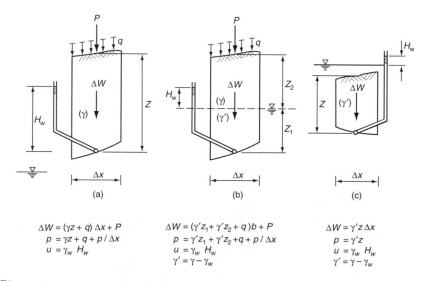

$$\Delta W = (\gamma z + q)\, \Delta x + P$$
$$p = \gamma z + q + p/\Delta x$$
$$u = \gamma_w\, H_w$$

$$\Delta W = (\gamma' z_1 + \gamma' z_2 + q)\, b + P$$
$$p = \gamma' z_1 + \gamma' z_2 + q + p/\Delta x$$
$$u = \gamma_w\, H_w$$
$$\gamma' = \gamma - \gamma_w$$

$$\Delta W = \gamma' z\, \Delta x$$
$$p = \gamma' z$$
$$u = \gamma_w\, H_w$$
$$\gamma' = \gamma - \gamma_w$$

Fig. 4.22 Summary sketch showing how vertical total stress p and pore water pressure u is calculated for each slice of a slope that is (a) dry, (b) partially submerged and (c) completely submerged

WORKED EXAMPLES

(1) To see how this works out in practice, consider the slope given in Fig. 4.23. For the slice shown, the vertical total stress at the base of the slice is:

$$p = 9 \times 20 = 180 \, \text{kPa}$$

The pore water pressure at the base of the slice is:

$$u = 5 \times 10 = 50 \, \text{kPa}$$

taking $\gamma_w = 10 \, \text{kN/m}^3$ (it is actually $9.81 \, \text{kN/m}^3$ for fresh water).

The vertical effective stress at the base of the slice is:

$$p - u = 130 \, \text{kPa}$$

Referred to the free water level, according to our rules, we have the equivalent stresses, p_e and u_e where the subscript 'e' stands for 'equivalent', as:

$$p_e = 4 \times 20 + 5 \times 10 = 130 \, \text{kPa}$$

$$u_e = 0 \, \text{kPa}$$

$$p_e - u_e = 130 \, \text{kPa}$$

If now we have a rapid draw-down of 2 m, then there is no change in effective stress but there is an increase in disturbing moment.

$$p_e = 6 \times 20 + 3 \times 10 = 150 \, \text{kPa}$$

$$u_e = 2 \times 10 = 20 \, \text{kPa}$$

$$p_e - u_e = 130 \, \text{kPa}$$

It can be seen that, before rapid draw-down, the value of p_e to be used in a stability analysis is 130 kPa, and after rapid draw-down p_e has increased to 150 kPa, resulting in a larger disturbing moment and a corresponding reduction in the factor of safety.

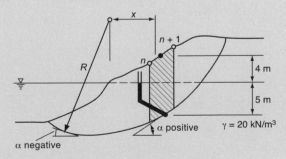

Fig. 4.23 Definition sketch for worked example

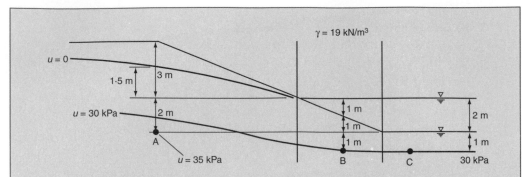

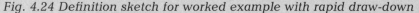

Fig. 4.24 Definition sketch for worked example with rapid draw-down

(2) In Fig. 4.24 we consider the stresses before and after rapid draw-down at the three points A, B and C shown. At point A there is no change in the water load directly above the point due to rapid draw-down. At point B, the change in water load is 1 m of water. At point C the change in water load is 2 m of water.

At point A:
before rapid draw-down

$$p_e = 3 \times 19 + 2 \times 9 = 75 \, \text{kPa}$$

$$u_e = 15 \, \text{kPa}$$

$$p_e - u_e = 60 \, \text{kPa}$$

after rapid draw-down

$$p_e = 5 \times 19 = 95 \, \text{kPa}$$

$$u_e = 35 \, \text{kPa}$$

$$p_e - u_e = 60 \, \text{kPa}$$

i.e. no change in effective stress.

At point B:
before rapid draw-down

$$p_e = 2 \times 19 = 18 \, \text{kPa}$$

$$u_e = 0 \, \text{kPa}$$

$$p_e - u_e = 18 \, \text{kPa}$$

after rapid draw-down

$$p_e = 1 \times 19 + 1 \times 9 = 28 \, \text{kPa}$$

$$u_e = 20 - 10 = 10 \, \text{kPa} \text{ (removal of 1 m of water)}$$

$$p_e - u_e = 18 \, \text{kPa}$$

At point C:
before rapid draw-down

$$p_e = 1 \times 9 = 9\,\text{kPa}$$

$$u_e = 0\,\text{kPa}$$

$$p_e - u_e = 9\,\text{kPa}$$

after rapid draw-down

$$p_e = 1 \times 9 = 9\,\text{kPa}$$

$$u_e = 10 - 10 = 0\,\text{kPa (removal of 2\,m of water)}$$

$$p_e - u_e = 9\,\text{kPa}$$

(3) Consider the example given in Fig. 4.25. Key parameters are as follows: $c' = 0$ and $5\,\text{kPa}$, $\phi' = 20°$, $\gamma = 19.5\,\text{kN/m}^3$. The trial slip circle is divided into six slices of equal widths. By scaling heights from the drawing, we have magnitudes of stresses given below.

Before rapid draw-down the stresses are as shown in Table 4.5.

After rapid draw-down the stresses are as shown in Table 4.6.

Before rapid draw-down the stability analysis proceeds as shown in Table 4.7.

After rapid draw-down the stability analysis proceeds as shown in Table 4.8.

For this trial circle, the factors of safety are shown in Table 4.9.

It can be seen that:

- rapid draw-down significantly reduces the factor of safety, and

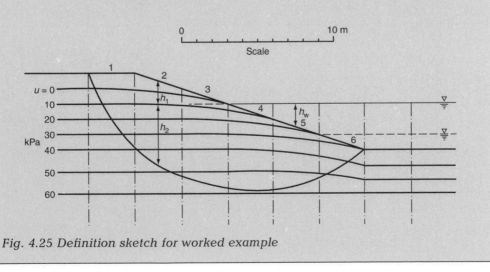

Fig. 4.25 Definition sketch for worked example

Table 4.5. Equivalent stresses for the submerged case

Slice	p_e: kPa	u_e: kPa	$p_e - u_e$: kPa
1	$2 \times 19.5 + 1 \times 9.5 = 48.5$	10	38.5
2	$1.5 \times 19.5 + 4 \times 9.5 = 67.3$	6	61.3
3	$0.5 \times 19.5 + 5.4 \times 9.5 = 61.1$	2	59.1
4	$5.3 \times 9.5 = 50.4$	1	49.4
5	$3.9 \times 9.5 = 37.1$	2	35.1
6	$1.6 \times 9.5 = 15.2$	3	12.2

Table 4.6 Equivalent stresses for rapid draw-down

Slice	p_e: kPa	u_e: kPa	$p_e - u_e$: kPa
1	$3 \times 19.5 = 58.5$	20	38.5
2	$3.5 \times 19.5 + 2 \times 9.5 = 87.3$	26	61.3
3	$2.5 \times 19.5 + 3.4 \times 9.5 = 81.1$	22	59.1
4	$1.5 \times 19.5 + 3.8 \times 9.5 = 65.4$	16	49.4
5	$0.5 \times 19.5 + 3.4 \times 9.5 = 42.1$	7	35.1
6	$1.6 \times 9.5 = 15.2$	3	12.2

Note there is no change in the vertical effective stress, $p - u$, for each slice!

Table 4.7 Stability analyses for submerged case

Slice	α: degree	$p \sin \alpha$: kPa	$c' = 5$ kPa			$c' = 0$		
			A $c' + p' \tan \phi'$	B m_α $F = 1.59$	A/B	A $p' \tan \phi$	B m_α $F = 1.20$	A/B
1	57	40.68	19.0	0.74	25.68	14.0	0.80	17.50
2	34	37.63	27.3	0.97	28.14	22.3	1.00	22.30
3	16	16.84	26.5	1.03	25.73	21.5	1.04	20.67
4	0	0	23.0	1.00	23.00	18.0	1.00	18.00
5	−15	−9.60	17.8	0.90	19.78	12.8	0.89	14.38
6	−32	−8.05	9.4	0.71	13.24	4.4	0.70	6.29
		$\sum 77.50$	$\sum 123.0$ $F_1 = 123/77.5$ $=1.59$		$\sum = 135.57$ $F_2 = 1.75$	$\sum 93.0$ $F_1 = 93/77.5$ $=1.20$		$\sum = 99.14$ $F_2 = 1.28$

- the factor of safety is sensitive to the value of c' used in the analysis – *it is therefore extremely important in practice to choose the value of c' to be used in the slope stability analysis with great care.* Remember the old saying 'no cee too low, no phi too high'!

Table 4.8 Stability analysis for rapid draw-down

Slice	α: degree	$p \sin \alpha$: kPa	$c' = 5\,\text{kPa}$			$c' = 0$		
			A $c' + p' \tan \phi'$	B m_α $F = 1.21$	A/B	A $p' \tan \phi$	B m_α $F = 0.92$	A/B
1	57	49.06	19.0	0.80	23.75	14.0	0.89	15.73
2	34	48.82	27.3	1.00	27.30	22.3	1.05	21.24
3	16	22.35	26.5	1.04	25.48	21.5	1.07	20.09
4	0	0	23.0	1.00	23.00	18.0	1.00	18.00
5	−15	−10.90	17.8	0.89	20.00	12.8	0.86	14.88
6	−32	−8.05	9.4	0.70	13.43	4.4	0.65	6.77
		$\sum 101.28$	$\sum 123.0$ $F_1 = 123/101.28$ $= 1.21$		$\sum = 132.96$ $F_2 = 1.31$	$\sum 93.0$ $F_1 = 93/101.28$ $= 0.92$		$\sum = 96.71$ $F_2 = 0.95$

Table 4.9 Summary of results for worked example depicted in Fig. 4.25

c': kPa	Before rapid draw-down	After rapid draw-down
5	1.75	1.31
0	1.28	0.95

(4) Consider the slope given in Fig. 4.26 representing a cut formed as part of a river control works. The steady state water level is given as the line ABC. Due to a sudden opening of a sluice gate, the water level in the waterway is subjected to a rapid draw-down at the end of which the water surface is represented by ABD, i.e. at rapid

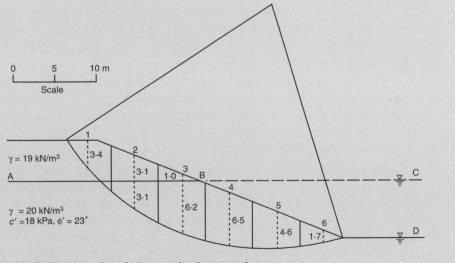

Fig. 4.26 Definition sketch for worked example

Table 4.10 Stability analysis for submerged case

Column No.	1	2	3	4	5	6	7
Slice No.	α: degree	p_e: kPa	u_e: kPa	$p\sin\alpha$	$c' + (p-u)\tan\phi'$	m_α ($F_1 = 1.9$)	5/6
1	46	66.3	0	47.7	46.1	0.85	54.3
2	34	91.4	0	51.1	56.8	0.96	59.2
3	22	81.5	0	30.5	52.6	1.00	52.6
4	20	65.0	0	11.3	45.6	1.03	44.3
5	−1	46.0	0	−0.8	37.5	1.00	37.5
6	−11	17.0	0	−3.2	25.2	0.94	26.8
SUM				136.6	263.8		274.7

draw-down, water is 'stranded' in the slope because there has not been sufficient time for drainage to take place. For the proposed slip circle, the stability of the slope is to be determined for both the prior steady state and for the condition of 'rapid draw-down'.

The technique is to sub-divide the proposed slipping mass into six slices of equal width. For this simple example, using five slices is too great an approximation, while using seven slices gives very little improvement over the accuracy obtained from six slices. The medians of each slice are dotted in and measured both above and below the water level as shown. The slope of the tangent to the circle at the base of each median gives the values of α. It may be noted that each radial line from the centre of the slip circle to the base of each median also makes an angle α with the vertical.

The data and computation may be presented conveniently in a tabular form as shown in Table 4.10. The steady state condition is analysed in Table 4.10. It can be seen that the vertical total stress for each slice is computed using submerged unit weight below the water level ABC. The consequence of this analytical device is that the pore water pressures must be calculated from the same line ABC. Consequently, column 3 in Table 4.10 contains only zeros. The computations may be easily carried out using a scientific calculator or a spreadsheet. Equation 10.16 contains the unknown factor of safety F on both sides of the equation. The iterative solution of this equation is carried out by first obtaining a rough approximation of F by dividing the sum of column 5 (i.e. effectively assuming $m_\alpha = 1$ throughout) by the sum of column 4, whence

$$F_1 = 263.8/136.6 = 1.93$$

The parameter $\tan\phi'/F = \tan 23°/1.93 = 0.22$ is then used to enter the curves of Fig. 4.13 (sometimes called 'Janbu's curves') to obtain

Table 4.11 Stability analysis for rapid draw-down

Column No.	1	2	3	4	5	6	7
Slice No.	α: degree	p_e: kPa	u_e: kPa	$p\sin\alpha$	$c' + (p-u)\tan\phi'$	m_α ($F_1 = 1.4$)	5/6
1	46	66.3	0	47.7	46.1	0.91	50.7
2	34	122.4	31	68.5	56.8	1.00	56.8
3	22	143.5	62	53.7	52.6	1.04	50.6
4	20	130.0	65	22.6	45.6	1.03	44.3
5	−1	92.0	46	−1.6	37.5	0.99	37.9
6	−11	34.0	17	−6.5	25.2	0.92	27.4
SUM				184.4	263.8		267.7

the value of m_α corresponding to each value of α. Each value of $c' + (p-u)\tan\phi'$ in column 5 is then divided by the corresponding m_α value and the result put in column 7. A more accurate value of the factor of safety is then obtained as $F_2 = 274.7/136.6 = 2.01$. Further iterations make little improvement.

The data and computations for the rapid draw-down case are given in Table 4.11. Here $F_1 = 263.8/184.4 = 1.43$, and the parameter $\tan\phi'/F = \tan 23°/1.43 = 0.3$, whence $F_2 = 267.7/184.4 = 1.45$.

In summary, $F = 2.01$ for the submerged case and $F = 1.45$ after rapid draw-down.

Summary of the steps required to analyse the stability of a slope by hand

(a) *Step 1: Scale drawing.* Prepare a scale drawing of a cross-section through the slope. Draw in a realistic proposed failure surface.

(b) *Step 2: Slice parameters.* Divide the sliding mass into slices, preferably of equal width. The slice width should be selected to take account of changes in soil properties, slope geometry and pore pressure distribution. Obtain the inclination with respect to the horizontal, α, of the tangent to the centre of the base of each slice (note that α is also the inclination to the vertical of the radial line joining the centre of the slip circle to the centre of the base of each slice).

(c) *Step 3: Weight parameters.* Find the total vertical stress acting at the centre of the base of each slice. This will be the slice height times the unit weight for a uniform soil. For the partly submerged case, submerged unit weight will be used below the water level to compensate for the restraining effect of water pressure on the submerged face of the slope.

(d)   *Step 4: Pore water pressure.* In a total stress analysis, the pore pressures are not known and not used. In an effective stress analysis, the pore pressures will be found from scaling off the drawing. For the case of steady seepage, a flow net will be used. For the case of a submerged slope, pore pressures will be measured from the level line of submergence as datum. For rapid draw-down, it is assumed that drainage (seepage) during rapid draw-down has not occurred and that a flow net has not been established and hence pore pressures are defined by the geometry of the slope.

(e)   *Step 5: Detailed calculations.* Calculate $p \sin \alpha$ for each slice and sum them in a table.

For total stress analysis, compute factor of safety $F = (\sum s_u / \cos \alpha) / \sum p \sin \alpha$. For effective stress analysis, evaluate pore pressures u or u_e, and evaluate the strength term for each slice, $c' + (p - u) \tan \phi'$, and sum in a table. Find the first trial

$$F_1 = \frac{\sum c' + (p - u) \tan \phi'}{\sum p \sin \alpha}$$

Now use this first trial F_1 to evaluate the parameter $\tan \phi' / F_1$ in order to enter Janbu's curves (Fig. 4.13) to find values of m_α to modify the $c' + (p - u) \tan \phi'$ terms and evaluate a revised value of factor of safety, F_2, whence the process can be iterated to give F_3.

Non-circular slip analysis

Overview

Non-circular slips occur by sliding on a soft layer or, for example, by sliding through the sloping clay core of an earth dam. While circular slips can move without internal distortion, with non-circular slips internal shear must take place. If internal shear forces are neglected then an artificially low factor of safety will result. The classic method for analysing non-circular slope stability is that of Janbu *et al.* (1956).

Janbu method

For a total stress analysis, refer to Fig. 4.27. Resolving parallel to the slope (where α is the inclination of the slope) gives:

$$\sum W \sin \alpha = \frac{\sum sl}{F}$$

$$F = \frac{\sum sl}{\sum W \sin \alpha}$$

$$\text{or } F = \frac{\sum sl / \cos \alpha}{\sum W \tan \alpha}$$

neglecting interslice forces.

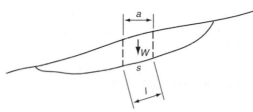

Fig. 4.27 Definition sketch for non-circular total stress analysis by method of slices, neglecting side forces

To take account of interslice forces use the Janbu's quasi-empirical correction factor f_0 whence

$$F = f_0 \frac{\sum sl/\cos\alpha}{\sum W\tan\alpha}$$

where f_0 is >1 and varies between 1 and 1.13 depending on geometry.

For an effective stress analysis for a non-circular failure surface, use the Janbu solution. Referring to Fig. 4.28

$$F = f_0 \frac{\sum\{[c' + (p-u)\tan\phi']a\}/n_\alpha}{\sum W\tan\alpha} \tag{4.15}$$

which, for equal slice widths, reduces to

$$F = f_0 \frac{\sum[c' + (p-u)\tan\phi']/n_\alpha}{\sum p\tan\alpha} \tag{4.16}$$

where $n_\alpha = \cos^2\alpha(1 + \tan\alpha\tan\phi'/F)$.

Computer programs

The Geotechnical and Geo-environmental Software Directory (GGSD)

This web site (www.ggsd.com) catalogues programs in the fields of geotechnical engineering, soil mechanics, rock mechanics, engineering geology, foundation engineering, hydrogeology, geo-environmental engineering, environmental engineering, data analysis and data visualization and lists worldwide suppliers and publishers of these programs. Programs are indexed by program name, program category, operating system and program status, and the Directory entries are listed by category. Program publishers and suppliers are indexed by organization name and by country, and the Directory entries are listed in alphabetical order. The Directories of programs, publishers and suppliers are cross-referenced. The Directory also gives links to other web sites featuring geotechnical, geo-environmental or related software. There is a free GGSD Newsletter. The GGSD web site for slope stability programs lists the following descriptions of programs, classified by program name, type (commercial, freeware or shareware), and operating system (DOS,

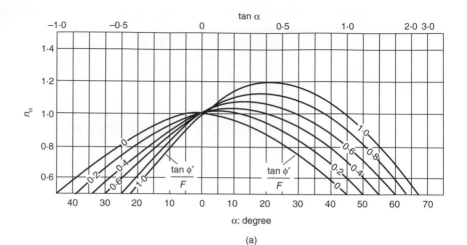

(a)

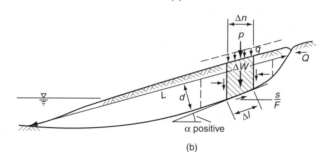

(b)

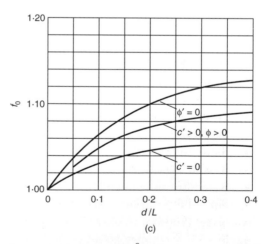

(c)

Fig. 4.28 (a) Janbu's curves for $n_\alpha = \cos^2 \alpha (1 + \tan \alpha \tan \phi'/F)$,
(b) definition sketch for non-circular effective stress analysis, (c) Janbu's
semi-empirical correction factor for influence of side forces.

Windows, Mac). Detailed up-to-date descriptions of each program and worldwide suppliers are given in the Geotechnical and Geoenvironmental Software Directory at www.ggsd.com. By mid-2000, this web site had received over one million hits and was at that time getting 140 000 hits per month.

Slope stability (soil) software index
- CHASM – Commercial – Win3x, Win95/98, WinNT, UNIX
- DLISA – Freeware – DOS
- ESAU – Commercial – DOS, UNIX, HP
- ESTAVEL – Commercial – DOS
- Galena – Commercial – Win95/98, WinNT
- GBSLOPE – Commercial – Win3x, Win95/98, WinNT
- Geo-Tec B – Commercial – Mac, PowerMac
- GEOSLOPE – Commercial – DOS
- GeoStar – Commercial – DOS
- GFA2D – Freeware – DOS, Win3x, Win95/98
- GGU-SLICE – Commercial – Win95/98, WinNT, Win2000
- GGU-SLOPE – Commercial – Win95/98, WinNT, Win2000
- GPS-PC – Commercial – DOS
- GSlope – Commercial – DOS
- GSTABL7 v. 1.14 – Commercial – Win3x, Win95/98, WinNT
- I.L.A. – Commercial – Win95/98, WinNT
- LISA – Freeware – DOS
- MPROSTAB – Commercial – DOS
- MSTAB – Commercial – DOS
- PCSTABL 5 M – Public domain – DOS
- Penta – Commercial – Win95/98, WinNT
- REAME – Commercial – DOS
- Sarma – Shareware – DOS
- SB-SLOPE – Commercial – DOS
- Slide – Commercial – DOS, Win3x, Win95/98, WinNT
- SLIP – Commercial – DOS
- SLOPBG – Commercial – DOS
- Slope (Geosolve) – Commercial – DOS
- Slope (Oasys) – Commercial – DOS, Win95/98, WinNT
- SLOPE 8R – Commercial – DOS
- Slope-W – Commercial – Win95/98, WinNT
- SLOPNC – Commercial – DOS
- Stabl for Windows – Commercial – Win95/98, WinNT, Win2000
- STABLE – Commercial – DOS, Win3x, Win95/98, WinNT
- STABLEPRO for Windows – Commercial – Win3x, Win95/98
- STEDwin 2.2 – Commercial – Win3x, Win95/98, WinNT

- SWASE – Commercial – DOS
- TSLOPE – Commercial – Web/Java, Java client, server calculation.
- TSLOPE3 – Commercial – DOS, UNIX
- TSTAB – Commercial – DOS, UNIX
- UTEXAS3 – Commercial – DOS
- WinStabl – Shareware – Win3x, Win95/98, WinNT
- XSTABL – Commercial – DOS

Slope stability (rock) software index
- ACCECALC – Commercial – Win95/98, WinNT
- CLU_STAR – Commercial – Win95/98, WinNT
- EzSlide – Freeware – Win95/98, WinNT
- Plane Failure Analysis – Freeware – Win95/98, WinNT
- RocFall – Commercial – Win95/98, WinNT
- ROCK3D – Commercial – Win95/98, WinNT
- ROCKPF – Commercial – DOS
- ROFEX – Commercial – Win3x, Win95/98
- ROTOMAP – Commercial – Win95/98, WinNT
- Slope Keyblock – Commercial – Win95/98, WinNT
- SLOPEPACK – Commercial – DOS
- SWARS – Commercial – DOS
- Swedge – Commercial – Win95/98, WinNT
- WEDGE – Commercial – DOS
- Wedge Failure Analysis – Freeware – Win95/98, WinNT

SLOPE/W Student Edition CD

The SLOPE/W Student Edition CD is produced by GEO-SLOPE International Ltd and is provided with this book. The best way to get going is to install the software and then follow the tutorial in Chapter 3 in the User's Guide. Once you have gone through this simple tutorial you will have a pretty good idea of how to operate the software. You should then be able to go through the Student Edition lessons. The complete User's Guide is also in the on-line help. You can get at the on-line help by pressing the F1 key when a command is highlighted, by looking at the table of contents or by searching indexed words (it works much like any Windows application on-line help system). Another way of getting an introduction to the software is to open the files included with the Student Edition. By opening these files in DEFINE and CONTOUR you can quickly see the definition environment and how you can view the results.

The technical overview of SLOPE/W can be downloaded from the GEO-SLOPE International Ltd web site in colour (www.geo-slope.com). The SLOPE/W package interfaces with the other GEO-SLOPE International Ltd modules for seepage (SEEP/W) and for stress distribution (SIGMA/W).

CHAPTER FIVE

Classic case records of slope failures

Overview

The measurement of the strength of clays has been discussed in Chapters 2 and 3. It has been shown that the measured strength, both in terms of total stress and effective stress, can vary between wide limits. The ratio of maximum measured strength to minimum measured strength may be of the order of three or more. It is vital, therefore, to know how the full-scale or operational strength mobilized in situ in a given situation compares with the strength measured in a reasonable and rational manner. This is why it is necessary to study relevant case records where failure has occurred in the field in order to determine by how much the mobilized in situ strength is different from the strength measured in a test.

Case records fall into two general limiting classifications as follows.

(a) *Undrained failure.* Undrained failure in the field where the time to failure is so short that insufficient time has elapsed for there to be a significant change in effective stress and hence strength (e.g. cuttings in the short term).

(b) *Drained failure.* Drained failure in the field where the time to failure is so long that sufficient time has elapsed for the pore water pressures to change and be in equilibrium with the controlling boundary conditions thus changing effective stresses and hence strength (e.g. long term cuttings and natural slopes).

An intermediate condition should also be considered in the stability of clay cuttings where failure may occur some time after the end of construction but before sufficient time has elapsed for complete equilibration of pore water pressures in the cut slope.

The following groups of case records are considered below:

(a) *The short term stability of clay cuttings* may be subdivided as occurring in:

(i)   soft to firm intact clays
(ii) stiff fissured clays.

(b)   *The long term stability of clay slopes* may be subdivided as:

(i)   first time slides in intact clays. An intact clay is one which has no structural discontinuities such as fissures, slickensides, bedding planes, slip surfaces, etc.

(ii)   first time slides in cuttings in stiff fissured clays

(iii)   natural slopes in stiff fissured clays

(iv)   slides on pre-existing slip surfaces.

Short term stability of cuttings

Soft to firm intact clays

Base failures of strutted excavations

Bjerrum and Eide (1956) analysed base failures of seven excavations in soft to firm intact Scandinavian clays. Although not cuttings in the usual sense, they nevertheless represent failures due to an unloading condition. The undrained shear strength was measured using the field shear vane apparatus and the analyses were based on bearing capacity factors, N_c, appropriate to the geometry of the plan of the excavation. The results of these calculations are shown in Table 5.1. It can be seen that the calculated factors of safety ranged from 0.82 to 1.16 with an average value of 0.96. It can be concluded that, for these clays and for these types of failure, a total stress analysis using the undrained shear strength obtained from the in situ vane, provides a reliable estimate of the factor of safety against base failure of a strutted excavation.

Congress Street, Chicago

Details of this slide have been published by Ireland (1954). Failure occurred during excavation of an open cut in Chicago for the Congress Street superhighway in 1952. The slide took the form of a simple rotational movement on a more or less circular slip surface, limited in depth by stiff clay of a Wisconsin age moraine. Apart from a cover of fill, placed in the late 19th century, and a relatively thin layer of sand and silt, the slide was located principally in soft to firm intact clay (typical index properties $w = 25\%$, $w_L = 32\%$, $w_P = 18\%$), the upper portion of which had been subjected to some desiccation before the sand was deposited. The clays above the moraine are probably tills laid down by the ice sheets under water.

The undrained shear strengths of the clays were measured by compression tests on Shelby-tube samples and corrected for sample disturbance by means of correlations established between the strengths of tube samples and of specimens from hand-cut blocks (Peck, 1940). Reasonable assumptions were made concerning the shear strength in the sand and fill layers, but these play a minor role in the analysis.

Table 5.1 Factors of safety for base failures of strutted excavations in saturated clay, after Bjerrum and Eide (1956)

Site	Dimension $B \times L$: m	Depth D: m	Surcharge: kPa	Unit weight: kN/m³	s_u: kPa	Sensitivity	B/L	D/B	N_c	Safety factor F
Pumping station, Fornebu, Oslo	5.0 × 5.0	3.0	0.0	17.5	7.5	50	1.0	0.60	7.2	1.03
Storehouse, Drammen	4.8 × ∞	2.4	15	19.0	12	5–10	0.0	0.5	5.9	1.16
Pier shaft, Gothenborg	Dia. 0.9	25.0	0.0	15.4	35	20–50	1.0	28.0	9.0	0.82
Sewage tank, Drammen	5.5 × 8.0	3.5	10	18.0	10	20	0.69	0.64	6.7	0.93
Test shaft (N), Ensjoveien, Oslo	dia. 1.5	7.0	0.0	18.5	12	140	1.0	4.7	9.0	0.84
Excavation, Grev Vedels pl., Oslo	5.8 × 8.1	4.5	10	18.0	14	5–10	0.72	0.78	7.0	1.08
'Kronibus shaft', Tyholt, Trondheim	2.7 × 4.4	19.7	0.0	18.0	35	40	0.61	7.3	8.5	0.84
										average 0.96

On the most critical circle, according to the $\phi_u = 0$ analysis, the calculated factor of safety was 1.11. As was to be expected, the critical circle lay further back within the slope than the actual slip surface (Skempton, 1948).

The Chicago clays probably show little variation in strength resulting from anisotropy. The presence of some cracks and joints, due to desiccation in the upper layer of clay, may mean that its laboratory strength exceeds the field strength. In the present case a reduction of, say, 20% would seem to be ample; and from figures given by Ireland this reduces the calculated factor of safety to about 1.07. Stress–strain curves of the softer clays show little drop in strength after the undrained peak (Peck, 1940) but there may be a reduction in mobilized strength due to progressive failure along that part of the slip surface lying within the stiff upper clay. However, a 30% reduction to allow for this effect only brings the calculated factor of safety down to 1.00.

The slide occurred while excavation was still taking place. The failure was therefore not delayed and, although the rate of shearing may well have been considerably slower than in the laboratory test, it is not thought that the difference between field and laboratory strengths, on this account, is likely to exceed 10%. Finally, the Chicago clays have a rather low sensitivity (around four) and the strengths as measured on hand-cut blocks are probably very close to the in situ values.

In conclusion, then, the calculated factors of safety range from about 1.1 with no corrections to about 0.9 with perhaps an over-generous allowance for such effects as progressive failure, rate of testing and fissures in the upper clay.

Porsmossen, Stockholm

On 16 July 1968 at 3 p.m. a landslide occurred during the driving of steel sheet piles at the toe of a cutting, shown in Fig. 5.1.

The soil conditions consist of a very soft glacial clay of maximum depth about 6 m, underlain by moraine. The soft clay has the following

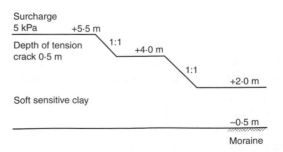

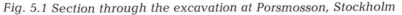

Fig. 5.1 Section through the excavation at Porsmosson, Stockholm

characteristics: $w = 40$ to 50%, $w_L = 40$ to 50%, $w_P = 20\%$, $s_u = 7.5$ to 15 kPa (in situ vane), sensitivity 10 to 50.

The calculated factor of safety was found to be 1.14, which can be considered as being in fairly reasonable agreement with the actual value of unity. If a correction is made as proposed by Bjerrum (1972), the calculated factor of safety drops to 1.08.

Stiff fissured clays

Bradwell

The classic case of a short term failure in a cutting in a stiff fissured clay occurred at Bradwell in 1957 and is described by Skempton and La Rochelle (1965). Two slides occurred in the deep excavation for a nuclear reactor at Bradwell, Essex, in the London clay. A plan of the excavation for Reactor No. 1 is shown in Fig. 5.2. Slide I took place on 24 April 1957, about five days after completion of this part of the excavation (Fig. 5.3) and Slide II followed at an adjacent section 19 days after excavation. The London clay is overlain by 2.7 m of soft post-glacial marsh clay and 3.5 m of fill, placed as the excavation proceeded. At Slide II the fill was 2.4 m thick. The weather during these operations was dry and continued so throughout the period leading to Slide II. Ground water level was located within the marsh clay from which there were small seepage flows on to the berm at the top of the London clay.

In Essex the London clay has been over-consolidated during its geological history by the removal by erosion of some 150 m of overlying sediments. It is a stiff fissured marine clay of Eocene age with horizontal bedding.

The first sign of instability at Slide I was the outward bulging and cracking of clay just above the toe of the slope. A few hours later a large wedge slipped out, to be followed after four hours by a more massive movement. Next day the main slide occurred, with a tension crack passing through the fill.

The undrained strength of the clay was measured on 38 mm diameter by 76 mm triaxial (vertical axis) specimens taken from borehole samples and hand-cut block samples, with a time to (peak) failure of the order 15 minutes. No significant difference could be found between the strengths of the two types of sample, and sufficient tests were made to establish firmly the variation of undrained strength with depth. Average values of the index properties of the London clay within the depth involved in the slides are: natural water content 33%, liquid limit 95%, plastic limit 30%, clay fraction 52%.

Five possible slip surfaces were analysed for Slide I (see Fig. 5.4) and of these the three most critical all showed the actual strength of the London clay mobilized during the failure to be only 56% ($\pm2\%$) of the average

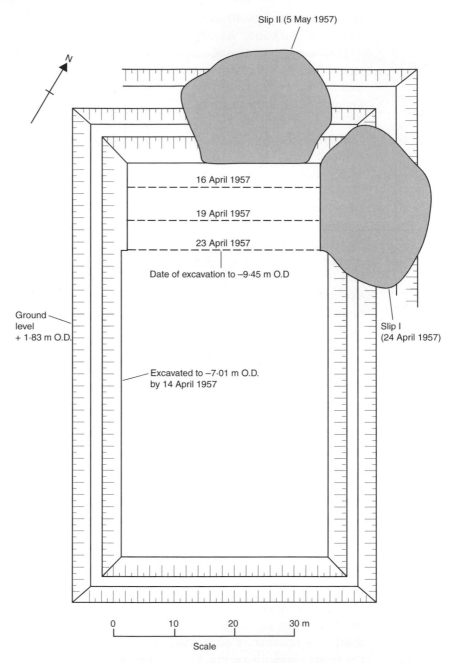

Fig. 5.2 Plan of excavation for Reactor No. 1 Bradwell, after Skempton and La Rochelle (1965)

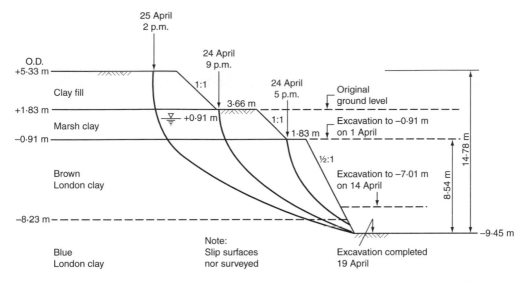

Fig. 5.3 Section through Reactor No. 1 Bradwell excavation, Slide I (east side) 1957 (slide occurred five days after completing excavation), after Skempton and La Rochelle (1965)

strength measured in the laboratory (Table 5.2). For Slide II the ratio was 52%. In other terms, using the laboratory strengths the calculated factors of safety were about 1.8 and 1.9 respectively for the two slides.

At the time of the preliminary investigations these results seemed to indicate surprisingly large discrepancies between laboratory and field

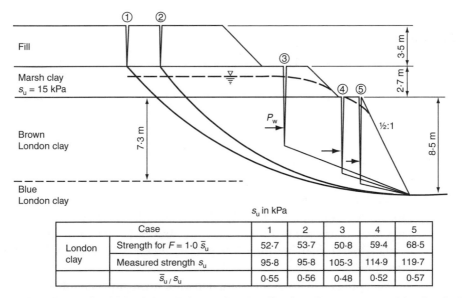

	Case	1	2	3	4	5
London clay	Strength for $F = 1.0\ \bar{s}_u$	52.7	53.7	50.8	59.4	68.5
	Measured strength s_u	95.8	95.8	105.3	114.9	119.7
	$\bar{s}_u / s_u$	0.55	0.56	0.48	0.52	0.57

Fig. 5.4 Stability calculations for slides at Bradwell, after Skempton and La Rochelle (1965)

Table 5.2 *Summary of data relating to the Bradwell excavations, after Skempton and Larochelle (1965)*

Excavation	Total depth: m	Time of slide: days		Stable for at least: (months)	Strength of London clay: kPa		$\bar{s}_u/s_u$
		Since start of excavation	Since completion of excavation		Required $\bar{s}_u$	Measured s_u	
Reactor No. 1, Slide I	14.8	23	5	–	54	96	0.56
Reactor No. 1, Slide II	13.7	34	19	–	50	96	0.52
Pump House	11.9	–	–	4.0	>40	101	>0.40
Turbine House	7.0	–	–	4.5	>33	86	>0.38

strengths, but the differences can now be readily explained. In the first place, tests with various times to failure showed that pore water migration takes place within the clay, towards the shear zone; and in five days this effect, together with a small rheological component, brings about a reduction in undrained strength of approximately 18% as compared with the strength measured in 15 minute tests (Skempton and La Rochelle 1965). In the second place it is now known, that the undrained strength of London clay en masse, or in very large triaxial specimens, is about 70% of the strength measured in 38 mm diameter by 76 mm triaxial specimens; due to the presence of fissures and other discontinuities which are by no means fully representing the small samples. There is every reason to suppose that the time and size effects work in conjunction, and if this is correct the ratio of the 5 day field strength to the 15 minute laboratory strength would be $(1 - 0.18) \times 0.70 = 0.57$.

The near coincidence of this figure with the results of stability calculations is obviously fortuitous to some extent. But it is considered that it establishes beyond reasonable doubt that the apparent discrepancy implied by a calculated factor of safety of 1.8 is not due to shortcomings of the $\phi = 0$ analysis, and can be explained in a quantitative manner by the highly misleading strengths obtained in conventional laboratory tests. Indeed, if some allowance were made for progressive failure and for the possibility of a slightly lower strength due to orientation effects, the conclusion could be reached that the field strength was actually a little greater than might be expected when all factors are taken into account.

The somewhat lower strength deduced from the delayed failure of Slide II is qualitatively perfectly rational. It amounts to a further reduction of 7%, and there is no difficulty in supposing that this could be the combined result of small additional pore water migration and progressive failure effects.

So far as other stiff fissured clays are concerned, the Bradwell results presumably give some indication of the problems to be encountered, but the quantitative pattern may differ considerably.

It is of interest to note that two slopes in the London clay at Bradwell were stable for about four months. In both cases the average shear stress in the London clay was calculated to be about 40% of the measured strength on 38 mm by 76 mm triaxial test specimens.

The investigations at Bradwell have been summarized by Skempton and La Rochelle (1965) as follows:

- In Slide I, which occurred five days after completion of excavation, the average shear stress, and hence the average shear strength, in the London clay at failure was 56% of the strength as measured in conventional (15 minute) undrained triaxial tests on small specimens.
- In Slide II, which occurred 19 days after excavation, the shear strength of the London clay in the ground was 52% of the measured strength.
- The slopes of the Pump House and Turbine House excavations remained stable for periods of at least 4 and 4.1 months respectively. In both cases the average shear stresses in the London clay were about 40% of the measured strength.
- Thus, based on experience at Bradwell, it would be unwise to design slopes for short term stability in the brown London clay with shear strengths exceeding 50% of the values measured in conventional undrained tests. However, strengths equal to 40% of the measured values would appear to guarantee stability for periods of several months.
- It may be added that at other sites in the brown London clay where long term slips have taken place, after periods ranging from 5 to 30 years, the strengths had fallen to values between 25% and 35% of the undrained strengths as measured before excavation (Skempton, 1948). In these cases softening and progressive failure had occurred in the clay.

Cut-off trench, Wraysbury

In 1966, a slide occurred into a cut-off trench taken down into the London clay. The slide occurred some three days after the end of construction of an unsupported trench some 3 m wide and 3 m deep taken down into the clay. A large proportion of the failure surface of the slip was approximately horizontal and it was decided therefore to include in the work an investigation of the variation of the undrained shear strength of the London clay with orientation of the failure surface.

Table 5.3 Comparison of undrained strengths for vertical and inclined triaxial test specimens of London clay from Wraysbury, after Simons (1962)

Axis of test	No. of tests	Water content w: %	Time to failure: min)	Undrained strength s_u: kPa	Corrected undrained strength s_u: kPa (w = 28%)	Strength ratio (compared with vertical strength)
Vertical	12	28.1	7	116	117	1.00
Horizontal	12	28.1	6	124	125	1.07
45°	12	28.2	6	82	84	0.72
56° to vertical	12	27.0	4	103	90	0.77

Undrained triaxial compression tests with a confining pressure of 207 kPa in were carried out in 38 mm diameter by 76 mm high specimens, with various inclinations of the axis, hand cut from blocks about 250 mm³, which in turn were hand-cut from an open excavation. The results of the tests are shown in Table 5.3.

In an attempt to eliminate some of the scatter, the results have been corrected to a common water content of 28%, for the purpose of comparison. It can be seen that the greatest strength was obtained for samples with their axes horizontal, and least for samples with their axes inclined at 45° to the vertical. Also, when the axis is inclined at 56° to the vertical (the failure surface in the test then being approximately horizontal referred to the field), the strength measured is 77% of that with the axis vertical. Thus for this particular slip in the field where the failure surface was mainly horizontal, estimates of the in situ strength based on samples with vertical axes, the usual method, would result in appreciable error.

Now it is well known that undrained strength measurements made on fissured clays are influenced by the size of the specimens tested. To investigate this, triaxial specimens of sizes 305 mm diameter by 610 mm, 152 mm diameter by 305 mm, 101 mm diameter by 203 mm, 38 mm diameter by 76 mm, and 13 mm diameter by 25 mm were tested under undrained conditions, with their axes vertical and a confining pressure of 200 kPa. The results obtained are shown in Table 5.4. Again, to reduce scatter, the measured strengths were corrected to a common moisture content of 28%. The main points to emerge are as follows:

- 305 mm diameter by 610 mm, 152 mm diameter by 305 mm, 102 mm diameter by 203 mm test specimens gave approximately the same measured strength, which was about 60% of that obtained from 38 mm diameter by 76 mm specimens taken by a U-4 sampler.
- 38 mm diameter by 76 mm test specimens from hand-cut blocks indicated a strength 43% higher than 38 mm diameter by 76 mm test

Table 5.4 Comparison of undrained strengths for different sizes of triaxial test specimens with respect to 38 mm × 76 mm triaxial tests from U-4 sampler for London clay from Wraysbury, after Simons (1967)

Size of test specimen: mm	No. of tests	Water content w: %	Time to failure: min	Undrained strength: s_u: kPa	Corrected undrained strength s_u: kPa ($w = 28\%$)	Strength ratio compared with 38 × 76 from U-4 sampler
305 × 610	5	28.2	63	49	51	0.62
152 × 305	9	27.1	110	51	46	0.56
102 × 203	11	27.7	175	48	46	0.57
38 × 76 (U-4)	36	26.9	8	93	82	1.00
38 × 76 (blocks)	12	28.1	7	116	117	1.43
13 × 25 (intact)	19	26.6	10	262	219	2.68

specimens from the U-4 sampler, showing the effect of the disturbance caused by the U-4 sampling.

- The highest measured strength was obtained for the 13 mm diameter by 25 mm intact specimens.

In addition to the laboratory shear tests, undrained direct shear tests were carried out in the field, shearing a block of clay 610 mm square. Table 5.5 compares the results obtained from an analysis of the slip, the 610 mm square in situ shear box, the 305 mm diameter by 610 mm vertical triaxial specimens and the 38 mm diameter by 76 mm vertical specimens from U-4 samples, the latter representing standard practice. Corrections have been made for the differing moisture contents and times to failure assuming purely undrained shear. The main points to note are as follows:

- The standard 38 mm diameter by 76 mm specimens give a strength 1.88 times that indicated by an analysis of the slip, and it should be mentioned that if no corrections are made for moisture content and time to failure, this ratio would be 3.1.
- The 305 mm diameter by 610 mm triaxial specimens show a strength 23% higher than that indicated by the slip analysis. As indicated previously, part of this difference is due to the differing inclinations of the failure surfaces.
- The 610 mm square in situ shear box tests give a strength 17% higher than that from the slip analysis.

Bearing in mind the approximate nature of the corrections made for moisture content and time to failure, and the possibility that slight restraint imposed by the shear box may have resulted in a higher

Table 5.5 Comparison of undrained strengths for different types of test with respect to slip strength for London clay from Wraysbury, after Simons (1967)

Size (in mm) and type of test specimen	Water content w: %	Time to failure T_f: min	Undrained strength s_u: kPa	Corrected undrained strength s_u: kPa ($w = 28\%$)	Corrected undrained strength s_u: kPa ($w = 28\%$) ($T_f = 4000$ min)	Strength ratio compared with slip
Slip	29.3	4000	30	35	35	1.00
610×610 shear box	28.1	71	48	48	41	1.17
305×610 triaxial vertical	28.2	63	49	51	43	1.23
38×76 triaxial vertical (U-4)	26.9	8	93	82	66	1.88

measured strength, reasonable agreement between the in situ shear box tests and the slip analysis is indicated.

To sum up, the standard undrained laboratory tests carried out on 38 mm diameter by 76 mm specimens taken from U-4 samples greatly overestimates the in situ strength of the London clay as indicated by an analysis of the end of construction slip. Much better agreement is obtained from the results of triaxial specimens 101 mm diameter by 203 mm high and larger, and 610 mm square in situ shear tests.

It can be concluded that provided the undrained shear strength of a clay is carefully measured, taking into consideration the factors outlined in Chapter 2, then the total stress analysis (the so-called $\phi = 0$ analysis) can be used to assess the end of construction stability of cuttings.

Long term stability of clay slopes

Overview

The long term stability condition is considered to cover the situation when the pore water pressures in a slope have reached an equilibrium value, i.e. they are no longer affected by construction operations. Under this class, therefore, falls the stability of natural slopes and of cuttings when sufficient time has elapsed for the excess pore water pressures set up during excavation to have dissipated and the water pressures in the slope are then governed by the prevailing ground water conditions. This is clearly a drained situation and it is obvious that an attempt to predict the stability of such slopes using the undrained shear strength as a basis for calculation (the $\phi = 0$ analysis) is bound in general to result in a completely unreliable calculated factor of safety. To illustrate this point, Table 5.6 (Bishop and Bjerrum, 1960), shows calculated factors of safety obtained by the $\phi = 0$

Table 5.6 Long-term failures in cuts and natural slopes analysed by the $\phi = 0$ analysis, after Bishop and Bjerrum (1960)

Locality	Type of slope	Data of clay					Safety factor, $\phi = 0$ analysis	Reference
		w: %	w_L: %	w_P: %	PI	Liquidity index $\dfrac{W - PL}{PI}$		
1. Over-consolidated, fissured clays								
Toddington	Cutting	14	65	27	38	−0.34	20	Cassel, 1948
Hook Norton	Cutting	22	63	33	30	−0.36	8	Cassel, 1948
Folkestone	Nat. slope	20	65	28	37	−0.22	14	Toms, 1953
Hullavington	Cutting	19	57	24	33	−0.18	21	Cassel, 1948
Salem, Virginia	Cutting	24	57	27	30	−0.10	3.2	Larew, 1952
Walthamstow	Cutting	–	–	–	–	–	3.8	Skempton, 1942
Sevenoaks	Cutting	–	–	–	–	–	5	Toms, 1948
Jackfield	Nat. slope	20	45	20	25	0.00	4	Henkel and Skempton, 1955
Park Village	Cutting	30	86	30	56	0.00	4	Skempton, 1948b
Kensal Green	Cutting	28	81	28	53	0.00	3.8	Skempton, 1948b
Mill Lane	Cutting	–	–	–	–	–	3.1	Skempton, 1948b
Bearpaw, Canada	Nat. slope	28	110	20	90	0.09	6.3	Peterson, 1952
English, Indiana	Cutting	24	50	20	30	0.13	5.0	Larew, 1952
SH 62, Indiana	Cutting	37	91	25	66	0.19	1.9	Larew, 1952
2. Over-consolidated, intact clays								
Tynemouth	Nat. slope	–	–	–	–	–	1.6	Imperial College
Frankton, N.Z.	Cutting	43	62	35	27	0.20	1.0	Murphy, 1951
Lodalen	Cutting	31	36	18	18	0.72	1.01	N.G.I.
3. Normally-consolidated clays								
Munkedal	Nat. slope	55	60	25	35	0.85	0.85	Cadling and Odenstad, 1950
Säve	Nat. slope	–	–	–	–	–	0.80	Cadling and Odenstad, 1950
Eau Brink cut	Cutting	63	55	29	26	1.02	1.02	Skempton, 1945
Drammen	Nat. slope	31	30	19	11	1.09	0.60	N.G.I.

analysis for long term slope failures in different clays. It can be seen that the factors of safety vary from 21, for over-consolidated clays, down to 0.6 for the normally-consolidated Drammen clay. In undrained tests, pore water pressures are set up during shear so that the effective stress at failure is quite different from the effective stress acting on the failure surface in the field, and therefore a completely different strength is obtained.

The following groups of problem will be considered separately:

- first time slides in intact clays
- first time slides in stiff fissured clays
- slides on pre-existing slip surfaces
- natural slopes in stiff fissured clays.

Much of the discussion here has been based on the work of Skempton (1964), Skempton and Hutchinson (1969) and Skempton (1977).

First time slides in intact clays

Drammen (Kjaernsli and Simons, 1962)

On 6 January 1955 a rotational slide occurred in the north bank of the Drammen River, at the town of Drammen in Norway (Fig. 5.5). The slide was located in a soft intact marine clay of post-glacial age, covered by about 3 m of sand and granular fill. The clay has occasional extremely thin seams of silt and fine sand. Its index properties are typically $w = 35\%$, $w_L = 35\%$, $w_P = 18\%$, clay fraction = 38%, and sensitivity of 8. Piezometer readings showed that the pore pressures were hydrostatic. Beneath the ground surface the clay is normally consolidated. Below the

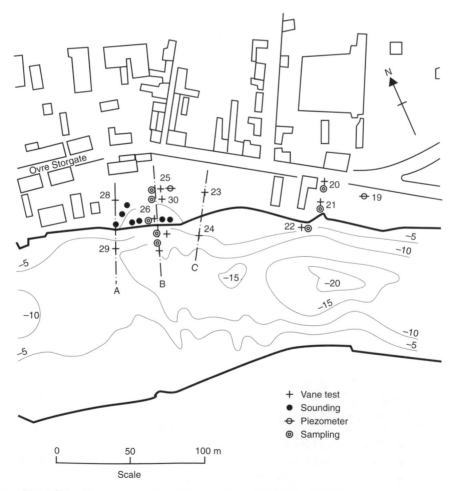

Fig. 5.5 Site plan, Drammen, after Kjaernsli and Simons (1962)

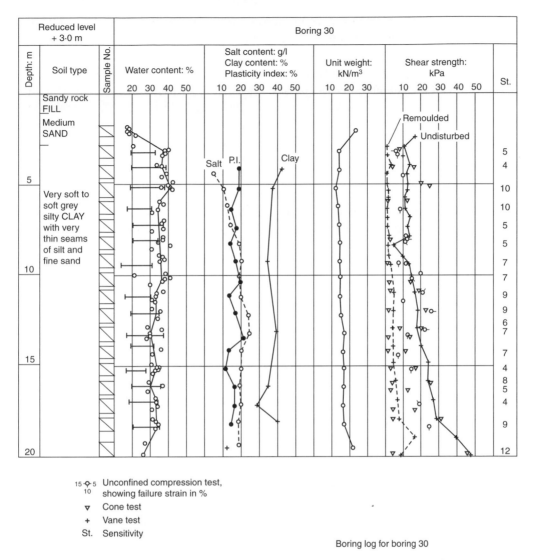

Fig. 5.6 Typical borehole log, Drammen, after Kjaernsli and Simons (1962)

slope the clay is very lightly over-consolidated as a result of removal of load by river erosion. A typical borehole is given in Fig. 5.6. Average values of the peak strength parameters are: $c' = 2\,kPa$, $\phi' = 32.5°$, as measured in drained triaxial tests on vertical axis specimens taken with a piston sampler in boreholes.

The main cause of the slide was a gradual steepening of the slope by erosion at the toe, but the failure was triggered by placing a small amount of fill at the edge of the bank. At the time of the slide the water level in the river was about 1 m below normal. There were no obvious signs of impending movement beforehand.

Stability analyses were carried out using Bishop's simplified effective stress method using peak parameters for three sections: A, immediately upstream of the slip; B, through the slip (the section being based on available information); and C, immediately downstream of the slip. The results of the calculations are given in Fig. 5.7. The minimum calculated factor of safety of 1.01 was obtained for a critical circle corresponding closely to the actual slip circle, as far as this could be determined from field observation. This suggests that the effects of rate of testing, strength anisotropy and progressive failure are negligible for the Drammen clay.

The results of total stress ($\phi = 0$) stability calculations based on the in situ shear vane and undrained compression strength tests on high quality samples from the NGI 54 mm diameter thin-walled stationary piston sampler are given in Fig. 5.8. The minimum calculated factor of safety was found to be 0.47, confirming that the undrained analysis cannot be used to assess the long term stability of a natural slope in the essentially normally consolidated soft clay at Drammen.

The results of piezometer observations at different depths at five locations are shown in Fig. 5.9 and indicate hydrostatic conditions. In the experience of the authors, this is a most unusual case as very often piezometer observations at different depths in a clay do not show hydrostatic conditions.

It should be noted that no residual shear strength determinations were carried out on the Drammen clay at the time the stability investigations were carried out. We believe, however, that the peak strength is close to the residual and that this is why progressive failure was not a factor in the failure.

Lodalen (Sevaldson, 1956)

A railway cutting, originally made in 1925, was widened in 1949. Five years later a slide occurred in the early morning of 6 October 1954. The sliding mass moved as an almost monolithic body, sinking about 5 m in the upper part and pushing forward about 10 m at the toe. Subsequent borings established the position of the slip surface at three points (Fig. 5.10) and, together with the back scarp, showed that the surface closely approximated a circular arc. Pore pressure measurements by piezometers revealed a small upward component of ground water flow, presumably influenced by artesian pressures in the underlying rock.

The clay, of post-glacial age, is lightly over-consolidated with an intact structure, and has an undrained shear strength ranging from 40 to 60 kPa. Average values of the index properties are: $w = 31\%$, $w_L = 36\%$, $w_P = 18\%$, clay fraction $= 40\%$, and sensitivity of 3. The peak strength parameters were determined in consolidated undrained triaxial tests, with pore pressure measurements, made on vertical axis specimens

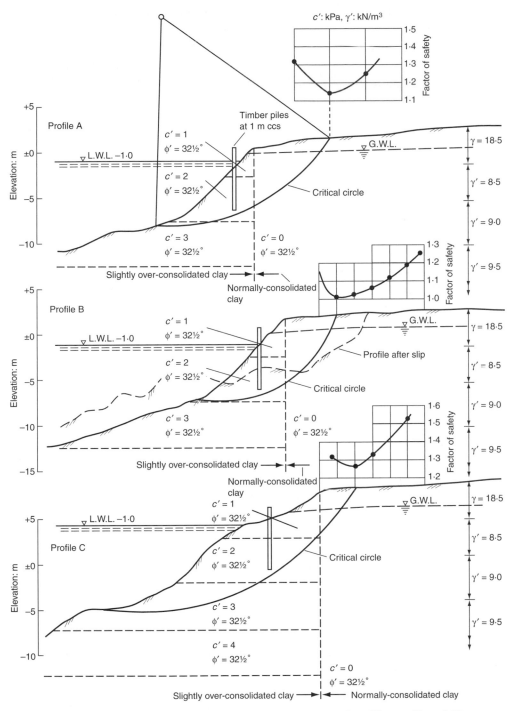

Fig. 5.7 Effective stress stability calculations, Drammen, after Kjaernsli and Simons (1962)

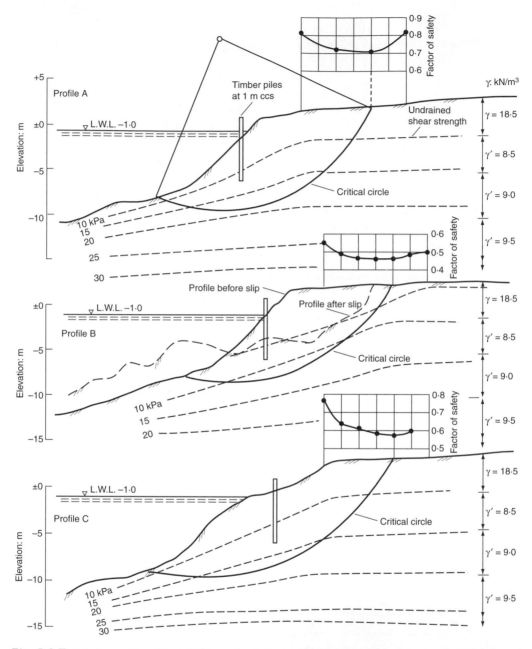

Fig. 5.8 Total stress ($\phi = 0$) stability calculations, Drammen, after Kjaernsli and Simons (1962)

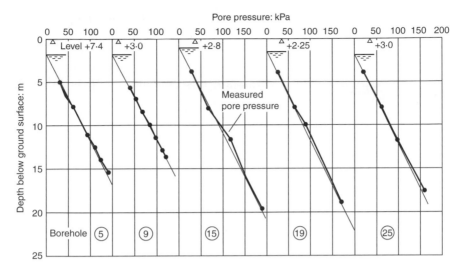

Fig. 5.9 Piezometer observations, Drammen, after Kjaernsli and Simons (1962)

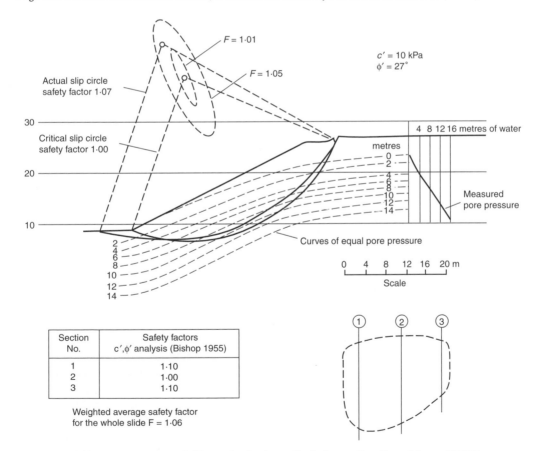

Section No.	Safety factors c′,φ′ analysis (Bishop 1955)
1	1·10
2	1·00
3	1·10

Weighted average safety factor
for the whole slide F = 1·06

Fig. 5.10 Effective stress stability calculations, Lodalen, after Sevaldson (1956)

135

taken by a thin-walled piston sampler from boreholes. The scatter of results was extraordinarily small, with average values: $c' = 10\,\text{kN/m}^2$, $\phi' = 27°$. The uppermost zone a few metres thick below original ground level constituted a 'drying crust' typical of Scandinavian clays, but the tension crack extended through this. Stability analyses using Bishop's method gave a minimum calculated factor of safety of 1.00. The corresponding critical slip surface differed slightly in position from the actual surface, and on the latter the calculated factor of safety was 1.07.

These figures are so close to unity that, as in the previous record from Drammen, the conclusion must be that the combined effect of various factors such as anisotropy and rate of testing is negligible.

Residual strength tests have not yet been made on the Lodalen clay, but as an upper limit we could safely assume: $c'_r = 0$, $\phi'_r = 27°$. With these parameters the factor of safety falls to 0.73. Thus it seems clear that progressive failure must have played a very small part as a cause of the slide. The delay of five years from excavation to failure is therefore probably associated with a slow decrease in effective stress following the removal of load from the slope in 1949.

The results of the total stress ($\phi = 0$) stability calculations based on the in situ shear vane and undrained compression strength tests on high-quality samples are given in Fig. 5.11. A minimum factor of safety of 0.93 was found with a weighted average factor of safety for the whole slip of 1.01. *It should be emphasized that this is a special case, Lodalen being a lightly over-consolidated clay, as in general a total stress analysis cannot be applied to a long term stability problem.* For normally consolidated clays, a $\phi = 0$ analysis will give a factor of safety much less than one for a long term failure (see Drammen above) and much greater than one for an over-consolidated clay (see Table 5.6). Lodalen, being a lightly over-consolidated clay, showed a factor of safety of about unity using a $\phi = 0$ analysis, and is not typical.

Selset (Skempton and Brown, 1961)

In the north Yorkshire Pennines, the River Lune, an upland tributary of the Tees, is eroding its valley through a thick deposit of clay till, probably of Weichselian age. At the section shown in Fig. 5.12, the river, when in flood, is cutting into the toe of a slope about 12.8 m high. When the site was first visited in 1955 clear evidence could be seen of a rotational landslide. Comparison of present topography with a map of 1856 showed that the rate of lateral movement of the river into the valley side was very slow. Piezometers indicated a flow pattern rather similar to that at Lodalen, with a component of upward flow from the underlying bedrock.

The till consisted of stones and boulders set in a sandy clay matrix ($w = 12\%$, $w_L = 26\%$, $w_P = 13\%$, clay fraction = 25%), forming a massive,

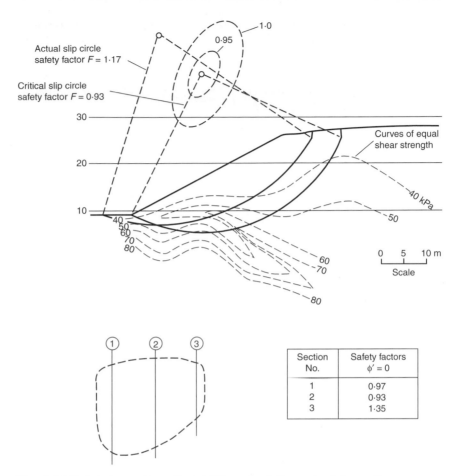

Section No.	Safety factors $\phi' = 0$
1	0·97
2	0·93
3	1·35

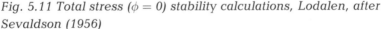

Fig. 5.11 Total stress ($\phi = 0$) stability calculations, Lodalen, after Sevaldson (1956)

stiff intact clay. Shear strength parameters of the matrix, measured in slow drained triaxial tests with a time to failure of up to two days, are: $c' = 8.6\,\text{kPa}$, $\phi' = 30°$ (peak); $c'_r = 0$, $\phi'_r = 28°$ (residual). The stress–strain curves were gently rounded at the peak.

Stability analyses by Bishop's method with peak strengths gave a minimum calculated factor of safety = 1.03 (typical result within a range from 0.99 to 1.14 controlled by two limiting assumptions concerning flow net patterns). In contrast, the factor of safety using residual parameters was 0.69 (see Fig. 5.13).

Thus, the actual strength around the slip surface at the time of failure must have been close to the peak strength as measured in the laboratory, and very much greater than the residual strength.

This conclusion is not unexpected. The strength of this type of clay would presumably be little influenced by rate of shearing and anisotropy

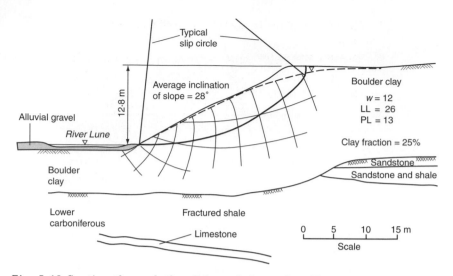

Fig. 5.12 Section through the slide at Selset, after Skempton and Brown (1961)

may well be insignificant, while any substantial reduction from peak strength by progressive failure is most unlikely with the non-brittle, almost flat-topped stress–strain curves.

Vaughan and Walbancke (1973) cite slope angles and heights of a number of other slopes in till under similar conditions, which give support to the conclusions derived from the detailed study of the Selset case.

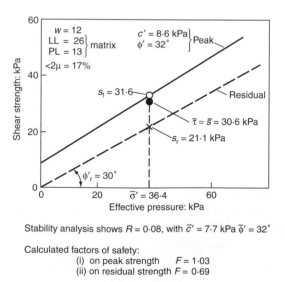

Stability analysis shows $R = 0.08$, with $\bar{c}' = 7.7$ kPa $\bar{\phi}' = 32°$

Calculated factors of safety:
    (i) on peak strength $F = 1.03$
    (ii) on residual strength $F = 0.69$

Fig. 5.13 Summary of data, Selset, after Skempton and Brown (1961)

First time slides in cuttings in stiff fissured clays

Overview

The development of railway and road systems in many countries required the construction of deep cuttings in stiff fissured clay. It quickly became clear that major slope instability problems were often involved. It was recognized that slope failures could develop with time, steep cuttings would fail during or shortly after construction, while with less steep slopes failure could occur months or years or decades after a cutting was made. It was also recognized that the failures were associated with a reduction in strength with time. For example, a cutting at Kensal Green in London which was widened in 1912 in order to construct a retaining wall, failed in 1941, the wall moving forward by about 0.45 m (see Fig. 5.14). Back-analysis of the failure indicated an undrained shear strength of 16 kPa, very much less than that measured on undisturbed samples. Early research concentrated on softening of the clay with time leading to reductions in undrained shear strength. The explanation of the softening process was assumed to be as given by Terzaghi in 1936, namely the infiltration of ground water into fissures opened as a consequence of lateral movements following stress release during excavation. Later on, it was felt that stability should be assessed in terms of effective stress and not total stress, and effective stress stability analyses of failures were carried out, assuming that equilibrium pore water pressures in the

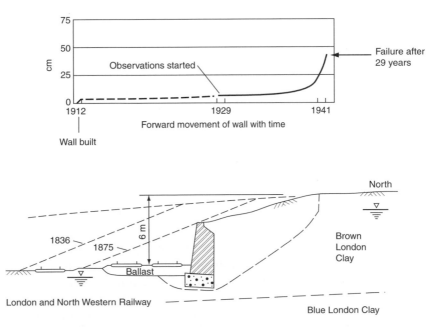

Fig. 5.14 Section through Kensal Green Wall, after Skempton (1977)

Fig. 5.15 Apparent reduction in c' with time, after De Lory (1957)

cut slopes would be achieved relatively quickly because of opening up of the fissures due to stress relief. Such analyses seemed to indicate that there was a reduction in c', the cohesion intercept with respect to effective stress, with time as indicated in Fig. 5.15.

It was thus necessary to:

- determine the in situ strength at failure in terms of effective stress
- find the best method of measuring or predicting this strength and
- find an explanation for the long delayed failures.

A major breakthrough occurred in 1977 when Professor Skempton gave a special lecture at the Ninth International Conference on Soil Mechanics and Foundation Engineering at Tokyo and the information that follows is based on that lecture.

Summary of Skempton's Tokyo Special Lecture (1977)

A cutting in blue London clay at Edgewarebury had been made in 1964, pore water pressures were measured in 1973 and were found to be negative. In 1973, r_u was -0.75 and in 1975, r_u was -0.62 ($r_u = \gamma_w h / \gamma z$ and r_u is the average r_u around a slip surface) – see Fig. 5.16.

A cutting at Potters Bar was constructed in the brown London clay to a depth of 11 m with a slope of 3H to 1V. The cutting was widened in 1956 on the west side only, with the east side unaltered apart from deepening by 1 m and the construction of a small toe wall. Piezometer readings on the west side in 1975 showed $r_u = 0.15$ after 19 years, and for the east side, $r_u = 0.32$ after 125 years (Fig. 5.17). The reduction in r_u with time

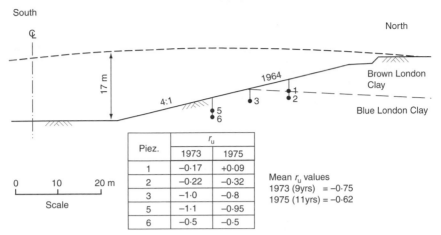

Piez.	r_u	
	1973	1975
1	−0·17	+0·09
2	−0·22	−0·32
3	−1·0	−0·8
5	−1·1	−0·95
6	−0·5	−0·5

Mean r_u values
1973 (9yrs) = −0·75
1975 (11yrs) = −0·62

Fig. 5.16 Edgewarebury cutting, after Skempton (1977)

is a very slow process in London clay, requiring about 40 to 50 years after construction of a cutting to reach equilibrium (Fig. 5.18). Data concerning first time slides in brown London clay are given in Table 5.7.

It can be seen from Table 5.7 that for slopes of $2\frac{3}{4}:1$ or steeper, failures can be expected in the intermediate term, while for slopes deeper than about 10 m, a slope of say 4 : 1 may be required for long term stability. The data for the long term slides shown in Table 5.7, have been back-analysed taking $F = 1.0$ and assuming $r_u = 0.30$. The results are shown by the solid points in Fig. 5.18. Effective stresses were also calculated for $r_u = 0.25$ and 0.35 and are indicated on Fig. 5.19.

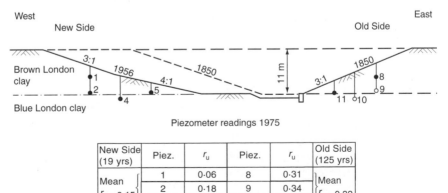

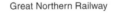

New Side (19 yrs)	Piez.	r_u	Piez.	r_u	Old Side (125 yrs)
Mean $r_u = 0.15$	1	0·06	8	0·31	Mean $r_u = 0.32$
	2	0·18	9	0·34	
	5	0·21	11	0·31	
	4	0·09	10	0·32	

Fig. 5.17 Potters Bar cutting, after Skempton (1977)

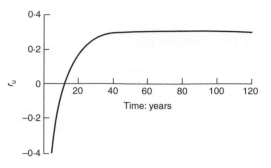

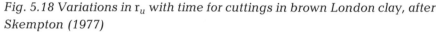

Fig. 5.18 Variations in r_u with time for cuttings in brown London clay, after Skempton (1977)

The best fit to the points is a line defined by the parameters:

$c' = 1 \, \text{kPa}$ and $\phi' = 20°$

and a lower limit is given by

$c' = 0 \, \text{kPa}$ and $\phi' = 20°$

Laboratory tests show that the peak strength parameters for brown London clay are:

$c' = 14 \, \text{kPa}$ and $\phi' = 20°$

for small samples, and

$c' = 7 \, \text{kPa}$ and $\phi' = 20°$

for large triaxial samples (250 mm diameter).

The residual parameters for brown London clay measured on natural slip surfaces gave the parameters:

$c'_r = 1.4 \, \text{kPa}$ and $\phi'_r = 13°$

which control the stability of natural slopes.

Table 5.7 First time slides in brown London Clay, after Skempton (1977)

	Site	Date of cutting	Date of slip	Time to failure: years	Height: m	Slope
Intermediate	New Cross	1838	1841	3	17.0	$1\frac{1}{2}$:1
	Kingsbury	1931	1947	16	6.0	$2\frac{1}{4}$:1
	St Helier	1930	1952	22	7.0	2:1
	Cuffley	1918	1953	35	7.2	$2\frac{3}{4}$:1
Long term	Sudbury Hill	1903	1949	46	7.0	3:1
	Crews Hill	1901	1956	47	6.2	$3\frac{1}{3}$:1
	Grange Hill	1902	1950	48	12.2	$3\frac{1}{4}$:1
	West Acton	1916	1966	50	4.9	3:1
	Hadley Wood widened	1850 1916	1947	c. 65	10.4	$3\frac{2}{3}$:1

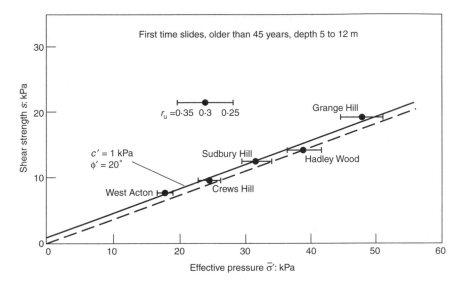

Fig. 5.19 Effective pressures for brown London clay, after Skempton (1977)

These parameters are shown in Fig. 5.20. The parameters that govern the stability of first time slides in the brown London clay are approximately equal to the 'fully softened' or 'critical state' condition which can be determined by measuring the strength of remoulded normally consolidated clay. Similar parameters are obtained from tests on joints and fissures.

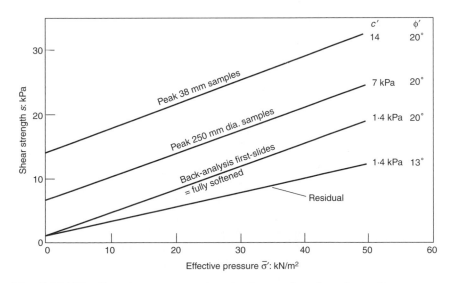

Fig. 5.20 Effective stress parameters for brown London clay, after Skempton (1977)

Table 5.8 Height–slope relationships for cuttings in brown London clay, after Chandler and Skempton (1974)

Height of slope: m	Slope gradient
12	$3\frac{1}{2}:1$
6	$3:1$
3	$2\frac{1}{2}:1$

In his Special Lecture at Tokyo, Professor Skempton drew the following conclusions.

- The shear strength parameters of the brown London clay relevant to first time slides are: $c' = 1\,\text{kPa}$ and $\phi' = 20°$.
- The peak strength even as measured on large samples, is considerably higher, so some progressive failure mechanism appears to be involved.
- The in situ strength is given approximately by the 'fully softened' value and also by the lower limit of strength measured on structural discontinuities (joints and fissures).
- The residual strength is much smaller than this and corresponds to the strength mobilized *after* a slip has occurred, with large displacements of the order of 1 m or 2 m.
- It is a characteristic feature of first time slides in London clay that they generally occur many years after a cutting has been excavated.
- The principal reason for this delay is the very slow rate of pore pressure equilibration; a process which in typical cuttings is not completed, for practical purposes, until 40 or 50 years after excavation.
- At equilibration, $r_u = 0.3$.

Chandler and Skempton (1974) carried out slip circle analyses to determine height–slope relationships for cuttings in the brown London clay based on the assumptions: $F = 1.0$, $r_u = 0.3$, $c' = 1.5\,\text{kPa}$, and $\phi' = 20°$. The results are given in Table 5.8.

Clearly, if a design factor of safety of greater than unity is required, and presumably this is always the case, then flatter slopes than those indicated in Table 5.8 will be required. The subsequent use of the slope, or the presence of nearby structures, will also influence the choice of factor of safety.

Chronology of beliefs relating to stability of cuttings in stiff fissured clay
Research has developed through a chronology of beliefs that delayed failures of cuttings in stiff fissured clays could be explained by

- reduction in undrained strength with time due to swelling (1940s), through

- apparent reduction of cohesion intercept with respect to effective stress, c', with time (assuming fully drained conditions, because of fissures opening up in a comparatively short time) (1950s), to
- slow rate of equilibration of pore water pressures (say over 50 years) (1970s to the present time); see long term stability models in Chapter 3.

Natural slopes in stiff fissured clays

Jackfield (Skempton, 1964)

In 1952, a landslide occurred at the village of Jackfield, Shropshire, on the River Severn, about 2 km downstream of Iron Bridge; destroying several houses and causing major dislocations in a railway and road. In this locality the Severn flows through a vee-shaped valley (the so-called Iron Bridge gorge) which has been eroded largely, if not entirely, since the retreat of the main ice sheet of the Last Glaciation. Erosion is indeed still continuing, and the sides of the valley are covered by a mosaic of landslides, of varying ages.

It is possible that previous landslides may have taken place along at least a part of the present slip surface, but the slope must have been more or less stable for a long time before 1950, when warnings of instability were observed in the form of a broken water main serving cottages near the river bank. Towards the end of 1951 further movement was noted, and by February 1952 the road was becoming dangerous. During the next month or two the landslide developed alarmingly. Six houses were completely broken up, gas mains had to be relaid above ground, the railway could be maintained only by daily adjustments to the track and a minor road along the river had to be closed to traffic. By this time the maximum downhill displacement totalled 18.3 m (see Fig. 5.21).

The strata, consisting of very stiff clays and mudstone, alternating with marl-breccia and occasional coal seams, dip gently in a south-easterly direction with the strike running roughly parallel to the section of the landslide. The slide, however, was confined wholly within the zone of weathered, fissured clay extending to a depth of 6–8 m below the surface (Fig. 5.22). The slip surface ran parallel to the slope (which is inclined at 10°), at an average depth of 5 m. The length of the sliding mass, measured up the slope, amounted to about 170 m and in the winter 1952/53 ground water level reached the surface at a number of points, although on average it was located at a depth of 0.6 m.

Analysis of the forces acting on the slip surface shows $\sigma'_n = 62$ kPa and $\tau = 19$ kPa. Drained shear tests on samples taken from depths between 5 m and 6 m, but not in the immediate vicinity of the slip plane, showed peak strength parameters of $c' = 11$ kPa and $\phi' = 25°$. When these tests were made, the significance of residual strengths was not clear. Fortunately, however, in most cases the observations were continued

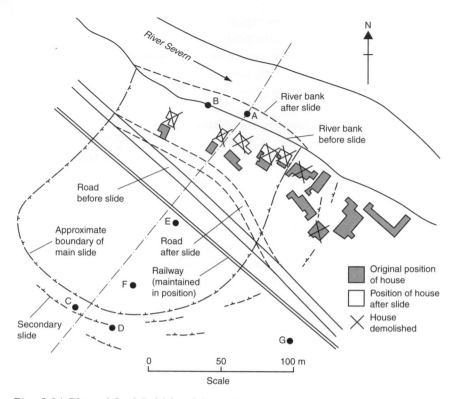

Fig. 5.21 Plan of Jackfield landslip, after Skempton (1964)

throughout the full travel of the shear box, and it is possible from the results to make an approximate estimate of the residual angle of shearing resistance, giving $\phi'_r = 19°$.

The peak and residual strengths corresponding to the average effective pressure of 62 kPa acting on the slip surface, are 40 kPa and 20 kPa respectively. But, as previously mentioned, the average shear stress (and hence the average shear strength) along the slip surface at the time of failure was 19 kPa.

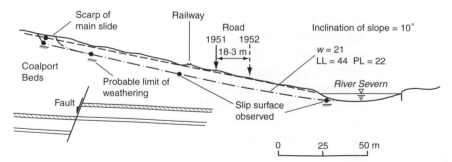

Fig. 5.22 Section through the Jackfield landslip, after Skempton (1964)

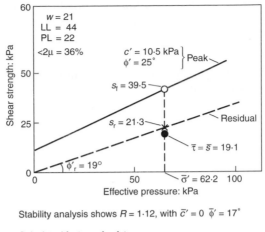

Stability analysis shows $R = 1\cdot12$, with $\bar{c}' = 0$ $\bar{\phi}' = 17°$

Calculated factors of safety:
       (i) on peak strength $F = 2\cdot06$
       (ii) on residual strength $F = 1\cdot11$

Fig. 5.23 Analysis of the Jackfield landslip, after Skempton (1964)

It is therefore clear that when the landslide took place the strength of the clay was closely equal to its residual value. In fact, taking $\phi'_r = 19°$, it is found that the residual factor $R = 1.12$ but, when the approximate nature of ϕ'_r is taken into account, it is doubtful if the value of R is significantly different from 1.0 (see Fig. 5.23).

Expressing the results in another way, had the peak strength been used in a stability analysis of the Jackfield slope, the calculated factor of safety would have been 2.06 (an error of more than 100%, since the true factor of safety was 1.0). On the other hand, using even the rather crude value of $\phi'_r = 19°$, the calculated factor of safety based on residual strength would differ by only 11% from the correct result.

Coastal cliffs (Hutchinson, 1967)

Under conditions of fairly strong marine erosion the cliffs are subject to rotational or compound sliding on deep or moderately deep slip surfaces. Where the rate of erosion is less severe the typical pattern is dominated by shallow slides and mudflows. The slopes of all these eroding cliffs are characteristically irregular, with average inclinations between about 15° and 30°.

At various places along the coast, sea defences have been constructed which prevent further erosion at the foot of the cliffs. If no further stabilization works are carried out, such as drainage or re-grading, the slopes then enter the phase of free degradation. Eight slopes in this category have been surveyed. Their inclinations range from 13° to 20° and they show clear evidence of instability, in the form of shallow rotational slides involving either the whole or part of the slope. These defended cliffs have been free from marine erosion for periods of about 30 to 150 years.

Where marshes have formed, particularly in estuaries, the sea has retreated from the old cliffs which, generally, have been left to flatten their slopes undisturbed by stabilization measures. Surveys of ten of these freely degrading abandoned cliffs show inclinations of 8.5° to 13°.

Slopes steeper than 9.5° are still unstable, and are characterized by successive shallow rotational slips, while the flatter slopes exhibit well marked undulations which almost certainly represent the subdued remains of quiescent successive slips. These slopes, at 8.5° to 9.5°, may be regarded as being in a transitional state, approximating the condition of final equilibrium.

Inland slopes

Observations by Skempton and De Lory (1957), greatly extended by Hutchinson (1967), have shown that many inland slopes in London clay or soliflucted London clay are unstable even though they are not currently subject to stream erosion. Two clearly differentiated types of instability can be noted: successive slips and transitional slab slides. Shallow, markedly non-circular slides also occur which may be a variant form of the slab-like movements. In addition the undulations, previously mentioned, are common.

Transitional slab slides are found at inclinations ranging from 8° to 10°. Their shape suggests that failure is taking place on pre-existing solifluction shears running parallel to the surface.

Successive slips have been observed on slopes inclined at angles between 9.5° and 12°, with one exception at 8.5°. These slopes are so similar to the abandoned cliffs in their form of instability and range of inclination that we consider them to be closely equivalent, and infer that postglacial erosion has removed the solifluction mantle leaving a slope essentially in the London clay.

Undulations occur at inclinations from 8.5° to 10.5°, the lower limit on these inland slopes being identical with that on the abandoned cliffs. It seems, then, that while the minimum unstable angle is 8°, this is almost certainly associated with renewed movements on solifluction slip surfaces; and the angle of ultimate stability of London clay itself is around 9°.

It has also been observed that the maximum stable slope in London clay is about 10°. An overlap of 1° or so can easily be accounted for by modest differences in the position of ground water level as between one site and another.

There is no great difficulty in deriving a quantitative explanation of these field observations provided it is assumed that the strength of the clay has fallen to its residual value. This indeed will be the case on a solifluction slip surface, and with $c_r' = 1.0$ kPa, $\phi_r' = 13°$, movement can take place at an inclination of $8°$ if ground water is near the surface of the slope.

The slightly steeper inclination of $9°$ for un-soliflucted London clay presumably reflects the presence of a series of curved but interlinking slip surfaces rather than a continuous planar shear.

Slides on pre-existing slip surfaces

Overview

Slip surfaces can be caused by landsliding, by solifluction and by tectonic shearing. Examples of recent movements along pre-existing slip surfaces in each of these categories are given below. The movements may be continuing post-failure displacements or they may result from re-activation (e.g. caused by excavation at the toe of a slope) but in both cases we are dealing with a condition of limiting equilibrium controlled by the residual strength along the slip surface. In this respect the present set of records differ sharply from the first time slides.

Sudbury Hill (Skempton, 1964)

After the slide in 1949 in the London clay cutting at Sudbury Hill no remedial measures were carried out and the slumped mass continued to move intermittently for several years. Small amounts of clay were removed from time to time at the toe (probably in the winter months) to prevent it encroaching on the railway track.

When the profile was surveyed in 1956 the overall displacement amounted to at least 2 m (see Fig. 5.24). The post-failure movements must therefore correspond to a factor of safety $= 1.0$ on residual strength.

Using the Morgenstern–Price analysis the average effective normal stress and shear stress along the slip surface are:

$$\sigma_n' = 29 \text{ kPa} \qquad \tau = 8 \text{ kPa}$$

If c_r' is made equal to zero the value of ϕ_r' corresponding to these stresses is $14.4°$ (calculations on a circular slip surface (Skempton, 1964) gave $\phi_r' = 15°$ if $c_r' = 0$). As mentioned earlier, tests on natural slip surfaces in brown London clay are best represented by $\phi_r' = 13°$ and a small cohesion intercept. The Sudbury Hill post-failure analysis shows $c_r' = 1.0$ Pa and $\phi_r' = 13°$.

Other classic slides of this type, which have been summarized by Skempton and Hutchinson (1969), are

- Folkestone Warren
- Walton's Wood
- Sevenoaks Weald
- River Beas Valley

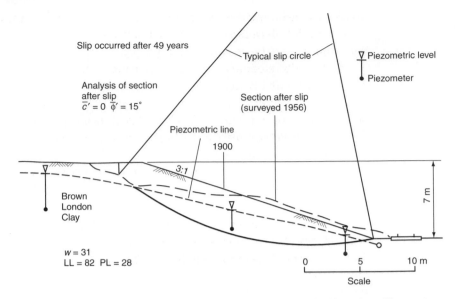

Fig. 5.24 Section through Sudbury Hill landslide (1949), after Skempton (1964)

All these cases confirm that, at failure, the mobilized strength was very close to the residual strength.

Lessons learned from case records of slope failures

- After a slide has taken place, the strength on the slip surface is then equal to the residual value. The residual strength is associated with strong orientation of the clay particles and is represented by an angle of shearing resistance ϕ'_r, which in most clays is considerably smaller than the value of ϕ' at peak strength.
- First time slides in slopes in non-fissured clays correspond to strengths only very slightly less than peak.
- First time slides in cuttings in fissured clays correspond to strengths well below peak, but generally above the residual and equate to the fully softened (or remoulded or critical state) value.
- Some form of progressive failure must be operative to take the clay past the peak. This is probably the result of a non-uniform ratio of stress to strength along the potential slip surface but also the fissures play an important role as stress concentrators and in leading to softening of the clay mass.
- London clay, and probably other stiff fissured clays, undergoes a loss in strength in cuttings tending towards the fully softened (or remoulded or critical state) value. Just before a first time slide occurs, there is a softened shear zone with many minor shears. It is possible that some

over-consolidated clays may exhibit a more marked reduction in strength before a first time slide takes place.

- In all clays, the residual strength will be reached after a continuous principal slip surface has developed and, in the field, this state appears to be attained typically after mass movements of the order of a metre or so.

- The angle of ultimate stability of clay slopes is probably controlled by the residual strength and this may result from successive slipping or solifluction movements.

- The residual strength obtains on pre-existing shear surfaces, whether these are the result of tectonic shearing or old landslides.

- First time slides in cuttings in London clay generally occur many years after excavation. The main reason for this delay is the very slow rate of pore pressure equilibration. Typically this is not complete until 40 or 50 years after the cutting has been excavated.

CHAPTER SIX
Stabilizing and investigating landslips

Remedial measures for stabilizing landslips

Overview

Many general reviews of methods of slope stabilization have been made. Some of the more recent are by Root (1953), Baker and Marshall (1958), Brawner (1959), Zaruba and Mencl (1982), Duncan (1971, 1976), Schweizir and Wright (1974), Smith (1974), and Broms (1969, 1975). Hutchinson (1978, 1982, 1983, 1984a, 1984b) has made outstanding contributions in this area and the authors have drawn freely on his work in preparing this chapter. Professor Hutchinson's kind permission to do this is gratefully acknowledged.

The main methods of slope stabilization are described in the sections that follow. They fall into the following categories:

- drainage
- changing the slope geometry
- earth retaining structures, including the use of anchors
- miscellaneous methods.

It may be advantageous on occasions to use a combination of some of these methods to obtain an optimum solution.

Drainage

Overview

Drainage can be a most effective form of slope stabilization provided that the drains are properly maintained but this rarely occurs in practice. Drainage channels should be inspected on a regular basis and kept free of debris and drainage layers should be carefully designed using the well known filter rules to prevent clogging up with time. Weep holes in walls should be cleared on a regular basis and a properly designed drainage blanket should be placed behind the wall.

A slope protected by drainage measures should be monitored by piezometers and other indicators to check that the drains are functioning properly.

It is common practice in earth dam engineering to monitor slopes on a regular basis to check that all is well. This is not usually the case with natural or man-made slopes, even if failure could result in damage to property or loss of life. In the latter cases, the authors believe that stabilization works should include a programme of monitoring and maintenance on a regular basis.

Surface drainage

Ponds formed on a slope or behind the crest should be drained and the water should be conducted away from the site and discharged at a safe location.

Tension cracks should be dealt with to prevent the build-up of water pressure in them. They are often sealed but this is likely only to be effective in the short term, as the seal is usually broken as a result of further down-slope movement. Drainage of tension cracks should be considered.

Cut-off trench drains are often placed behind the crest of a motorway cutting to intercept ground water flow. The location of such cut-off trenches should be chosen with care to avoid the trenches acting as potential tension cracks in subsequent landslides. It is advisable to cover the bottom of a cut-off trench with an impermeable flexible lining to minimize the risk of a trench acting as a source of water at another point in the cutting.

An interesting situation developed in a long slope in the London clay at Whitstable. Although the slope was long and fairly uniform, instability was only evident at one location. Careful examination revealed that the ground surface at the top of the slope behind the failing masses sloped almost imperceptibly towards the unstable section thus leading surface water directly to the slip. There is no doubt that this exacerbated an already critical situation.

Erosion control

Toe erosion   If the toe of a landslide is located under water, in the sea or a river for example, it is essential to prevent erosion at this critical point. Rock armour, cribwalls or gabions, can supply effective protection.

Surface erosion   This can be controlled by establishing vegetation on the slope, using berms to lead water safely away, or providing shallow herringbone drainage systems. In parts of the world where rainfall is high, e.g. in Hong Kong and Malaysia, a thin layer of soil cement is sometimes used to provide not only erosion control, but also to prevent rainfall from entering the slope. These soil cement layers require drainage behind them to prevent water pressure building up. Short lengths of pipe are

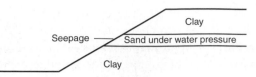

Fig. 6.1 Seepage erosion

often inserted through the soil cement layer for this purpose although the effectiveness of such drainage pipes may be doubtful.

Seepage erosion   A potentially dangerous situation is illustrated in Fig. 6.1 and is not uncommon. If left untreated, the superincumbent strata can be undermined and the underlying clay stratum may be softened or eroded leading to stability problems. Seepage erosion may be controlled by placing inverted filters over the area of discharge (Terzaghi and Peck, 1967), taking care that the materials used satisfy the well known filter criteria, or by intercepting the seepage at some distance back from the face with wells or sand drains.

Deep drainage trenches
Case study: University of Surrey, Guildford UK   An interesting case record where deep drainage trenches were used to stabilize an unstable slope in London clay arose when the University of Surrey was constructed directly over part of an old landslip at Stag Hill in Guildford, Surrey, in the UK (Simons, 1977).

That such a suitable site as this could be found undeveloped so close to a city like Guildford, is due entirely to a geotechnical reason and that is that a landslip had taken place on the slopes of Stag Hill probably some time during the 19th century and, following this, the site was blighted and not developed.

Boreholes and trial pits were put down, the distribution of pore water pressure throughout the slopes, i.e. at different depths and locations, was measured, a comprehensive laboratory investigation was carried out on undisturbed samples recovered from boreholes and trial pits, and stability calculations in terms of effective stress were performed. Figure 6.2 shows the site in 1952 before construction, and Fig 6.3 shows the site during construction in 1969. In both aerial photos, the extent of the slip can be clearly seen.

The site: topography and geology: The site is shown on the site plan of Fig. 6.4 and the slopes on which the main part of the university is built are remarkably uniform varying from about 8° to 10°. The slopes become gentler moving further to either the east or the west.

The whole site is underlain by London Clay and, according to geological records, this is followed by the Woolwich and Reading beds, and then the

Fig. 6.2 Aerial view of the University of Surrey site in 1952 before construction of the university commenced. Guildford Cathedral is partly built. Slip scarps can be seen to the left of the Cathedral boundary hedgerow. Note the concentration of vegetation in the slip scarps

Chalk, but none of the boreholes put down on the site penetrated through the London Clay. The London clay is an over-consolidated jointed fissured clay of Eocene age.

Towards the top of Stag Hill, undisturbed brown London Clay overlying the blue London Clay, extends to the ground surface. Close to the toe of the slip and further north, the undisturbed clay is overlain by up to 6 m of brown and yellow mottled sandy and silty clay. It is probable that this upper zone is a redeposited mixture of London Clay and the more sandy Claygate beds which overlie the London Clay.

Confirmation that this upper zone was redeposited was obtained from a trial pit more than half-way up the slope where fragments of chalk were found at a depth of 1.5 m.

Fissures and joints were observed throughout the London Clay, and in the redeposited material in the trial pits, and in the samples taken from greater depth.

Fig. 6.3 Aerial view of the University of Surrey site in 1969. The scarp and toe of the slip can be seen in the right hand corner of the photograph

Piezometers: To establish the pore water pressure distribution in the clay, piezometers were installed. These consisted of a porous pot surrounded by a plug of sand in a borehole with a small diameter pipe or tube being connected to the pot and taken vertically up the borehole to the ground surface, the space above the sand plug being backfilled with a bentonite–cement grout, to prevent leakage down the borehole, thus ensuring that the measurements reflect correctly the pore water pressure around the sand plug. The dimensions of the sand plug and of the internal diameter of the connecting tube were so chosen with respect to the permeability of the clay, that a suitably short response time was achieved, i.e. that a change in pore pressure was followed reasonably quickly by a corresponding change in water level in the tube.

In October and November 1965, 19 piezometers were installed (the 100 and 200 series), and records were taken, but most of these piezometers were soon lost during construction of the Phase 1 buildings.

In December 1966, six additional piezometers were installed, A, B, C and D on a line down the slope west of Wates House and through what

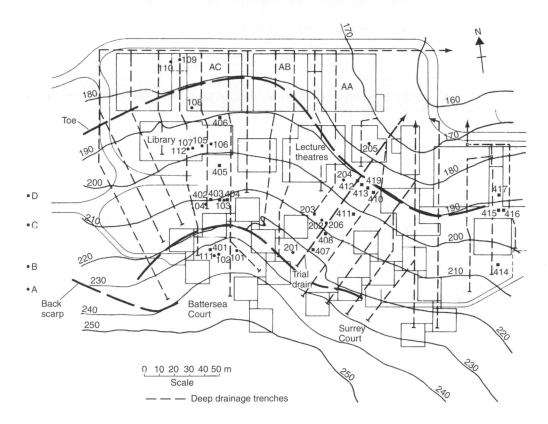

Fig. 6.4 Site plan of the University of Surrey (contours in feet above sea level)

is now the Leggett Building, and 201 and 204 were reinstalled. Piezo-meters A, C, B and D were in ground not affected by drainage measures at that time, so that the readings here could be compared with those affected by the drainage system.

In March 1969, 17 new piezometers were installed (the 400 series) and regular records have been taken since then. At the present time, of the 42 piezometers originally installed, only six appear now to be functioning properly and this kind of wastage is by no means unusual. Loss of piezo-meters due to subsequent building and to vandalism is only too common.

It was decided that the only practicable and economic way of stabilizing the slope would be to reduce the water pressures in the slopes perma-nently by installing a system of deep, gravel-filled, drainage trenches. This system had successfully been proved previously on London Clay slopes. Gregory (1844) described the use of gravel-filled counterfort drainage trenches to stabilize cuttings in the London Clay on the line of the London and Croydon Railway. A number of recommendations were then made for Stag Hill as follows.

- A trench drainage system to be installed with drains nominally 5 m deep at 15 m centres.
- Foundations for heavy buildings to be piled transferring loads below the level of the slip surface.
- Low rise buildings applying a gross foundation pressure of not more than 37 kPa at 1.5 m below existing ground level, could be constructed without piling. Taking account of the weight of excavated soil this means that the net increase in load on the ground at foundation level was about 8.6 kPa.

In December 1965, a trial slope drain was constructed, and between March and June 1966 the main Phase I slope drainage system was constructed, bored cast in situ piling was then installed and construction of the buildings followed.

In September 1965, the total cost of making the site available for the proposed University development was estimated as follows:

Land price	£350,000
Stabilizing drainage system	£153,000
Extra foundations	£100,000
Total	£603,000

This gave a cost per acre of £7,100 that compared with a cost of £30,000 per acre quoted for another new university being developed in an urban environment.

So, in spite of the fact that additional costs were required so that the University of Surrey could be safely constructed on an existing landslip, there was no doubt whatsoever that the development should proceed, and in fact the university was fortunate in acquiring so appropriate and valuable a location for its development.

Results of piezometer observations   Because of space limitations, only the main points arising from the observations can be discussed.

(a) *Initial distribution of pore water pressure with depth.* There is evidence that a classical under-drainage system exists on the site. Figure 6.5 shows the results of pore water pressure observations made on piezometers prior to the introduction of the drainage system, and it is clear that the deeper a piezometer is installed, the lower the piezometric level. This is due to under-drainage into the underlying chalk, resulting from wells having been sunk in the chalk for water supply purposes, which has led to a progressive under-drainage in the London Basin. Figure 6.6 shows the reduction in water head in the chalk for a point located in Hyde Park (Wilson and Grace, 1942). This has resulted in a settlement in the London

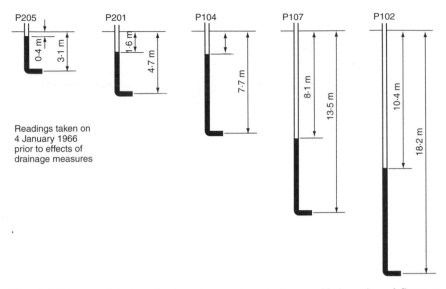

Fig. 6.5 Piezometric levels showing under-drainage, University of Surrey

clay, which varies depending on the thickness of the London Clay at any point, and the amount of water level reduction, and has a maximum value of about 0.3 m for the period 1865 to 1975.

This is of crucial importance in slope stability problems, since the actual distribution of pore water pressure with depth will govern the factor of safety of the slope and also the depth of the failure surface. In cases where the effects of under-drainage are apparent in the London Clay then the depth to the failure surface will be significantly

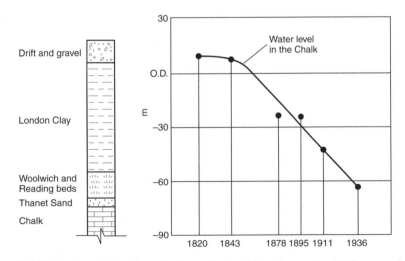

Fig. 6.6 Water levels in the chalk in Central London, after Wilson and Grace (1942)

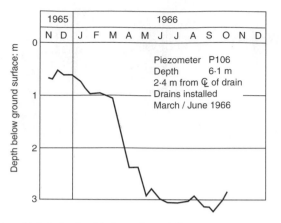

Fig. 6.7 Effects of trial drain, University of Surrey

less than if a hydrostatic pore water pressure distribution is assumed. Similar distributions of pore water pressure with depth in the London clay, at other locations, have been reported by Butler (1972). It is therefore essential when investigating the stability of a London clay slope that piezometers be installed at varying depths, and of course at different locations, so that a reliable knowledge of the pore water distribution throughout the slope is obtained.

(b) *Effect of drain installation – initially.* The effect of the trial drain installed in December 1965 is shown in Fig. 6.7 and a significant draw-down in water pressure can be seen. It should be pointed out, however, that this draw-down is also partly due to the seasonal drop in water pressure, which must be expected during the spring–summer period. Furthermore, the construction of the trench drains involves cutting deep slots into the London clay thereby for a short time reducing the horizontal total stress on the sides of the trench to zero. This temporary stress reduction would also lead to a reduction in pore water pressure.

Most of the piezometers were then lost during building construction, and it is necessary to turn to the piezometers installed in 1969 for further information. Figure 6.8 shows the observations for piezometer C, in an undrained area, and for piezometer 404 installed some 1.5 m from a drain, at the same surface ground level. It can be seen that the drainage system has depressed the pore water pressures in the slope, in particular in the winter months where a reduction in head of about 2.7 m has been achieved. If the average depth of the slip surface is 6.1 m this means an increase in the factor of safety against sliding of some 42%, which is appreciable! There are, therefore, no grounds for concern as to the safety of the university

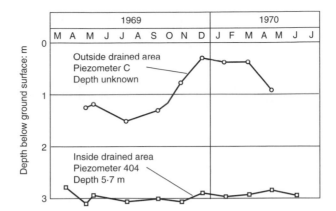

Fig. 6.8 Comparison of piezometers C and P404, University of Surrey

buildings. Figure 6.9 shows the variation in pore water pressure for piezometer 404 over a number of years.

Figure 6.10 indicates the distribution of pore water pressure across two drainage trenches and although the draw-down towards the drains is perhaps somewhat less than might have been hoped for, the beneficial effects of the drainage system are obvious. A word of warning must, however, be inserted here. While deep drains have been shown to lower the water pressures throughout the slopes here at Guildford, it does not necessarily follow that a similar reduction can be expected for all London clay slopes.

Buckthorp *et al.* (1974) report a case where deep drains did not have the desired effect and this was because the site had previously been wooded, and the trees were removed immediately prior to the drains being installed. The destruction of the root system offset any reduction in pore water pressure which the drains otherwise would have had.

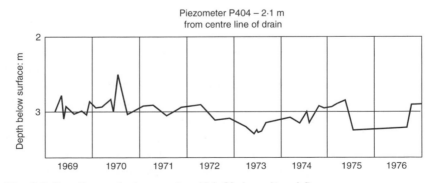

Fig. 6.9 Readings of piezometer 404, University of Surrey

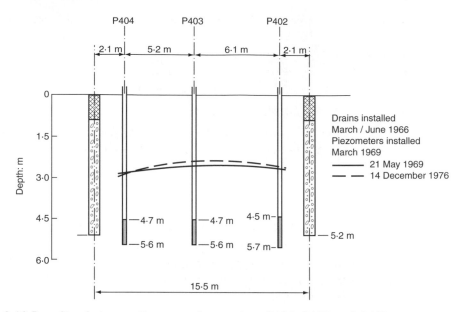

Fig. 6.10 Results of observations on piezometers P404, P403 and P402

Weeks (1969) described the effects of counterfort drains at Sevenoaks, near London, where an average reduction in pore water pressure head of 2 m was achieved at the depth of the slip surface.

The results of the pore water pressure observations at the University of Surrey site confirm the original design decisions and show that major structures can be safely founded directly on an old landslip, provided a proper drainage system is installed and provided heavy building loads are taken down below the level of the slip surface. Due attention must, however, be paid to the actual pore water pressure distribution throughout the slope, and to the effects of removal of vegetation.

Influence of barometric pressure It is interesting to note that when measuring pore water pressures on the Stag Hill site, the observations reflected changes in barometric pressure. Köhler and Schulze (2000) measured barometric variations in pore pressure in a cut constructed in the 1920s in Lias clay close to Lühnde (Germany) that they instrumented in 1997. From inclinometer readings they found that slope movements could be correlated with major barometric pressure drops (up to 5 kPa below mean barometric pressure).

Three-dimensional design considerations The three-dimensional trench problem is highly complex, taking into account several factors including:

- non-homogeneous and anisotropic permeability varying with effective stress,
- partial clogging of the sides of the trenches (smear)
- variable inflow of ground water from upslope
- intermittent rainfall, involving saturated and partially saturated flow.

A simplified solution based on two-dimensional flow has been presented by Hutchinson (1978) and by Bromhead (1984) and provides a basis for the design of a trench drainage system. Drain widths on 0.5 m to 0.9 m, depths of 3.5 m to 5.0 m, and spacings of 10 m to 20 m are common.

Bored piles and counterfort drains

Case study: Gypsy Hill, London UK   The use of bored piles together with counterfort drains to stabilize a major landslip in a clay slope at Gypsy Hill, South London, has been described by Allison *et al.* (1991). Counterfort drains are trench drains that also act as a structural element strengthening or 'buttressing' the slope.

In June 1980, a row of Victorian detached houses in Victoria Crescent on the west side of Gypsy Hill in South London was demolished to make way for a new housing development. The rear gardens of the houses sloped upwards, away from the houses, at a steep angle and were retained at the toe of the slope by brick walls up to 2.6 m in height. The slope angle immediately behind the walls was about 19°.

The height of the retaining wall was reduced to approximately 0.5 m to improve daylighting to the rear of the new properties. This was achieved by providing a new reinforced concrete wall at the toe of the slope, and regrading the rear gardens by steepening them over a distance of some 20 m to join the existing slope line near to the upper site boundary.

Work proceeded on the new three-storey townhouses and the roof construction was nearing completion when, in February 1984, after a period of heavy rain, major downhill slip movement took place in the rear garden slope. Saturated soil flowed over the top of the new reinforced concrete retaining wall and piled up against the rear wall of the new buildings.

The movement took place across practically the full width of the site, and extended beyond the upper site boundary, causing severe disruption to an access road and partly exposing the foundations of a three-storey block of flats. The area of the slip within the site boundary is shown in Fig. 6.11.

Site investigations revealed that the slip had taken place on a pre-existing shear plane at a depth of about 4 m to 5 m below the ground surface. This shear plane was found to be the interface between relatively

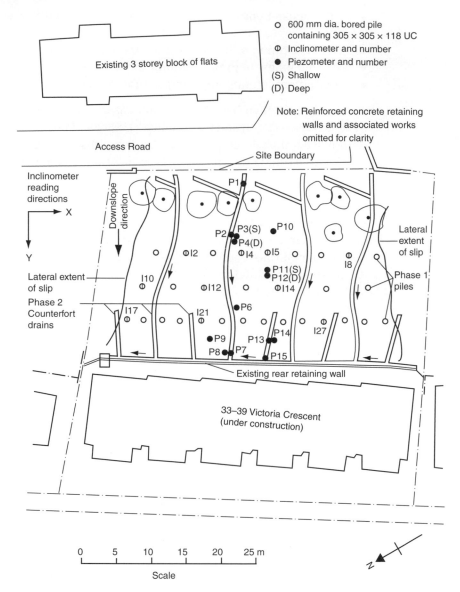

Fig. 6.11 Site plan, Gypsy Hill, after Allison et al. (1991)

intact London clay and an overlying mantle of colluvium. Piezometers installed immediately after the slip indicated a ground water surface approximately parallel with the slope, about 0.5 m below ground level. The movement was occurring on an approximately planar surface with an average slope of about 9°.

In the absence of remedial measures, continued downhill movement of material on the slope would result in the following.

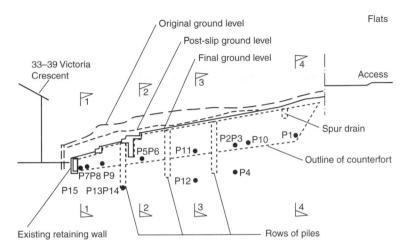

Fig. 6.12 Section through site, Gypsy Hill, after Allison et al. (1991)

- A build-up of saturated soil against the rear of the new properties, leading to penetration of door and window openings and structural damage to the walls.
- Uphill regressive movements leading to the undermining of the foundations of the existing three-storey block of flats to the upslope side of the site boundary.

The need for rapid action was a matter of over-riding importance. After considering various options, it was decided to install 30 No. 600 mm diameter concrete bored piles about 10 m long in 3 rows, 9 piles in the top row, 7 piles in the middle row, and 14 piles in the bottom row as shown in Fig. 6.12. Each pile was reinforced by a 305 × 305 × 118 kg steel universal column. The piles were designed following Viggiani (1980). Inclinometers were installed in 10 piles to monitor performance.

It was recognized that the factor of safety would be greatly reduced if the ground water should subsequently rise to near the ground surface. Accordingly, a system of counterfort drains was adopted and designed following Bromhead (1986). The system comprised four counterfort drains running the full length of the slope. The drains were located with an average spacing of 4.5 m, a minimum drain depth of 3.5 m, and a drain width of 0.5 m.

Inclinometer readings showed that pile deformations stabilized shortly after pile installation and prior to construction of the counterfort drains. Only very minor movements were recorded subsequently. The piles in the upper row had a maximum deformation of 37 mm, while the deformations of the piles in the middle and lower rows were up to 15 mm.

165

Twelve piezometers were installed in the slope and piezometer readings for the deeper section of the slip zone indicated that the average level of the water table was at least 2.5 m below ground level. For the purpose of slope stability calculations, the average draw-down was assumed to be 2.0 m.

This is an important case record showing that bored cast-in-place concrete piles, reinforced with heavy structural steel sections, proved to be a satisfactory means of arresting an active shallow landslide within a short time period, on a site with very limited access. The provision of slope drainage in the form of counterfort drains resulted in a significant lowering of the piezometric levels and greatly enhanced the stability of the slope. The installation of instrumentation, comprising inclinometers embedded in the piles and also piezometers installed within and adjacent to the counterfort drains, provided useful data which generally confirmed the original design assumptions and provided confidence in the satisfactory performance of the remedial works.

Reference may also be made to works by Hong and Park (2000), Ergun (2000) and Nichol and Lowman (2000).

Pumping from wells

An interesting example of slope stabilization involving pumping from wells is given by Pilot *et al.* (1985). The building of the A8 autoroute in south-eastern France near the Italian border involved the construction of a large cutting with a maximum depth of 40 m. The ground conditions are marl, which when intact is very hard but is highly sensitive to the action of water, in contact with which it readily decomposes forming a plastic material in which the slip surfaces encountered during the works were found.

Drainage was provided by installing eight pumping wells 125 mm in diameter and up to 15 m long, each with a pump in the bottom of the well. The operation of the pumps is discontinuous and is controlled by the water levels in the wells which were taken down to the underlying intact marls. Additional drainage measures included sub-horizontal drains up to 80 m long with a slope of 20%, and some run-off ditches.

It was reported that all of the arrangements worked well, and piezometers indicated that the ground water table was lowered practically to the bottom of the decomposed marl.

If a clay slope is underlain by a more permeable material, for example the chalk or a sand layer, and the water head in such a permeable layer is below the ground water table in the clay, the stability of the clay slope can be improved by installing a system of vertical sand drains through the clay layer and terminating in the underlying material. This is by no means an uncommon situation that can be used with advantage to provide an economical and effective method of lowering ground water pressures.

Reference may also be made to the works of Maddison and Jones (2000) and to Maddison *et al.* (2000).

Horizontal drains

Overview These are usually drilled into a slope on a slightly rising gradient and provided with perforated or porous liners (Root, 1958; Tong and Maher, 1975). They are often 60 m to 100 m in length, thus providing drainage deep into the slope. They are quick to install and can bring about an improvement in slope stability in a short period of time which can be important if nearby structures are endangered. In cold climates, it may be necessary to protect the drain outlets from freezing.

Case study: Lyme Regis, Dorset, UK Lyme Regis is situated on an actively eroding stretch of the West Dorset coast and has therefore always faced considerable challenges from coastal erosion and landslipping. Although the earliest known coast protection structure, the Cobb, dates back to the 13th century, many of the town's sea walls are relatively recent. It was not until around 1860 that the Marine Parade sea wall was successfully completed and the East Cliff sea wall was built only in the 1950s. Prior to the construction of these defences, the coastline would have been actively retreating in a similar way to the unprotected parts of the coast today, and there is strong historical evidence to indicate that a large part of the original mediaeval town has been lost to the sea.

Problems arising from coastal landsliding have been particularly serious during the 20th century. Some fifteen individual properties have been destroyed and many more severely damaged. There have been several major sea wall breaches along the main frontage, frequent substantial damage to Cobb Road and the complete loss of the main coastal road to Charmouth.

The Lyme Regis Environmental Improvements were initiated by West Dorset District Council in the early 1990s, with the principal aim of implementing engineering works to help ensure that the integrity of the town's coast protection is maintained in the long term and to reduce the damage and disruption caused by coastal landsliding.

The construction work for Phase 1, which comprises a new sea wall and rock armour adjacent to the mouth of the River Lim, was completed in 1995. Since then, West Dorset District Council has been carrying out a series of preliminary studies to gain information for the conceptual design of economic and environmentally acceptable coast protection works for the remaining areas of the town.

167

The coast protection works, when implemented, are likely to comprise a combination of the following broad elements:

- slope drainage
- slope strengthening
- provision of new foreshore structures
- strengthening and refurbishment of existing sea walls
- beach replenishment.

The stability of the landslides at Lyme Regis has been shown by both ongoing monitoring and the preliminary stability analyses to be very sensitive to ground water levels and seasonal ground water variations (Fort *et al.*, 2000). By reducing ground water levels down to or below dry 'summer' levels, the stability of the active landslides should be significantly improved. This has lead to preliminary proposals for the installation of a network of sub-horizontal drain arrays positioned at various elevations within the landslide systems at both Lyme Regis town and East Cliff. These would be a technically feasible, environmentally acceptable and cost effective technique of reducing ground water levels, thus improving stability and reducing ground movements. They have major potential advantages over conventional trench drains in that they have a relatively low construction and environmental impact, can be installed at significantly greater depths below existing ground surface, and can be designed to allow for longer term maintenance.

In essence the drains should satisfy a number of basic requirements to enable them to perform successfully, as follows:

- The size should be adequate to carry the maximum water flow without disturbance to the adjacent ground or development of excessive outflow pressures.
- There should be no significant loss of flow by re-infiltration into the ground along the drain length.
- Any liner or pipe should be sufficiently strong and rigid to be easily installed to the design length and orientation, and capable of supporting the borehole without significant deformation or collapse. In addition the liner or pipe should be able to accommodate some ground movement without failure.
- The slotted or perforated length of any liner should be formed so as to prevent soil ingress, or it should be provided with an appropriate filter.
- In the long term the drain should function satisfactorily without clogging and with the minimum of maintenance.

The ground conditions at Lyme Regis comprise an interbedded sequence of mudstones and clays with layers of limestone. Landslide deposits comprising largely cohesive materials of variable thickness mantle the in situ

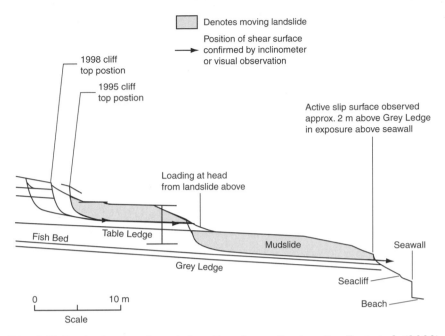

Fig. 6.13 Typical ground model section, Lyme Regis, after Fort et al. *(2000)*

geology. Figure 6.13 shows a typical ground model section based on observations (Fort *et al.*, 2000). Measured ground water levels within the landslide materials are close to the ground surface. Pressures recorded within the in situ geology are generally less than hydrostatic, suggesting under-drainage.

A preliminary proposal is to conduct trials that are located at two positions within Lyme Regis town to test the application within two areas dissimilar in topography, elevation, geology and landslide system. They would also be positioned so that it would form part of the future stabilization works at Lyme Regis. Possible drain array locations and configuration are shown in Fig. 6.14. Installation of the drains would be carried out by drilling plant of the type shown in Fig. 6.15.

Changing the slope geometry

Regrading a slope may be used to increase stability and this may be carried out as follows:

- the slope is made flatter
- a toe berm is constructed (see Fig. 6.16)
- material is removed from the top of the slope to reduce the overall height.

Site conditions often dictate the most advantageous solution.

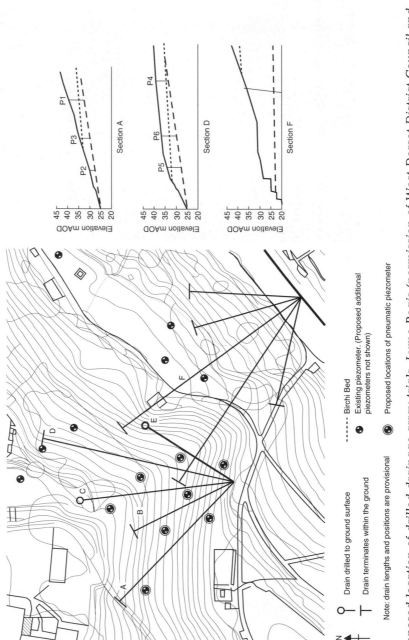

Fig. 6.14 Proposed location of drilled drainage array trials, Lyme Regis (permission of West Dorset District Council and High-Point Rendel)

Fig. 6.15 Drilling of sub-horizontal drain arrays, Robin Hood's Bay, Scarborough, UK (permission of Mr Keith Cole, West Dorset District Council)

When flattening a slope, the possibility of reactivating the old deep seated slip surface should be considered, as illustrated in Fig. 6.17 (Hutchinson, 1978, 1983, 1984).

When considering cut and fill solutions, the neutral line theory of Hutchinson (1978) is helpful. Figure 6.18 illustrates a simple example. A fill placed on a slope will have two effects.

(a) It changes the net disturbing moment which, depending on the location of the fill with respect to the centre of rotation of a potential slip surface, may have a positive or a negative effect.

(b) It will increase the effective stress acting on a slip surface but only after dissipation of excess pore water pressures has occurred. In a saturated clay slope, any fill placed will not result in an increase in effective stress in the short term.

Removal of soil

Toe, berm preferably free draining

Fig. 6.16 Cut and fill to improve slope stability

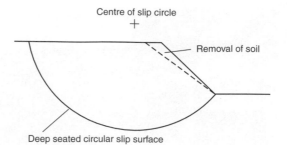

Fig. 6.17 Flattening a slope may lead to reactivation of a deep-seated slide, after Hutchinson (1983, 1984b)

Conversely, if a clay slope is regraded to a flatter angle, the possibility of a reduction in effective stress and hence strength with time as the clay swells should be taken into account.

Great care must be taken to ensure the correct positioning of a toe berm. It must not, of course, initiate a local slide and may not be beneficial with translational slides where there may be a potential for an over-ride slide to occur (see Fig. 6.19).

A large-scale example of the use of a toe berm to improve slope stability is found at Folkstone Warren on the south coast of England. Here a massive landslip occurred during the First World War that affected the transport of men and equipment to the Western Front. Extensive filling was placed on the foreshore as a toe load (Viner-Brady, 1955). Although well positioned, the toe berm improved the overall factor of safety by only 3.5% to 5%.

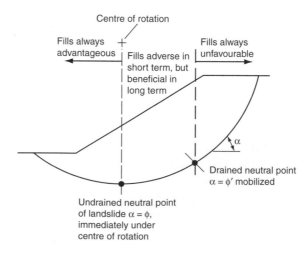

Fig. 6.18 The neutral line theory, after Hutchinson (1978)

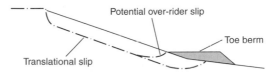

Fig. 6.19 Translation slip stabilized by a toe berm, illustrating a potential over-rider slip

Earth retaining structures

Drainage and changing the slope profile are usually the first choice for stability improvement schemes. If, however, space is restricted, then earth retaining structures, if properly engineered, can play a useful role.

Retaining walls of various types have been used in slope stabilization works, e.g. gravity structures including cribwalls and gabions, as well as reinforced concrete walls, steel sheet pile walls, and large diameter bored piles. They can be free standing or anchored. The use of steep reinforced slopes has greatly increased in recent years and reinforced earth structures can be found in many parts of the world. Soil nailing can be a viable solution.

The design of such structures is outside the scope of this Short Course book. Reference can be made to works by Clayton *et al.* (1991), Hoek and Bray (1977), Ingold (1982), Littlejohn *et al.* (1977), Grant *et al.* (2000), Martin and Kelly (2000) and Jamaludin and Hussein (2000).

Miscellaneous methods

Electro-osmosis

Overview L. Casagrande introduced electro-osmosis into civil engineering in the mid-1930s and its application has been demonstrated on a number of occasions (Casagrande, 1947, 1952, 1953). In most cases, electro-osmosis was used in silty soils to produce a temporary stabilization, for instance, during excavation. A famous case was the stabilization of slopes for the construction of U-boat pens in the very soft Trondheim clayey silt (Casagrande, 1947).

In the electro-osmosis system, direct current is made to flow from anodes which are usually steel rods driven into the soil, to cathodes which can also be steel rods or slotted pipes. The electrical gradient produces a flow of pore water through the soil from anode to cathode where it can be removed, for example, by pumping from pervious cathodes. The increase in shear strength obtained was primarily due to the increase in effective stress due to the flow of water. When the current is switched off, the permanent increase in shear strength is small.

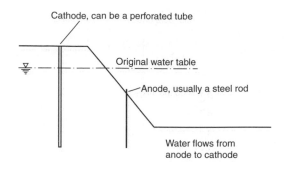

Fig. 6.20 Simplified electro-osmosis installation

Figure 6.20 shows a typical arrangement to stabilize the face of a cutting. The flow of water is directed away from the face of the cutting, from anode to cathode.

Case study: Ås, Norway At one stage it was considered that the use of electro-osmosis to increase the shear strength of clays was economically impractical. There are now cases where electro-osmosis has been successfully applied to that end, see e.g. Casagrande (1953) and Bjerrum *et al.* (1967) who describe an important and extremely interesting case where electro-osmosis was used to obtain a *permanent* increase in shear strength of a Norwegian quick clay so that two excavations for a sewage treatment plant to a depth of about 4.5 m could be safely carried out.

The site is located at Ås, about 30 km south of Oslo on the eastern side of the Oslofjord. The ground conditions are shown in Fig. 6.21 and a cross-section across the site is shown in Fig. 6.22. The undrained shear strength near the base of the excavation was as low as 6 kPa. It was clear that the danger of a bottom heave failure (Bjerrum and Eide, 1956) was great and it was decided to use electro-osmosis to increase the shear strength of the quick clay.

The electro-osmosis installation used reinforcing bars of 19 mm diameter, 10 m long, as electrodes that were pushed 9.6 m into the ground in 10 rows 2.2 m apart and at a spacing of about 0.6 m in each row. In total, 190 bars were used and the area of the installation was 200 m^2.

No pumping devices were installed since it was expected that the water expelled at the cathodes would be capable of forcing its way up through the clay and this is in fact what happened. Small trenches were dug along the cathodes to allow the expelled water to drain away, and in such a way, the water was collected and the discharge measured.

The installation was monitored by 14 piezometers, one precision settlement gauge, and 11 Borros settlement gauges that measured the settlement at various depths. Measurements of voltage applied and current

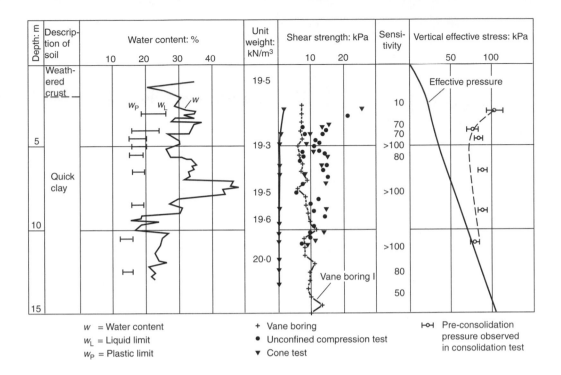

Fig. 6.21 Ground conditions of the electro-osmosis installation, Ås, after Bjerrum et al. (1967)

consumed are shown in Fig. 6.23 which also gives the observed settlement at three gauges.

After the electro-osmosis was complete, the cathodes were easy to remove and, not being corroded, were later used as reinforcing steel. The adhesion developed between the soil and the anodes prevented removal of the anodes. During excavation, however, some pieces of the anodes were recovered and it was found that corrosion of the anodes amounted to 37% of the original weight.

The total amount of electricity consumed was 30 000 kWh and the total weight of electrode steel was 4.5 tonnes. Altogether, 100 m³ of water was expelled and a volume of about 2000 m³ of clay was stabilized, i.e. 17 kWh of electric power and 2.25 kg of steel were necessary to stabilize 1 m³ of clay.

The settlement observations given in Fig. 6.24 show that the ground surface within the treated area amounted to about 500 mm for the period of treatment of some four months. At a distance of only 2 m from the outside rows of electrodes the settlement was negligible.

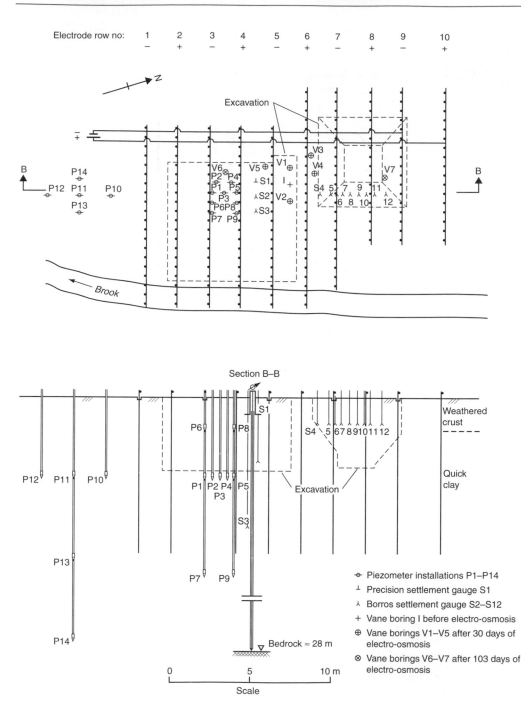

Fig. 6.22 Plan and cross-section of the electro-osmosis installation, Ås, after Bjerrum et al. (1967)

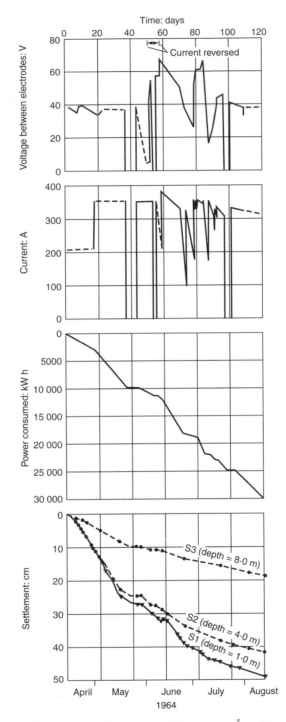

Fig. 6.23 Electricity consumption and settlements, Ås, after Bjerrum et al. *(1967)*

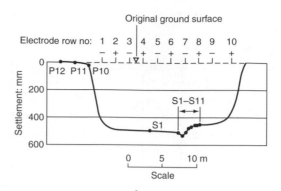

Fig. 6.24 Settlement observations, Ås, after Bjerrum et al. (1967)

A total of 14 conventional NGI-type piezometers were installed. These consist of a porous bronze tip 300 mm long, with an outside diameter of 30 mm, fixed to a 6 mm inside diameter copper tube from the tip to the top of the pipes. Copper tubing was used instead of nylon in anticipation of negative pore pressures. It would be expected that electro-osmosis would radically alter the pore pressure conditions. As water is removed from the clay next to the anodes, suctions will develop in these areas while at the cathodes the pore pressure will increase. It quickly became obvious that a production of gas took place in the piezometers, which would have made the readings unreliable. Bubbles rising from the holes left by some borings performed along the cathodes indicated that the gas formation was considerable. Ignition of the gas by a match flame demonstrated the presence of hydrogen. Bjerrum *et al.* (1967) point out that piezometers in electro-osmotic installations should consist of non-conducting materials such as ceramic filters and plastic tubing to minimize problems of gas formation due to electrolysis.

The following main conclusions may be drawn from this extremely important case record.

- Electro-osmosis is a useful engineering procedure for improving the engineering properties of a limited volume within a large deposit off soft clay.
- Due to electro-osmosis, the undrained shear strength increased from an initial value less than 10 kPa to a maximum value of 110 kPa next to the anodes.
- During the period of treatment an unexpected temporary reduction in undrained shear strength was observed, in spite of the fact that water was being removed from the clay and surface settlement was occurring. This reduction in strength occurred not only within the treated clay volume but also to a large depth in the clay beneath the treated area. Bjerrum *et al.* (1967) believed that the reduction in strength

within the treated clay volume was due to shear deformations but no explanation can be offered for the reduction observed in the clay beneath the treated volume.

- This case record emphasizes the necessity for carrying out well planned control measures including observations of settlement, pore water pressure and, in particular, undrained shear strength at various times during the electro-osmosis treatment. Failure to do so could result in disaster.
- The response of soil to electro-osmosis is not fully understood but clearly changes in a soils fundamental physical–chemical properties occur leading to surprising results.

Another unexpected result was reported by Eide and Eggestad (1963) in connection with an excavation carried out in Oslo for the foundations for the new headquarters of the Norwegian Telecommunications Administration. Electro-osmosis was used to stabilize the very soft highly sensitive clay. It was noted that significant increases in undrained shear strength occurred together with a settlement of the ground surface, even though the pore pressures showed a slight increase!

Grouting

Injection grouting as a treatment for stabilizing clay embankment slips has been known for over 50 years and has found particular application to railway embankments.

Controlled hydro-fracture systems are used. A grid of injection points is set up with the base of the points below the slip plane, and a quantity of grout is specified for injection into the various points based upon experience, the local conditions, and depth of injection. More than one level of injection may be required. A 30% sand–cement grout that has been aerated by 15% to 25% to increase the viscosity and flow properties has been found to be suitable (Ayres, 1985) and at a cost of one third of alternative conventional methods. Ayres' paper gives two case records where grouting was successful. Grouting has also been successfully used to stabilize cutting slopes in stiff clays (Ayres, 1961). If done without care, however, grouting can increase pore pressures and trigger a slide.

Vegetation

The beneficial effects of vegetation on slope stability develop gradually over a period of time that may be weeks or months in the case of grasses and herbaceous vegetation or several years for shrubs and trees. The adverse effects of sudden removal of vegetation on slope stability are well known and have lead to slope instability problems on a number of occasions, e.g. Gray (1977) and Wu (1976).

179

The beneficial effects include

- reduction in pore water pressures caused by evaporation and tran-spiration
- reinforcement of the soil by root systems
- anchoring, arching and buttressing by tap roots.

Adverse effects include

- down-slope wind loading from large trees
- additional down-slope forces caused by the weight of large trees.

It should be noted that vegetation will result in a reduction of surface run-off and a corresponding increase in water infiltration but the overall effect on pore water pressures is beneficial, i.e. it reduces them.

While it is difficult to assess with any precision the degree of benefit that vegetation can have on slope stability, attempts have been made to quantify such effects, e.g. see works by Coppin and Richards (1990), Wu (1994) and Operstein and Frydman (2000).

It would seem sensible for design purposes to ensure that any slope has a factor of safety above unity without taking into account the effects of vege-tation, and then to consider that any vegetation is an added insurance.

Species of trees which may be suitable for improving the stability of slopes because of their extensive root systems, would be willow, poplar, oak, elm, horse chestnut, ash, lime and sycamore maples.

Freezing

This is an expensive technique and is rarely used to stabilize slopes. A dis-advantage is that the whole process of drilling holes, installing the plant, and freezing the ground may take several months. Freezing certain types of ground can cause severe heaving.

Ground freezing involves sinking boreholes in the area to be treated at 1.0 m to 1.5 m centres. Concentric tubes are inserted in the boreholes and refrigerated brine is pumped down the inner tubes and rises up the annu-lar space between the inner and outer tubes or casings. It then returns to the refrigeration plant via the return ring main. Usually it takes from six weeks to four months to freeze the ground. A well known application is in temporarily stabilizing a flow of silt during construction of the Grand Coulee Dam (Gordon, 1937). A number of examples of ground freezing in civil engineering works have been given by Harris (1985). A general review of the technique is provided by Sanger (1968).

Combining stabilizing methods

When using more than one method of improving slope stability, the possi-bility of progressive failure should be considered. If the various methods

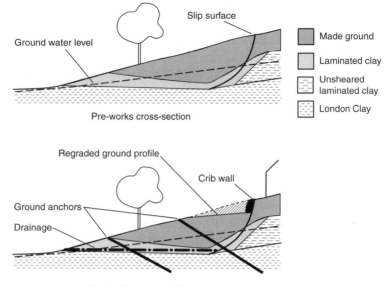

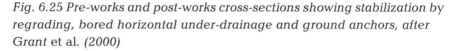

Fig. 6.25 Pre-works and post-works cross-sections showing stabilization by regrading, bored horizontal under-drainage and ground anchors, after Grant et al. (2000)

employed invoke different stress–strain relationships from the soil and from the materials and/or techniques involved, one method may be approaching failure before the another technique can make a significant contribution to the overall stability. In this way, the design intention, of each of the methods increasing stability at the same time, might not eventuate.

Grant *et al.* (2000) used a combination of regrading, bored horizontal under-drainage, and ground anchors to give an acceptable factor of safety of a slope that had been oversteepened and had a history of instability (Fig. 6.25). A particular feature of the works was the way in which the separate and combined actions of each of the methods were considered. The overall stabilization scheme was designed to give a factor of safety greater than 1.5 assuming the worst likely conditions. The contributions of the methods of stabilization, however, were balanced in such a way that even if any two of them should fail to perform as intended, then the slope would still remain with a minimum likely factor of safety of at least 1.1 under worst case conditions.

Investigating landslips

Overview

The remedial methods for stabilizing landslips outlined in this chapter require that a proper and thorough site investigation be carried out in advance. The techniques available for site investigation generally are well known and have been described in some detail by Clayton *et al.* (1995), for example, and need not be repeated here. There are some particular aspects of site investigation, however, that need to be emphasized. These aspects are covered in the following sections and include

- surface mapping
- distribution of pore water pressure
- location of slip surfaces
- early warning systems.

Surface mapping

The desk study and walk-over survey are two essential components of any ground investigation but are of particular importance when slopes are being assessed. They can provide a great deal of information at little cost. The site itself and the following sources of information should be consulted:

- topographic maps
- stereo aerial photography
- geological maps and publications
- geotechnical and engineering geology journals
- previous ground investigation reports
- well records
- meteorological records
- newspapers.

The value of aerial photographs cannot be overemphasized. Aerial photos taken at different times of the year and at different times of the day can be most revealing and can indicate down-slope movements that may not be obvious to the naked eye at ground level (e.g. see Figs. 6.2 and 6.3).

When carrying out the walk-over survey, the following features should be looked for:

- tension cracks
- toe bulges
- lateral ridges
- uneven ground
- graben features
- back-tilted blocks
- marshy ground
- streams
- springs and ponds
- cracked roads
- inclined trees

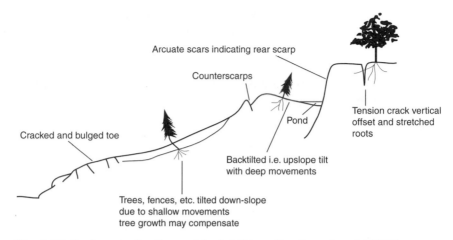

Fig. 6.26 Surface indications of instability, after Bromhead (1979)

- breaks and cracks in walls
- distorted fences
- drainage paths, both natural and man-made.

Some of these features are indicated in Fig. 6.26.

Walk-over surveys, carried out during and immediately after heavy rainfall, can be most revealing and can show adverse drainage conditions that may not be apparent under dry weather conditions. Weep-holes in walls should be carefully studied to see whether they are functional or blocked up, as is often the case.

Distribution of pore water pressures
Overview
An absolutely essential part of an investigation into the stability of a slope is the determination of the distribution of pore water pressures, the importance of which cannot be overemphasized.

The basic principle of operation of all piezometers used for measuring pore water pressures is that of a water-filled porous element placed in the ground so that the soil water is in continuity with the water in the porous element – this may be saturated at the time of installation or the ground water may be allowed to flow directly into the porous element.

Hanna (1973) has defined the requirements of any piezometer as:

- to record accurately the pore water pressure in the ground
- to cause as little interference to the natural soil as possible
- to be able to respond quickly to changes in ground water conditions
- to be rugged and remain stable for long periods of time
- to be able to be read continuously or intermittently if required.

Because some flow of water from the spoil into the piezometer is required for any piezometer to be able to record pore water pressure changes, a time lag exists between a change in pressure and the recording of that pressure change by the piezometer.

Standpipes and standpipe (Casagrande) piezometers

The simplest form of pore pressure measuring device is the observation well or standpipe (see Fig. 6.27). This consists of an open-ended tube that is perforated near the base, and is inserted in a borehole. The space between the tube perforations and the wall of the borehole is normally packed with sand or fine gravel, and the top of the hole is then sealed with well tamped puddle clay or bentonite grout or concrete to prevent the ingress of surface water.

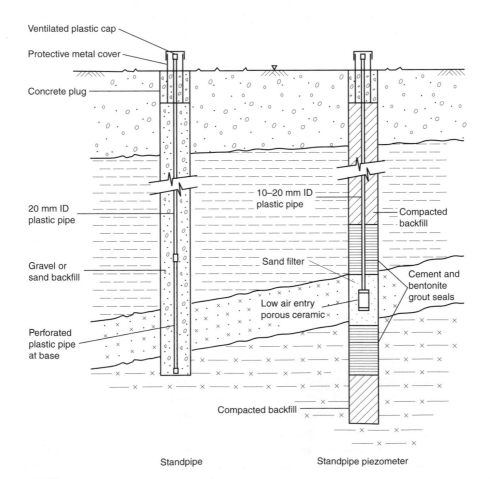

Fig. 6.27 Standpipe and standpipe piezometer

Measurements of water level in the standpipe are made by lowering an electrical 'dipmeter' down the open standpipe, giving a visual or audible signal when the water level is met.

The standpipe is very simple to install but, unfortunately, suffers from considerable disadvantages. First, no attempt can be made to measure pore water pressure at a particular level, and it has to be assumed that a simple ground water regime exists with no upward or downward flow between strata of differing permeability. A second major disadvantage is due to the considerable length of time required for equalization of the level of the water in the standpipe with that in the ground in soils of low permeability.

To overcome uncertainties connected with the standpipe the most common practice is to attempt to determine the water pressure over a limited depth by sealing off a section of the borehole. The system commonly used is termed a standpipe piezometer or a 'Casagrande' piezometer and is illustrated in Fig. 6.27. The sand filter is sealed above and below with grout, often a well mixed cement/bentonite seal in proportions of 1 : 1, and the mix should be as stiff as is compatible with tremie pipe placement at the base of the hole or by bentonite balls or pellets. The cement/bentonite seal is typically 2 m long. Vaughan (1969) has examined the problems of sealing piezometers installed in boreholes, when the grout seal is extended up to the ground surface, and concluded that for a typical installation the permeability of the seal can be significantly higher than that of soil surrounding the piezometer tip without serious errors arising. This illustrates the value of backfilling the entire hole with grout rather than using relatively short seals.

In order to speed equalization between ground water pressures and the level in the standpipe, it is important to ensure that the sand filter is saturated. The time required for equalization of pressures may be computed if the 'flexibility' of the piezometer system and the permeability and compressibility of the soil is known. Solutions for the time lag have been presented by Hvorslev (1956) for incompressible soils, and by Gibson (1963, 1966) for compressible soils.

Porous metallic tips are in extensive use and the Geonor piezometer tip is shown in Fig. 6.28. It comprises a 30 mm diameter filter cylinder connected to a central shaft by top and bottom pieces. The bottom piece is cone-shaped, the top piece being threaded to steel tubes. The piezometer element is fitted with a small diameter plastic riser pipe, usually nylon. In soft ground the piezometer can be installed by driving; pre-boring may be required in harder ground.

It is possible to measure the pore water pressure at different depths at any one location by using more than one porous elements carefully isolated from each other. This is not easy to achieve in practice and because

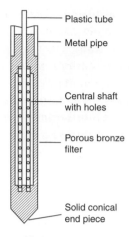

Fig. 6.28 Geonor piezometer tip

of this, when measurements of pore water pressure are required at different depths, separate boreholes are often used with only one filter element in any one borehole.

Pneumatic piezometers

The pneumatic piezometer tip consists of a ceramic porous stone, behind which is mounted an air-activated cell (Fig. 6.29). The tip is connected to instruments at the surface via twin nylon tubes and these are connected in

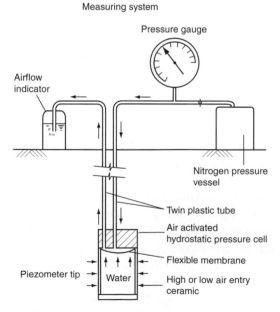

Fig. 6.29 Pneumatic piezometer equipment

turn to a flow indicator and a compressed air supply and pressure measuring apparatus. When the pore water pressures are measured, air or nitrogen is admitted to one line but is prevented from flowing up the other line by a blocking diaphragm in the tip. When the air pressure reaches the pore water pressure, the diaphragm is freed away from the inlet and the outlet tubes in the tip; air returns up the vent line and a visible signal of this is given by air bubbles in the air flow indicator. When the return air ceases to flow, the pressure in the feed line is equal to the pore water pressure.

Typically, the amount of water displaced by the diaphragm is very small ($<100\,mm^3$) and the total fluid volume of the tip is low, and therefore the time required for equalization between the ground water pressure and the air line pressure is very small.

The pneumatic piezometer is considerably more expensive than a simple standpipe piezometer and the system requires a more sophisticated read-out unit.

It has been suggested that in the long term some pneumatic piezometers become unreliable and, probably because of this, their application to slope stability problems has been limited.

Electrical piezometers

The principle of the electrical piezometer is that of a diaphragm deflected by the water pressure acting against one face, the deflection of the diaphragm being proportional to the applied pressure. This pressure is measured by means of various electrical transducers. Such devices have very small time lags and are sensitive but, unfortunately, their long term reliability is open to question. One of the authors once spent an entire summer installing vibrating wire type piezometers in a major earth dam only to have the instrument hut struck by lightning in the following winter. Because the hut did not have a lightning conductor, the electrical piezometers were destroyed! Fortunately, we had also installed hydraulic piezometers which survived the lightning strike and these functioned successfully – a classic reminder that not only do instrument huts require lightning conductors, but also that we always need backup systems.

Hydraulic piezometers

The closed hydraulic piezometer was developed at the Building Research Station in the UK and is widely used in unsaturated earth fill applications. When using the closed hydraulic piezometer, measurements of water pressure are made at a point remote from the piezometer tip. The tubing connecting the tip to the measuring device must be filled with a relatively incompressible fluid. To achieve this, two tubes connect the tip to the measuring point at the ground surface and then are flushed with de-aired

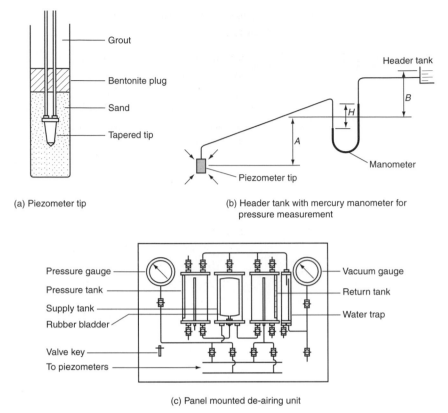

(a) Piezometer tip

(b) Header tank with mercury manometer for pressure measurement

(c) Panel mounted de-airing unit

Fig. 6.30 Twin tube hydraulic piezometer equipment

water, or an antifreeze solution in cold climates, before taking readings (Fig. 6.30).

When planning a piezometer installation, thought must be given to protecting the piezometers from vandalism. This has cost implications. Furthermore, provision must be made for piezometers to be read on a regular basis. It is important that the readings are taken during and immediately after periods of heavy rain so that critical pore water pressures are obtained.

Location of slip surfaces

An important part of the investigation of a landslide is the determination of the depth and three-dimensional shape of the slip surfaces that characterize it. Knowledge of these features are needed:

- in order to enable piezometers, inclinometers and other instrumentation to be correctly placed
- to provide a guide for sub-surface sampling techniques
- to help with the carrying out of back-analysis

- to permit the design of appropriate stabilization or control measures and long term monitoring.

It is helpful to locate the slip surface or slip surfaces before deciding on the position and depths of piezometers but the possible existence of an undiscovered slip should not be overlooked. Particularly if major structures could be involved in a landslip or there is a possibility of loss of life, it would be prudent when assessing the stability of a slope to ensure that it is stable no matter where a slip surface could be located and with the design using the residual strength. This would require knowing the distribution of pore water pressure with depth and this could mean at any one location measuring the water pressure at two or three depth intervals.

The first approach to locating slip surfaces is to observe them in test pits or in borehole samples. Strict safety precautions should be observed when inspecting test pits, bearing in mind that the inspection may be time consuming and collapse of the pit sides may occur. In addition, it is easy to miss a slip surface if the inspection is carried out by staff that are not experienced. The location of slip surfaces from borehole samples requires, ideally, continuous sampling or two boreholes located close together with sampling depths chosen so that the complete depth range is covered. It is also important to remember that multiple slips may well exist and to make quite sure that the investigations locate the lowest of these.

Inclinometers are frequently used to monitor movements in earth fills and slopes. Most of these instruments consist of a pendulum-activated electrical transducer enclosed in a waterproof 'torpedo'. The torpedo or probe is lowered down a near vertical guide casing installed in the ground. The inclination of the casing from the vertical is measured at pre-determined intervals along the casing, often 0.5 m, and the profile of the shape of the casing is obtained by integration of the observed slope values from a fixed point either by taking the casing into rock or some other stable material at depth where it can be assumed that there is no translation, or using the top of the casing as the datum, fixing its position by survey methods.

It is convenient to use a casing with diametrically opposite grooves, as shown in Fig. 6.31, down which the probe can run on wheels.

A crude type of inclinometer is a flexible tube installed in a borehole, down which steel rods of increasing length are lowered in turn and the rod length which is just able to pass a given point gives a measure of the curvature of the tubing in the vicinity of that point. Alternatively, a rod can be lowered to the bottom of the tube and then pulled up at intervals. If slip movements sufficient to flex the tube have occurred, the rod will jam in the lower part of the flexure.

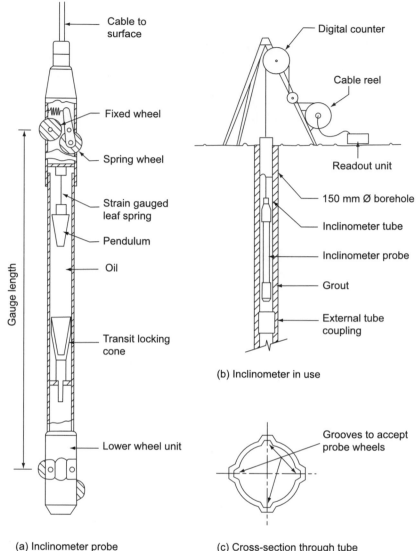

Fig. 6.31 Inclinometer equipment

In a test to failure in a trial bank on soft clay near Bangkok, Eide and Holmberg (1972) pushed 10 m long brittle wooden sticks into the ground (inside a metal tube that was withdrawn after installation) adjoining the proposed bank, on lines following cross-sections of the anticipated foundation failure. When this occurred, the sticks were broken at the level of the slip surface. By pulling out the upper broken-off portions of the sticks, the depth to the slip surface at each point could be determined.

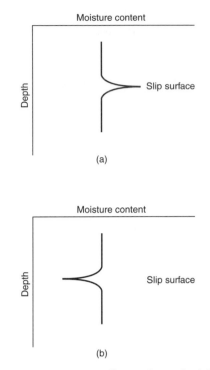

Fig. 6.32 Moisture content cusps at slip surfaces in (a) over-consolidated clay and (b) normally consolidated clay

A check on the location of a slip surface may be made if cusps in the distribution of moisture content with depth can be detected. In a dilatant soil that expands on shearing, for example an over-consolidated clay, an increase in the moisture content would be expected and such increases have been reported by Henkel (1957), Skempton (1964) and D'Appolonia *et al.* (1967). In a contractant soil, for example like a normally consolidated clay, local decreases in the moisture content in the vicinity of the slip surface would be expected (see Fig. 6.32).

Other methods, perhaps used less frequently, for detecting slip surfaces have been described in detail by Hutchinson (1983).

Early warning systems

While early warning systems are highly desirable, they generally become neglected or damaged as a result of vandalism. Piezometer and inclinometer readings can give early warning of impending instability, but the readings need to be taken on a regular basis and then assessed by an experienced geotechnical engineer or engineering geologist.

Vertical and horizontal deformation observations taken on fixed points, and measurements of changes in the width of tension cracks can be most helpful. Ranging rods placed in straight lines across a slope can give visual indication of down-slope movements, but are subject to vandalism.

If early warning systems are installed on a slope, it is absolutely essential that they are maintained and read and assessed on a regular basis and provision must be made for the cost involved.

PART 2
Rock Slopes

On 7 April 2000 at 11.00 a.m., Adrian Adams photographed a rock fall at Charmouth, Dorset, England. The tower or column of Belemnite Marl of Lower Jurassic age had become detached from the main cliff face for many years, before it suddenly came thundering down. A sequence of photographs showing the event is given below.

CHAPTER SEVEN

Discontinuities in rock: description and presentation for stability analysis

Description of rock discontinuities

In soils, slope failures range from being circular in relatively homogeneous materials to being non-circular or planar in layered soils. In rock, however, most slope failures are controlled by ever-present discontinuities such as joints, faults and fractures. These discontinuities are planes of weakness across which there is little or no tensile strength. Slope failures will be feasible or not and will propagate depending on the extent, pattern and types of discontinuity present in the rock mass. The assessment of rock slopes for their susceptibility to failure must, therefore, incorporate a description system (see below) and a means for presenting orientation data in a form that can be used directly in stability analysis (this will be seen later to be the hemispherical projection).

BS 5930: 1999, *Code of practice for site investigations*, includes the following types of discontinuity.

- *Joint*: a discontinuity in the body of rock along which there has been no visible displacement.
- *Fault*: a fracture or fracture zone along which there has been recognizable displacement.
- *Bedding fracture*: a fracture along the bedding (bedding is a surface parallel to the plane of deposition).
- *Cleavage fracture*: a fracture along a cleavage (cleavage is a set of parallel planes of weakness often associated with mineral realignment).
- *Induced fracture*: a discontinuity of non-geological origin, e.g. brought about by coring, blasting, ripping, etc. They are often characterized by rough fresh (i.e. no discoloration or surface mineral coatings) surfaces.
- *Incipient fracture*: a discontinuity which retains some tensile strength, which may not be fully developed or which may be partially cemented. Many incipient fractures are along bedding or cleavage.

Of these, joints and bedding fractures are the most common and in most cases these form a distinct pattern of parallel or sub-parallel sets. The orientation of these sets can in combination with the excavated or natural face of the rock, bring about one or more failure mechanisms that involve free-fall, sliding or rotation of rock blocks. In the special case of randomly orientated discontinuities failure is likely to occur on a circular plane in a similar manner to soils. Clearly the description of the discontinuities in rock is critical to the identification of potential failure mechanisms in fractured rock masses. The important characteristics of discontinuities which must be recorded can be placed under the headings of the following checklist:

- orientation
- spacing (one dimension)
- block size and shape (spacing in three dimensions)
- persistence
- roughness
- wall strength
- wall coating
- aperture and infilling
- seepage
- discontinuity sets.

Most of these characteristics are illustrated in Fig. 7.1. Table 7.1 summarises the nature and significance and these characteristics.

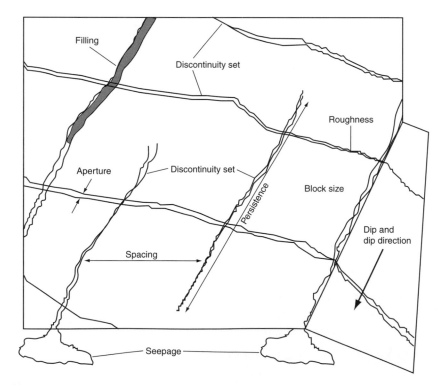

Fig. 7.1 Observations and measurements required for the description of discontinuities

Table 7.1 Description of discontinuities

Property	Measurement	Measurement method	Limitations	Uses
Orientation (Figs. 7.1 and 7.2)	The convention dip direction/dip should be used, e.g. 026/86 (Note in order to avoid confusing dip and dip direction the dip direction should always be reported as 3 digits, e.g. 006 and the dip as 2 digits, e.g. 09)	Geological compass (Fig. 7.3a,b) or a compass (Fig. 7.3c) and a clinometer (Fig. 7.3d)	Orientation of discontinuities cannot be determined accurately from core samples unless the core has been orientated (Barr, 1977) or the fractures have been correlated with those in the borehole using a device such as a borehole impression packer (Hinds, 1974; Barr and Hocking, 1976)	Orientation is one of the critical parameters that determines failure mechanism and is used in kinematic feasibility analysis
Discontinuity sets (Fig. 7.1)	The number of sets should be identified and average orientation of each set determined	Hemispherical projection		Permits identification of likely failure mechanisms
Spacing (one dimension) (Fig. 7.1)	Spacing should be measured for each joint set; the convention is to measure spacing perpendicular to the discontinuities. In cores with steeply dipping discontinuities, it may only be possible to measure spacing along the core axis; if so, this should be stated. For scanline surveys spacing is generally measured along the line of the tape	Surveyor's measuring tape (30 or 100 m) (Fig. 7.3e)	The use of vertical boreholes will introduce directional bias into discontinuity spacing measurements. It is recommended that inclined boreholes be used to reduce the bias. Scanlines should be in three near orthogonal directions to reduce bias	Spacing is used to deduce degree of structural anisotropy and block size and shape
Block size and shape (spacing in three dimensions)	The spacing of discontinuities in three dimensions may be described with reference to the size and shape of the rock blocks bounded by discontinuities (see Table 7.2). Rock blocks may be approximately equi-dimensional, tabular or columnar in shape. The three-dimensional characteristics of discontinuities cannot be measured effectively from a borehole/core	Surveyor's measuring tape (30 or 100 m) (Fig. 7.3e)	See above	Used as aid in determining the likely failure mechanism, e.g. whether toppling is more likely than sliding

Table 7.1 Continued

Property	Measurement	Measurement method	Limitations	Uses
Block size and shape (spacing in three dimensions)	Often block size and shape is determined from one-dimensional spacing measurements			
Persistence (Fig. 7.1)	Descriptive terminology can be applied to sets; actual measurements of trace lengths are preferred for individual discontinuities A tape is essential where actual measurements are required	Surveyor's measuring tape (10 or 30 m) (Fig. 7.3e)	Very limited information is available from cores Measurements are scale dependent and the size of the exposure should be recorded.	Aids the interpretation of the relative importance of discontinuities, e.g. a single highly persistent discontinuity is likely to be more important than a set of low persistence discontinuities
Termination	The nature of the discontinuity termination should be recorded in the context of the size of the exposure. A discontinuity may start and end within the exposure	Visual recognition	Measurements are scale dependent and the size of the exposure should be recorded	Aids the interpretation of the relative importance of discontinuities
Roughness (Fig. 7.1)	Descriptions of roughness should be made at three scales where possible (ISRM, 1978) *Large scale* (tens of metres) reported as measured wavelength and amplitude *Intermediate scale* (several metres) is divided into stepped, undulating or planar. Wavelength and amplitude may also be used *Small scale* (several centimetres) is divided into rough, smooth or striated is superimposed on the intermediate scale. Wavelength and amplitude may also be used Small-scale roughness may be measured using a profiling tool or converted to a Joint Roughness Coefficient (Barton, 1971) using published roughness profiles or a field tilt test (Barton, 1976; Barton *et al*, 1985)	Surveyor's measuring tape (30 m), straight edge, scale rule, profiling tool (Fig. 7.3f), profile charts (ISRM, 1978), geological compass or inclinometer (e.g. Dr Dollar's indicator— Fig. 7.3d)	It may not be possible to measure roughness at the large and intermediate scales due to the size of the exposure. Only small-scale roughness can be measured using core samples	Roughness is important in the determination of the shear strength of discontinuities. Intermediate and large-scale roughness will have the greatest influence on slope stability

Parameter	Description	Measurement	Limitation	Significance
	If the roughness has a preferred orientation (e.g. stepped, striated) the orientation should be recorded A tape, scale rule and profiling tool are required for these measurements An individual joint may be described as wavy (wavelength 12 m, amplitude 1 m), stepped (wavelength 2 m, amplitude 0.2 m) or smooth A tape, scale rule and profiling tool are required for these measurements			Provides indication of degree of asperity breakage that will occur during shearing across the discontinuity
Wall strength	The unconfined compressive strength of the discontinuity walls is determined from index tests such as the Schmidt hammer (Barton and Chouby, 1977)	Schmidt hammer	Index tests such as the Schmidt can be unreliable for the accurate determination of strength	
Wall coating	Any mineral coating discontinuity walls should be described. Particular attention should be given to low friction minerals such as chlorite. Wherever possible the extent of the coating should be recorded and the set or sets with which it is associated	Visual recognition	Estimating the extent of mineral coatings can be difficult particularly if exposures are limited	The presence of low-friction mineral coating will significantly reduce the basic friction angle (i.e. the friction angle for smooth discontinuities). The presence of such coating will mean that the shear strength of the discontinuities affected must be measured rather than estimated
Wall weathering and alteration	A factual description of the weathering/alteration of discontinuity walls should be made (Approach 1; BS 5930: 1999). The manifestation of weathering/alteration of discontinuities is most likely to be by increasing apertures (through dissolution) and/or discoloration and strength reduction (through chemical decomposition). Aperture and wall strength are described separately and hence the most important aspect of weathering/alteration that must be	Visual recognition		The weathered/alteration stage of the discontinuity walls will affect the shear strength and compressibility of the discontinuities

Table 7.1 Continued

Property	Measurement	Measurement method	Limitations	Uses
Wall weathering and alteration	described here is discoloration. The extent and amount of penetration into the rock of the discoloration should be reported			
Aperture and infilling (Fig. 7.1)	Where possible, measurements of aperture should be reported. Full description of rock, soil or mineral infill should be provided. The observer should comment on whether the reported apertures are present in the intact rock mass, or a consequence of geomorphological/weathering agencies, or whether due to engineering activities or creation of the exposure. The thickness and type of infill should be reported using standard terms, e.g. 1 mm surface film of calcite, 10 mm cemented breccia, stiff brown sandy clay	Scale rule, feeler gauges	Aperture is difficult/ impossible to determine from core samples. Aperture may be measured in the borehole itself using a device such as a borehole impression packer (Barr and Hocking, 1976) Infill thickness and type cannot be determined effectively from core samples since it is likely to have been washed out/contaminated by the drill flush fluid	Where the thickness of infill is equal or greater than the roughness amplitude (small/ intermediate scale) the strength of the infill will have a dominant effect on the shear strength of the discontinuity The presence of infill and small apertures will reduce the permeability of the discontinuities
Seepage (Fig. 7.1)	Evidence of seepage associated with each discontinuity set should be recorded. Where running water is observed issuing from a discontinuity attempts should be made to measure the rate of seepage using vessel of known volume and watch with a second hand	Visual recognition, vessel of known volume and stop watch	Seepage conditions associated with individual discontinuities cannot be determined from core samples or the borehole	Seepage observations will give an indication of drainage characteristics of particular discontinuity sets and hence the potential for the build-up of joint water pressure and the consequences this has on slope stability

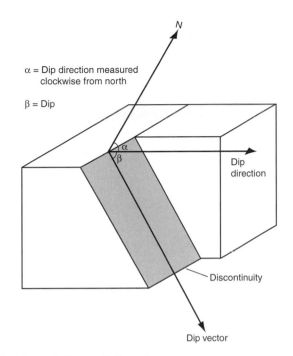

Fig. 7.2 Definition of dip and direction

Orientation

The orientation of discontinuities in a rock mass is of paramount importance to design in rock slope engineering. The majority of discontinuity
surfaces are irregular, resulting in a significant amount of scatter of
measurements being made over a small area. To reduce this scatter it is
recommended that a 200 mm diameter aluminium measuring plate be
placed on the discontinuity surface before any measurement is made. In
many cases, there may not be enough of the discontinuity surface exposed
to allow the use of such a plate. If the exposure cannot be enlarged then a
smaller plate must be used. A suitable combined compass and clinometer
(geological compass) to which measuring plates can be attached is shown
in Hoek and Bray (1981). The most common types of geological compass is
the Silva compass (type 15T) and the Clar type compass (see Fig. 7.3a,b).
These devices combine a compass and an inclinometer. This allows the
measurement of both dip and dip direction using the same instrument.
Discontinuity orientations may also be measured using a digital compass
and a Dr Dollar's clinometer (Fig. 7.3c,d).

Many compasses have the capacity to correct for differences between
magnetic north and true north. It is recommended that this adjustment
is always set to zero; corrections can be made later during processing or
plotting (Priest, 1993). It should also be noted that compass needles

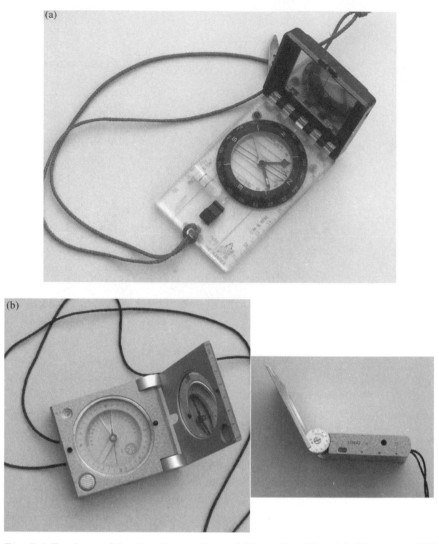

Fig. 7.3 Tools used for the description of discontinuities: (a) Silva type 15T geological type compass; (b) Clar type compass (manufactured by Carl Zeiss Jena Ltd)

balanced for magnetic inclination in the northern hemisphere will be severely out of balance in the southern hemisphere. Furthermore, electronic (digital) compasses set up for use in the northern hemisphere should not be used in the southern hemisphere.

Through careful use of the conventional geological compass and practice it is possible achieve a resolution of less than 30 seconds in dip and dip direction on readily accessible discontinuities (Priest, 1993). However, Ewan and West (1981) conclude that different operators measuring the

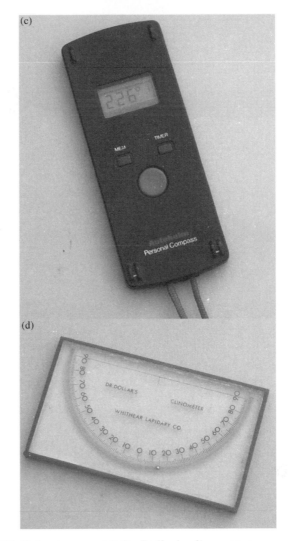

Fig. 7.3 (c) Digital compass; (d) Dr Dollar's clinometer

orientation of the same feature have a maximum error of ±10° for dip direction and ±5° for dip angle.

Discontinuity sets

A discontinuity set represents a series of near parallel discontinuities (refer to Fig. 7.1). The appearance of the rock mass together with its mechanical behaviour will be strongly influenced by the number of sets of discontinuities that intersect one another. The number of sets tends to control the degree of overbreak in excavations and structural anisotropy of the rock mass. The more discontinuity sets there are, the more isotropic

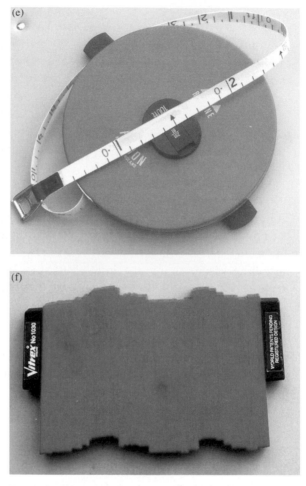

Fig. 7.3 (e) Surveyor's measuring tape; (f) profiling tool, length 140 mm

the rock mass becomes. This may be modified, however, by the spacing associated with one or more of the discontinuity sets. The number of sets also affects the degree to which the rock mass can deform without failure of intact rock.

A number of sets may be identified by direct observation of the exposure. The total number of sets present in the rock mass, however, is normally determined from a statistical analysis of the discontinuity orientation data which makes use of hemispherical projection methods (Matherson, 1983; Priest, 1985, 1993) which are described later.

Spacing

Discontinuity spacing is a fundamental measure of the degree of fracturing of a rock mass and hence it forms one of the principal parameters in the engineering classification of rock masses. In particular, for tunnelling this property has been used in the classification for support requirements (Barton *et al.*, 1974; Bieniawski, 1976) and for foundation settlement predictions on rock (Ward *et al.*, 1968). The spacing of adjacent discontinuities largely controls the size of individual blocks of intact rock. In exceptional cases, a close spacing may change the mode of failure of the rock mass from translational to circular. In such cases where the joints are extremely closely spaced the rock mass will tend to behave like a granular soil and joint orientation is likely to be of little consequence.

Discontinuity spacing may be considered as the distance between one discontinuity and another. More specifically ISRM (1978) defines discontinuity spacing as the perpendicular distance between adjacent discontinuities. It is easier when collecting spacing data in the field to adopt the former more general definition. For example, a random sample of discontinuity spacing values may be obtained from a linear scanline survey (described later). Such a survey provides a list of the distances along the scanline to the points where it is intersected by the discontinuities which have been sampled. Subtraction of consecutive intersection distances provides the discontinuity spacing data. Perpendicular discontinuity spacing data may be determined during data processing in the office. It is more meaningful if such measurements are made for discontinuities of the same type (e.g. the same discontinuity set).

Priest (1993) defines three different types of discontinuity spacings.

- *Total spacing*: the spacing between a pair of immediately adjacent discontinuities, measured along a line of general, but specified, location and orientation.
- *Set spacing*: the spacing between a pair of immediately adjacent discontinuities from a particular discontinuity set, measured along a line of any specified location and orientation.
- *Normal set spacing*: the set spacing when measured along a line that is parallel to the mean normal to the set.

The mean and range of spacings between discontinuities for each set should be measured and recorded. Ideally these measurements should be made along three mutually perpendicular axes in order to allow for sampling bias. Where discontinuity sets are readily identifiable in the field the normal set spacing of each set may be recorded in terms of the maximum, minimum and modal (most frequent) or mean spacing. A comprehensive treatment of the statistical analysis of discontinuity spacing and frequency is given by Priest (1993).

Discontinuity spacing data are best presented in the form of histograms. Histograms may be produced for individual sets of discontinuities or for all

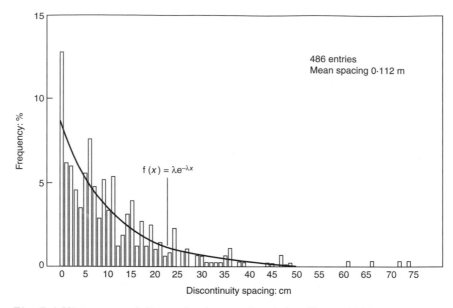

Fig. 7.4 Histogram of discontinuity spacing (after Yenn, 1992)

discontinuities intersecting a scanline. If the discontinuities in a particular set exhibit a regular spacing they will give rise to a normal distribution and a mean spacing may be easily determined. In many cases fractures are clustered or randomly spaced giving rise to a negative exponential distribution (Priest and Hudson, 1976). Examples of joint frequency distributions measured from scanlines in sandstone and mudstone are given in Fig. 7.4. The histograms show a close agreement with the negative exponential distribution expressed as:

$$f(x) = \lambda e^{-\lambda x} \tag{7.1}$$

where λ is the mean discontinuity frequency per metre. By fitting a negative exponential distribution to the spacing data the mean spacing may be determined from $1/\lambda$.

Priest and Hudson (1976) established the following relationship between Rock Quality Designation (RQD) and the mean discontinuity frequency per metre (λ):

$$RQD = 100e^{-0.1\lambda}(0.1\lambda + 1) \tag{7.2}$$

where RQD is a parameter normally derived from drillcore (Deere, 1964) and is commonly used in the classification of rock masses.

Block size and shape

Block size and shape are important indicators of rock mass behaviour in slope stability. Rock masses containing tabular or columnar-shaped

Table 7.2 Block size and shape (BS 5930: 1999)

First term (size)	Maximum dimension
Very large	>2 m
Large	600 mm–2 m
Medium	200 mm–600 mm
Small	60 mm–200 mm
Very small	<60 mm

Second term (shape)	Nature of block
Blocky	Equi-dimensional
Tabular	One dimension considerably smaller than the other two
Columnar	One dimension considerably larger than the other two

blocks may be more prone to toppling rather than sliding. A small block size may result in ravelling or circular failure brought about by the rock mass behaving like a granular soil.

The block size is determined from the discontinuity spacing in one dimension, number of sets and persistence. The number of sets and the orientation of discontinuities will determine the shape of the resulting blocks. Since natural fractures are seldom consistently parallel, however, regular geometric shapes such as cubes, rhombohedrons and tetrahedrons rarely occur.

BS 5930: 1999 recommends the descriptive terms for block size and shape given in Table 7.2. A more quantitative approach to block size description is given by ISRM (1978).

Other descriptive terms which give an impression of the block size and shape include:

- *massive*: few fractures or very wide spacing
- *irregular*: wide variations of block size and shape
- *crushed*: heavily jointed to give medium gravel size lumps of rock.

Persistence and termination

Persistence refers to the discontinuity trace length as observed in an exposure. It is one of the most important factors in discontinuity description, but unfortunately it is one of the most difficult to quantify. One of the common problems that arises is the measurement of the persistence of major joints which are continuous beyond the confines of the rock exposure. It is recommended that the maximum trace length should be measured, and comment made on the data sheet to indicate whether the total trace length is visible and whether the discontinuity terminates in solid rock or against another discontinuity. Clearly, persistence is very much scale dependent and any measurements of persistence should be accompanied by the dimensions of the exposure from which the measurements were made. Matherson (1983) considers persistence to be a fundamental feature in quantifying the relative importance of discontinuities in a rock mass.

Wall roughness

The wall roughness of a discontinuity is a potentially important component of its shear strength, particularly in the case of undisplaced and interlocked features. In terms of shear strength, the importance of wall roughness decreases as aperture, or infilling thickness or the degree of displacement increases. In cases where adjacent walls are not fully interlocked or mated the wall roughness will directly influence the degree of contact which in turn effect the compressibility of the discontinuity.

In general, the roughness of a discontinuity can be characterized by the following.

- *Intermediate and large-scale roughness* (Table 7.1). First-order wall asperities which appear as undulations of the plane and would be unlikely to shear off during movement. This will affect the initial direction of shear displacement relative to the mean discontinuity plane.
- *Small-scale roughness* (Table 7.1). Second-order asperities of the plane which, because they are sufficiently small, may be sheared off during movement. If the wall strength is sufficiently high to prevent damage these second-order asperities will result in dilatant shear behaviour. In general this unevenness affects the shear strength that would normally be measured in a laboratory or medium-scale in situ shear test.

Intermediate and large-scale roughness may be measured by means of a surveyor's measuring tape or rule placed on the exposed discontinuity surface in a direction normal to the trend of the waves. The orientation of the measuring tape, together with the mean wavelength and maximum amplitude should be recorded. In some cases it may be necessary to assess the roughness in three dimensions in which case a compass and disc inclinometer are recommended (Hoek and Bray, 1981). In many cases, however, the measurement of intermediate and large-scale roughness is made difficult or impossible by the extent to which the discontinuities are exposed. Small-scale roughness may be assessed by profiling the discontinuity surface. Short profiles (<150 mm) can be measured using a profiling tool (Fig. 7.3f) that is obtainable in most home improvement stores. Longer profiles may be measured using a 2 m rule as described by ISRM (1978). A number of quantitative techniques for measuring roughness are described in detail by ISRM (1978).

As mentioned in Table 7.1 the orientation of roughness features such as steps and undulations in relation to the direction of sliding can have a significant effect on shearing resistance. If the direction of sliding is parallel to the steps or undulations these features will have a negligible effect on shearing resistance. If the direction of sliding is perpendicular to the roughness features the shearing resistance is increased significantly. Clearly, the orientation of such roughness features should be recorded wherever possible.

Wall strength

Wall strength refers to the equivalent compression strength of the adjacent walls of a discontinuity. This may be lower than the intact strength of the rock owing to weathering or alteration of the walls. The relatively thin 'skin' of wall rock that affects shear strength and compressibility can be tested by means of simple index tests. Barton and Choubey (1977) explain how the Schmidt hammer index test can be used to estimate wall strength σ_d from the following empirical expression

$$\log_{10} \sigma_d \approx 0.88\gamma R + 1.01 \tag{7.3}$$

where γ is the unit weight of the rock material (MN/m^3), R is the rebound number for an L-type Schmidt hammer and σ_d ranges from 20 MPa to approximately 300 MPa. The apparent uniaxial compressive strength can be estimated from scratch and geological hammer tests (Table 7.3)

It is recommended that such tests be carried out on freshly broken rock surfaces such that the estimated wall strength may be directly compared with that of the intact rock. It is likely that the intact strength may be measured in the laboratory as part of the investigation and this will provide a means of calibrating these somewhat crude field measurements.

Aperture, infilling and wall coatings

Aperture is the perpendicular distance separating the adjacent rock walls of an open discontinuity, in which the intervening space is air or water filled. Discontinuities that have been filled (for example, with clay) also come under this category if the filling material has been washed out locally.

Large apertures may result from shear displacement of discontinuities having a high degree of roughness and waviness, from tensile opening resulting from stress relief, from outwash and dissolution. Steep or vertical discontinuities that have opened in tension as a result of valley formation or glacial retreat may have extremely wide apertures measurable in tens of centimetres.

In most sub-surface rock masses, apertures may be closed (i.e. <0.5 mm). Unless discontinuities are exceptionally smooth and planar it

Table 7.3 Rock material strength

Term	Unconfined compressive strength: MPa	Field estimation of hardness
Very strong	>100	Very hard rock, more than one blow of geological hammer required to break specimen
Strong	50–100	Hard rock, hand-held specimen can be broken with single blow of geological hammer
Moderate strong	12.5–50	Soft rock, 5 mm indentations with sharp end of pick
Moderately weak	5.0–12.5	Too hard to cut by hand into a triaxial specimen
Weak	1.25–5.0	Very soft rock, material crumbles under firm blows with the sharp end of a geological pick
Very weak rock or hard soil	0.60–1.25	Brittle or tough, may be broken in the hand with difficulty
Very stiff	0.30–0.60*	Soil can be indented by the fingernail
Stiff	0.15–0.30*	Soil cannot be moulded in fingers
Firm	0.08–0.15*	Soil can be moulded only by strong pressure of fingers
Soft	0.04–0.08*	Soil easily moulded with fingers
Very soft	<0.04*	Soil exudes between fingers when squeezed in the hand

* The unconfined compressive strengths for soils given above are double the undrained shear strengths

will not be of great significance to shear strength that a 'closed' feature is 0.1 mm wide or 1.0 mm wide. Such a range of widths however may have a greater significance with respect to the compressibility of the rock mass.

Large apertures may be measured with a tape of suitable length. The measurement of small apertures may require a feeler gauge. Details of measurement techniques may be found in ISRM (1978).

Infilling refers to material that separates the adjacent rock walls of a discontinuity and that is usually weaker than the parent rock. Typical filling materials are sand, silt, clay, breccia, gouge and mylonite. Filling may include thin mineral coatings and healed discontinuities. Mineral coatings such as chlorite can result in a significant reduction in shearing resistance of discontinuities (Hencher and Richards, 1989). In general, if the filling is weaker and more compressible than the parent rock its presence may have a significant effect on the engineering performance of the rock mass. The drainage characteristics of the filling material will not only affect the hydraulic conductivity of the rock mass but also the long- and short-term mechanical behaviour of the discontinuities since the infill may behave as a soil.

ISRM (1978) suggests that the principal factors affecting the physical behaviour of infilled discontinuities are as follows:

- mineralogy of filling material
- grading or particle size
- over-consolidation ratio (OCR)
- water content and permeability
- previous shear displacement
- wall roughness
- width of infill
- fracturing or crushing of wall rock.

If the thickness of the infill exceeds the maximum amplitude of the roughness the properties of the infill will control the mechanical behaviour of the discontinuity. Clearly the wall roughness and the thickness or width of infill must be recorded in the field. An engineering description of the infill material should be made in the field and suitable samples taken for laboratory tests. The infill should be carefully inspected in the field to see whether there is any evidence of previous movement (for example, slickensides) since this is likely to reduce the shearing resistance of the fracture significantly.

Seepage

Water seepage through rock masses results mainly from flow through discontinuities ('secondary permeability') unless the rock material is sufficiently permeable that it accounts for a significant proportion of the flow. Generally it should be noted whether a discontinuity is dry, damp or wet or has water flowing continuously from it. In the latter case the rate of flow should be estimated. Of course such observations are dependent upon the position of the water table and the prevailing weather conditions. It is important to note whether flow is associated with a particular set of discontinuities.

Methods for collecting discontinuity data

The method used in collecting discontinuity data will depend largely on the degree of access to the rock mass. Surface exposures may be available but may not be representative of the rock mass at the depth of interest, owing to weathering agencies (for example, stress relief causing reduction in joint spacing and an increase in aperture). In rock slope design, the necessary rock mass information may be obtained from surface exposures if available. Drillhole information may be used to supplement the data obtained from surface exposures. Where surface exposures are not available, or are considered to be unrepresentative of the rock mass, drillholes alone may be the only source of data. Table 7.4 shows how the quality of

Table 7.4 Quality of information from different types of discontinuity survey and access to the rock mass (based on Geological Society of London Working Party Report on the Description of Rock Masses (1977))

Type of information	Direct measurement (surface exposure, trial adit or shaft)	Surface photography	Drillhole core	Orientated drillhole core	Drillhole camera	Drillhole impression packer	Geophysics acoustic methods
Location	Good	Good	Good	Good	Good	Good	Medium
Type of discontinuity	Good	Medium	Good	Good	Good	Poor	Poor
Description of rock material	Good	Poor	Good	Good	Poor	None	None
Orientation: dip	Good	Medium	Medium/Poor	Good	Poor	Good	Poor
Orientation: dip direction	Good	Medium	Poor	Medium	Medium	Medium	Poor
Spacing	Good	Good	Medium	Medium	Medium	Medium	Poor
Persistence	Good	Good	Poor	Poor	Poor	Poor	Poor
Wall roughness: waviness	Good	Medium/Poor	Poor	Medium	Poor	Poor	Poor
Wall roughness: roughness	Good	Medium/Poor	Medium	Medium	Poor	Poor	Poor
Wall strength	Good	None	Medium	Medium	None	None	None
Aperture	Good	Poor	Poor	Poor	Medium	Medium	Medium/Poor
Infill: nature	Good	Poor	Medium	Medium	Poor	Poor	None
Infill: thickness	Good	Poor	Medium/Poor	Medium/Poor	Medium	Poor	Poor
Seepage	Good	Medium	None	None	Medium	None	Poor
Number of sets	Good	Good/Medium	Poor	Medium	Medium	Medium	None
Block size	Good	Good/Medium	Poor	Medium	Medium	Medium	Poor

Key to Table 7.4

Good: feature measured reliability

Medium: feature measured but not easily and often with poor reliability

Poor: feature difficult to measure, often measurement is inferred

None: impossible to identify feature

discontinuity data is affected by the type of access to the rock mass and the type of survey method used.

It may be seen from Table 7.4 that access to the rock mass via a drillhole suffers from a number of disadvantages. First, a drillhole permits only a small volume of the rock mass to be viewed such that the persistence of discontinuities cannot be adequately assessed. Second, the orientation of the core must be known before any fracture orientation measurements can be made. Also, drillholes are prone to directional biasing of discontinuity data unless they are drilled with different orientations. For example, if only vertical drillholes are employed any vertical or near vertical sets of discontinuities may be missed altogether or a false impression may be given with respect to the frequency of these fractures. The only way to overcome this problem is to drill inclined holes at a number of different orientations. Terzaghi (1965) and Priest (1993) discuss methods of dealing with directional biasing.

It is impossible to measure aperture from drillhole cores since it is inevitable that the sticks of core will have moved relative to one another during and after sampling. The only way of measuring aperture in this case is by inspection of the drillhole wall. This is achieved using a borehole impression packer or a borehole television camera.

Discontinuity infill may be washed out or contaminated by the drilling fluid such that it becomes difficult to assess its thickness or properties adequately. Mineral coatings on joint walls may be observed in core samples. It is impossible, however, to assess the degree of coverage from such a small sample.

Surface exposures offer a much larger expanse of rock for examination and permit direct observation and measurement of discontinuities. They can be just as prone to directional biasing as the drillhole if only a single orientation of exposed face is available. Where surface exposures have revealed one or more discontinuity sets drillholes and drillhole cores may be used effectively to check whether these persist at depth.

Discontinuity surveys

The rock mass contains a considerable amount of geometrical information which must be collected and interpreted. Clearly an irregular highly fractured rock face presents a somewhat daunting challenge to anyone who wishes to quantify the rock structure or discontinuity network in an unbiased manner. It is important, therefore, to ensure that measurement systems are based upon objective but flexible sampling strategies linked to rigorous data analysis (Priest, 1993). Typically between 1000 and 2000 discontinuities should be sampled to provide an adequate characterization of a site (Priest and Hudson, 1976). This number is generally made up from samples between 150 and 350 discontinuities taken at between

Table 7.5 Terminology and checklist for rock discontinuity description (BS 5930: 1999)

Spacing	Orientation	Persistence	Type of termination	Roughness	Wall strength	Aperture	Filling	Seepage	No. of sets
Extremely wide >6 m	Dip amount only in cores	Discontinuous	Cannot normally be described	Small scale (cm) and intermede scale (m) Stepped — Rough / Smooth / Striated	Schmidt hammer	Cannot normally be described in cores	Clean	Cannot be described in cores	Cannot be described in cores
Very wide 2 to 6 m		Continuous in cores					Surface staining (colour)		
Wide 600 mm to 2 m				Undulating — Rough / Smooth / Striated			Soil infilling (describe in accordance with 41)		
Medium 200 to 600 mm		Very high >20 mm	Termination		Point load test	Very open >10 mm		Moisture on rock surfaces	Record spacing and orientation of sets to each other and all details for each set
Close 60 to 200 mm	Take No. of readings, of dip direction/dip, e.g. 015/08°	High 10 to 20 mm	x (outside exposure)	Planar — Rough / Smooth / Striated	Other index tests	Open 2.4 to 10 mm	Mineral coatings (e.g. calcite, chlorite, gypsum, etc.)	Dripping water Water flow measured per time unit on an individual discontinuity or set of discontinuities	
Very close 20 to 60 mm		Medium 3 to 10 m	r (within rock)	Large scale (dm) Waviness Curvature Straightness		Moderately open 0.5 to 2.5 mm			
Extremely close <20 mm	Report as ranges and on stereonet if appropriate	Low 1 to 3 m	d (against discontinuity)		Visual assessment	Tight 0.1 to 0.5 mm Very tight <0.1 mm	Other – specify	Small flow 0.05 to 0.5 l/s Medium flow 0.5 to 5.0 l/s Strong flow >5 l/s	
Take number of readings; state min., average and max.		Very low <1 m	Record also size of exposure	Measure amplitude and wavelength feature		Take number of readings; state min., average and max.	Record width and continuity of infill		

5 and 15 sample locations chosen to represent the main zones based on geological structure and lithology. In some cases the extent of the site or the exposed rock makes such large numbers of measurements impractical or impossible. In such cases the minimum sample of 200 discontinuities should be taken.

The method of collecting discontinuity data will vary according to the type of access to the rock mass. The two broad sampling strategies that can normally be adopted involve either the logging of drillhole core or the examination of an exposed rock face.

When using drillholes, a detailed fracture log of the core is required together with an inspection of the drillhole wall. In the case of exposed rock faces above or below ground the most widely used sampling methods include scanline sampling (ISRM, 1978; Priest and Hudson, 1981; Priest, 1993) and window sampling (Pahl, 1981; Priest, 1993).

It is often useful to employ descriptive terms for many of the features of discontinuity, particularly where direct measurements are difficult or impractical due to time constraints. The descriptive terms for most of the key features recommended by BS 5930: 1999 are shown in Table 7.5.

Fracture logging of drillhole core

A common problem in logging fractures in core samples or in a man-made rock face is identification of artificial fractures resulting from the drilling process or by the creation of the face (blasting and stress relief). These fractures are normally excluded from the log, unless a conscious decision is made to the contrary which should be clearly stated on the log. A degree of judgement is therefore required. Artificial and natural fractures can often be distinguished from each other by observing the freshness, brightness, staining and erosion of the fracture surface. For example, in the chalk natural fractures often exhibit manganese spots or dendritic patterns and relatively smooth surfaces, whereas artificial fractures are clean and rough.

Every natural fracture which cuts the core should be described in the following manner.

- The position of the fracture in the core sample should be recorded. A pictorial log of the fractures cutting the drillhole may be made from this information.
- The angle the fracture makes with the core axis should be noted. Where the orientation of the core is known the dip and dip direction of the fracture should be determined.
- The roughness of the fracture surfaces should be noted.
- If any infill is present, its thickness and nature should be described.
- The presence of any mineral coatings should be noted.

215

- Where possible or practical the compressive strength of the fracture surface should be determined using a Schmidt hammer. A qualitative assessment of the wall strength may be made by indenting the wall and the side of the core using the point of a knife, pick or other sharp implement.
- The average spacing of discontinuity sets identified from the fracture log may be determined from the recorded positions of the relevant fractures. The general fracture state of the rock mass may be assessed from the determination of the total and solid core recovery, the fracture index and Rock Quality Designation (RQD).

Scanline sampling

Although many different techniques have been described for sampling discontinuities in rock exposures (Muller, 1959; Pacher, 1959; Da Silveria *et al.*, 1966; Knill, 1971) the line or scanline approach is preferred (Piteau, 1970; Broadbent and Rippere, 1970) on the basis that it is indiscriminate (all discontinuities whether large or small should be recorded) and provides more detail on discontinuity spacing (Priest and Hudson, 1976, 1981) and attitude than other methods. There is no universally accepted standard for scanline sampling.

In practice, a scanline survey is carried out by fixing a measuring tape to the rock face by short lengths of wire attached to masonry nails hammered into the rock. The nails should be spaced at approximately 3 m intervals along the tape which must be kept as taut and as straight as possible. The face orientation and the scanline orientation should be recorded along with other information, such as the location, date and the name of the surveyor. Where practicable the face and scanline, including a scale and appropriate label, should be photographed before commencing the sampling process. In cases where the face is irregular it will be necessary to take photographs from several viewpoints. A simple way to provide a scale is to attach clearly visible markers at say 1 m intervals along the length of the scanline. Care should be taken to minimize distortion of the face on the photographs.

Once the scanline is established the surveyor works systematically along the tape recording the position and condition of every discontinuity that intersects it. The features that are commonly recorded include the following.

- *Intersection distance.* This is the distance in metres (rounded to the nearest cm) along the scanline to the intersection point with the discontinuity. Where the face is irregular it will be necessary to project the plane of fractures not in contact with the tape on to the tape such that their position can be accurately recorded. In highly irregular faces this method can lead to significant errors in the determination of joint spacing. Ideally a clean, approximately planar rock face should be selected for scanline sampling.

- *Orientation.* This is the dip direction and dip of the discontinuities.
- *Semi-trace length.* This the distance from the intersection point on the scanline to the end of the discontinuity trace. The distance may be measured directly, estimated by eye or scaled from a photograph of the rock face, when it becomes available. There will be two semi-trace lengths associated with each discontinuity: one above and one below for a horizontal scanline; one to the left and one to the right for an inclined for vertical scanline.
- *Termination.* It is helpful to record the nature of termination of each semi-trace (see Table 7.5).
- *Roughness.* A profile of small-scale roughness may be made in the manner described earlier or the Joint Roughness Coefficient (JRC) (Barton, 1973) may be estimated visually. Intermediate and large-scale roughness may be described as in Table 7.5 and amplitude and wavelength can be measured where necessary.

Other features such as type of discontinuity, nature of infill, aperture, water flow, slickensides, are generally reported in a comments column on the logging sheets. An example of a typical logging sheet is shown in Fig. 7.5.

Further scanlines should be set up on a second rock face, approximately at right angles to the first, to minimize the orientation sampling bias. Corrections have been devised to compensate for directional bias (Terzaghi, 1965; Robertson, 1970) but these will not aid the identification of joints sets which intersect the scanline at low angles ($<10°$).

The length of the scanline should be at least fifty times the mean discontinuity spacing (Priest and Hudson, 1976) in order to estimate the frequency of discontinuities to a reasonable degree of precision. Priest (1993) recommends that the scanline should contain between 150 and 350 discontinuities. Often a compromise must be sought owing to restrictions to size of face or access to parts of the face.

Window sampling

Window sampling provides an area-based alternative to the linear sampling techniques outlined above which reduces the sampling bias for discontinuity orientation and size. The measurement techniques are essentially the same as for scanline sampling except that all discontinuity traces which cross the defined area of the rock face are measured.

The sampling window may be defined by setting up a rectangle of measuring tapes pinned to the rock face. In order to minimize sampling

Job No. **LNI/45**   Location   Rock type **Dolerite**   Sheet 1 of

Date **15/12/2000**   Observer **MCM**   Photo No. **LNI/45/12 looking SW**

Face orientation (dip direction/dip) **034/55**   Scan line orientation (dip direction/dip) **124/00**

Scanline length (m) **25**   Start point (description) **SE end working to NW**

Description of face (size, rough/smooth etc.) **Lateral extent 30 m, height 10 m. Slope rough with scree covering bottom 2 m**

Discontinuity

Chainage: m	Type	Dip direction	Dip	Set (if known)	Persistence	Roughness	Aperture/ infill	Comments
0·254	Joint	260	70	-	r. medium	Stepped (wavelength 1 m, amplitude 0·2 m)	Clean	Discontinuity damp, no measurable seepage, occasional calcite coating

Fig. 7.5 Typical scanline logging sheet (see Table 7.5 for an explanation of terms used)

bias effects it is recommended that the window should be as large as possible, with each side of a length such that it intersects between 30 and 100 discontinuities (Priest, 1993). Where possible, two windows of similar size should be set up on mutually perpendicular faces.

In general, area sampling provides a poor framework within which to collect orientation, frequency and surface geometry data for individual discontinuities. The window is likely to contain a large number of relatively small discontinuities, making it difficult to keep track of which discontinuities have been measured. The process of window sampling is generally more laborious than scanline sampling when attempting to apply the same rigorous sampling regime.

Face sampling

At the preliminary stage of an investigation it is often necessary simply to establish the number of discontinuity sets, the average orientation of each set and the relative importance of each set. Matherson (1983) suggests that for simple rock slope stability assessments at the preliminary site investigation phase it is sufficient to observe persistence, aperture and infilling in addition to orientation. In such cases it may be expedient to adopt a sampling strategy that is less rigorous than those outlined above.

Face sampling involves recording the orientation, persistence, roughness, aperture, infilling and seepage of a representative number of discontinuities in the exposed face. A suitable geological compass should be used for the orientation measurements and the classification recommended by BS 5930: 1999 should be used for the other parameters (see Table 7.5). The size of the sample must be large enough to ensure statistical reliability. A minimum of 200 measurements per face is recommended. Such large sample sizes may preclude the measurement of all the parameters listed above for each fracture. An evaluation of the importance of each discontinuity, however, should be made. This is normally based on persistence.

The grouping of data collected from a very large exposure or from a number of different locations may obscure discontinuity patterns. Ideally the area studied should be divided into units, domains or structural regions (Piteau, 1973). Data from each face should be assessed separately. If it can be established that each face shows a similar discontinuity pattern it may then be possible to group the data from a number of faces. If different rock types are present in the survey area the discontinuities observed in each type should be placed in separate groups. Collection of data in domains allows individual or grouped assessment. This is likely to be of major importance where alignment, design or rock type change along a proposed route.

Presentation of discontinuity data for stability analysis: the stereonet

Overview

Discontinuity data should be presented in a form that allows ease of assimilation and is amenable to rapid assessment. Discontinuities may be shown on maps, scale drawings of exposures or block diagrams which may be used to indicate the spatial distribution and interrelationships of these features. Such methods, although very useful, do not allow the quantitative assessment of orientation and spacing which are perhaps the most important aspects of discontinuities within a rock mass. In many cases, it is not possible to recognize all the discontinuity sets or assess the dominant orientation of some discontinuity sets in the field. Such factors may only be assessed by statistical analysis of discontinuity orientation data.

Discontinuity orientation data are easier to visualize and analyse if presented graphically. The most commonly used method of presenting orientation data is the hemispherical projection. This method allows the distribution of dip and dip direction to be examined simultaneously and provides a rapid visual assessment of the data as well as being readily amenable to statistical analysis. Although this method is used extensively by geologists, it is little understood by engineers, since it bears no recognizable relationship to more conventional engineering drawing methods. The basis of the method and its classic geological applications are described by Phillips (1971). Rock engineering applications are described in detail by Goodman (1976), Hoek and Brown (1980), Priest (1980, 1985, 1993), Hoek and Bray (1981) and Matherson (1983).

Hemispherical projection methods for display and analysis of discontinuity data

The hemispherical projection is a graphical method for the presentation and analysis of the orientation of planar and linear features in two dimensions (i.e. on a sheet of paper). All forms of hemispherical projection use an imaginary sphere as a basis for converting three-dimensional data into a two-dimensional form. The sphere is arranged such that linear features or planar features (e.g. discontinuities) found in a rock mass pass through its centre. The intersection of such lines or planes with the surface of the sphere are projected on to the equatorial plane. The intersection with either the upper or lower hemisphere is used in the projection process, hence the name hemispherical projection. In most cases where a horizontal plane of projection is being used, the intersection with the lower hemisphere is considered. Such projections are referred to as lower hemisphere projections.

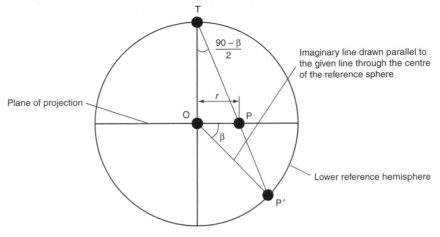

Radius of reference sphere = R

Fig. 7.6 Equal angle projection

There are two commonly used methods of projection:

- *Equal angle projection*. This projection accurately preserves the angular relationships between features.
- *Equal area projection*. This projection preserves the spatial distribution of features.

Equal angle projection

Figure 7.6 shows a vertical section through the imaginary sphere and the construction of the equal angle projection. If a line of dip direction α and dip β intersects the reference sphere in Fig. 7.6 at P′, the projection this point is achieved by drawing a straight line from P′ to a point T which is on the surface of the sphere vertically above the centre O. The projection of P′ occurs at P where the line P′T passes through the plane of projection. The relationship between the radial distance r (=OP), the radius of the reference sphere R and the dip of the line β is given by:

$$r = R \tan \frac{90 - \beta}{2} \tag{7.4}$$

Equal area projection

Figure 7.7 shows a vertical section through the imaginary sphere and the construction of the equal area projection. If a line of dip direction α and dip β intersects the reference sphere in Fig. 7.7 at P′, the projection this point is achieved by rotating the chord joining P′ to a point B vertically below the centre of the sphere (O) about point B until it is horizontal. The projection of P′ occurs at P″ where the arc drawn out by BP′ meets the horizontal

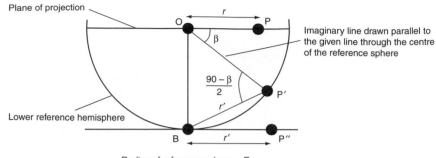

Plane of projection

O

r

P

β

Imaginary line drawn parallel to the given line through the centre of the reference sphere

$\dfrac{90 - \beta}{2}$

P′

r′

Lower reference hemisphere

B

r′

P″

Radius of reference sphere = R

Fig. 7.7 Equal area projection

plane of projection. The relationship between the radial distance r' (=BP″), the radius of the reference sphere R and the dip of the line β is given by:

$$r' = 2R\cos\frac{90 + \beta}{2} \qquad (7.5)$$

When $\beta = 0°$, $r' = 2R\cos 45° = R\sqrt{2}$. This means that the radius of the resultant projection is larger than the radius of the reference sphere by a factor of $\sqrt{2}$. The point P″ is therefore transferred to point P which is located at a radial distance $r = r'/\sqrt{2}$ from the centre of the reference sphere and hence

$$r = R\sqrt{2}\cos\frac{90 + \beta}{2} \qquad (7.6)$$

The equal area projection method is commonly used in rock engineering because it permits the statistical analysis of discontinuity data.

Plotting lines and planes using hemispherical projection

A line will always be projected as a point. A vertical line will plot as a single point at the centre of the reference circle and a horizontal line will plot as two diametrically opposed points on the circumference of the reference circle.

A plane presents a more complex problem. To help understand the projection of a plane (e.g. a discontinuity) it may be thought of as being made up of a series of lines radiating out (in the plane of the discontinuity) from the centre of the reference sphere as shown in Fig. 7.8(a). One of these lines (Ob) will have the same dip and dip direction as the plane and this line will form a line of symmetry. Each side of this line there will be a series of lines with dips ranging from the true dip of the plane to 0° and dip directions ranging from that of the plane to a direction perpendicular to it. The line of symmetry will plot on the plane of projection closest to the centre of the reference circle. The other lines will plot progressively

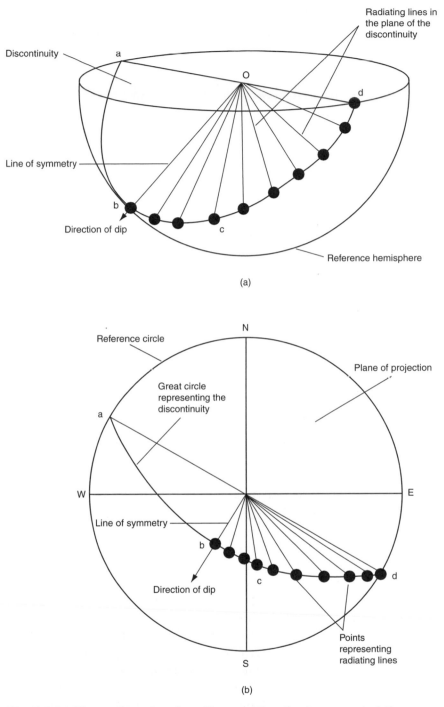

Discontinuity

Radiating lines in the plane of the discontinuity

a

O

d

Line of symmetry

b

Direction of dip

c

Reference hemisphere

(a)

Reference circle

N

Plane of projection

Great circle representing the discontinuity

a

W

E

Line of symmetry

b

Direction of dip

c

d

Points representing radiating lines

S

(b)

Fig. 7.8 (a) Illustration of a plane/lines cutting the lower part of the reference sphere. (b) Projection of plane/lines

223

further away until in the extreme the horizontal line (aOd), which is perpendicular to the dip direction of the plane, will plot on the circumference of the reference circle. The locus of the projection of all of these lines will trace out a great circle on the reference circle (Fig. 7.8(b)).

A vertical plane will plot as a straight line passing through the centre of the reference circle. A horizontal plane will plot directly over the circumference of the reference circle.

To aid plotting lines and planes and the measurement of angular relationships between these features (e.g. the orientation of the line of intersection between two inclined planes) a special type of graph paper has been produced called a 'stereonet'. An equal area stereonet is shown in Fig. 7.9. It is made up of a series of great circles representing planes dipping due East and due West at dip angles ranging from zero at the edge of the net to 90° at the centre at 2° intervals. The projection of lines and planes are plotted on tracing paper which is overlaid on the

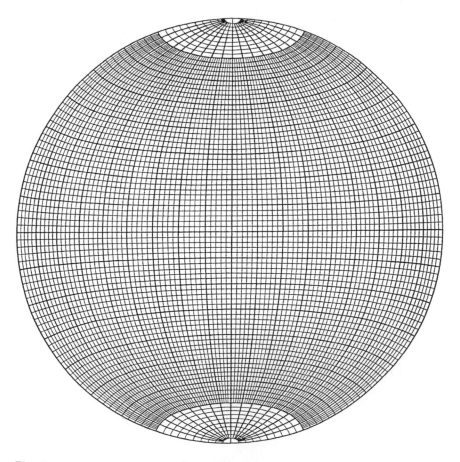

Fig. 7.9 Example of an equal area stereonet

stereonet. Since only E and W dipping planes are shown on the stereonet it is necessary to rotate the tracing paper about the centre of the stereonet in order to produce projections of lines and planes at other orientations.

Plotting projections of lines and planes using a stereonet
Preparing the stereonet for use
Take the stereonet and fix it (using draughting tape) to a stiff board (A4 size). Using a drawing pin make a hole through the centre of the stereonet. Remove the drawing pin and insert it through the back of the board and fix it in place with some tape. You are now ready to use your stereonet.

Plotting lines and planes using a stereonet
All data are plotted on tracing paper that is placed over the stereonet.

WORKED EXAMPLE

Problem 1

Plot the projection of a line dipping at 40° towards 120°.

- Place the tracing paper over the stereonet and pierce it with the point of the drawing pin.
- Mark the North Point with a long tick and write 'N' above it (Fig. 7.10(a)).
- Count 120° clockwise from the North Point and place a small tick over the edge of the stereonet (Fig. 7.10(a)).
- Rotate the tracing paper until the tick drawn is over either the East Point or the West Point. The East Point is more convenient in this case (Fig. 7.10(b)).
- Count 40° inwards from the circumference of the stereonet and mark the point with a dot (Fig. 7.10(b)).
- Rotate the tracing paper such the large tick representing the North Point coincides with the North Point of the stereonet (Fig. 7.10(c)).

The dot you have drawn represents the projection of the line.

Problem 2

Using the same piece of tracing paper as for Problem 1, plot the projection of a plane dipping at 50° towards 300°.

- Count 300° clockwise from the North Point and place a small tick over the edge of the stereonet (Fig. 7.11(a)).
- Rotate the tracing paper until the tick drawn is over either the East Point or the West Point. The West Point is more convenient in this case (Fig. 7.11(b)).

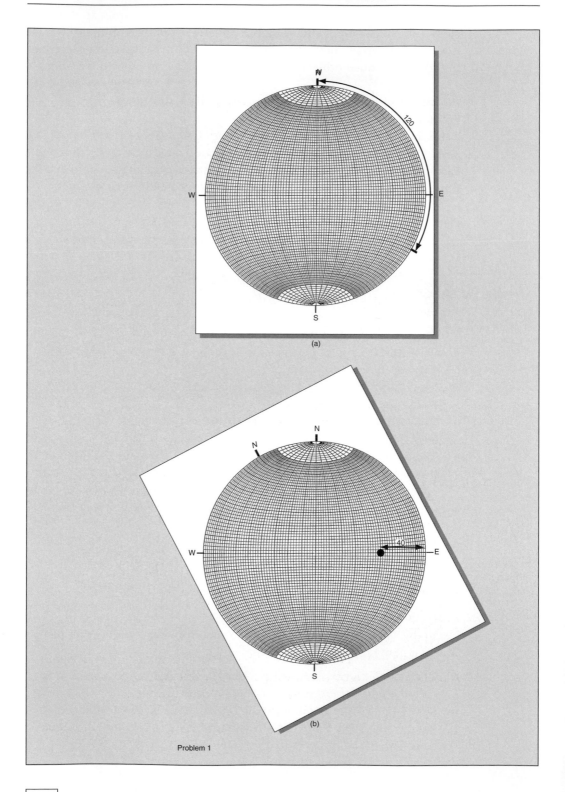

(a)

(b)

Problem 1

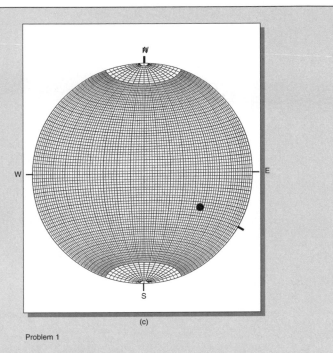

(c)

Problem 1

Fig. 7.10 Solution to Problem 1

- Count 50° inwards from the circumference of the stereonet and mark the point with a dot (Fig. 7.11(b)).
- Using the gridlines on the net trace out a great circle which passes through the dot drawn (Fig. 7.11(b)).
- Rotate the tracing paper such the large tick representing the North Point coincides with the North Point of the stereonet (Fig. 7.11(c)).

The great circle you have traced represents the projection of the plane. Now rotate the tracing paper back to the position it was in Problem 2 when the great circle was drawn. You will note that the point you plotted in Problem 1 is on the E–W line. If you measure the angular distance (by counting squares on the stereonet) along the E–W line between the great circle and the projection of the line in Problem 1 you should find the answer to be 90°. This indicates that the line is perpendicular to the plane.

It is convenient particularly for the purposes of statistical analysis to represent a plane by a line which is perpendicular to the plane. This line is called a pole. If a plane has dip and dip direction of β and α respectively then its pole will have dip and dip direction $90 - \beta$ and $180 + \alpha$ respectively. To plot the pole to the plane in the above problem we must first calculate the dip and dip direction of the pole. The dip and

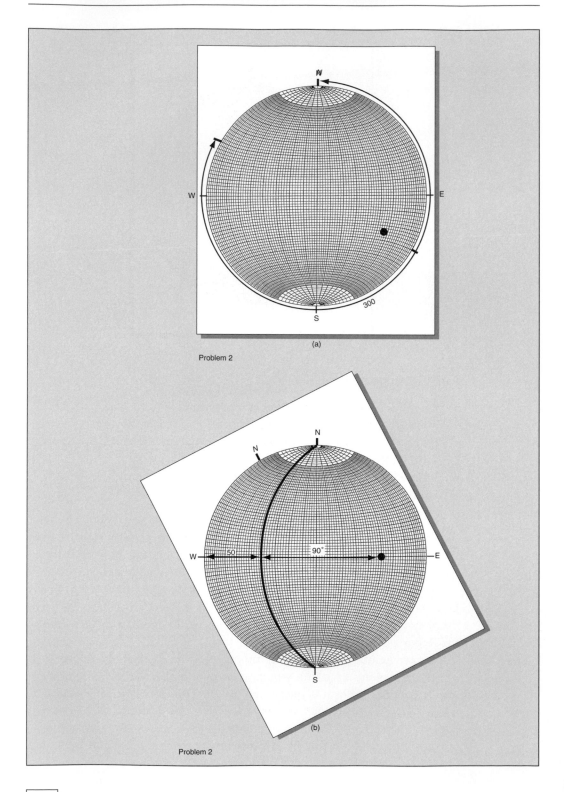

(a)

Problem 2

(b)

Problem 2

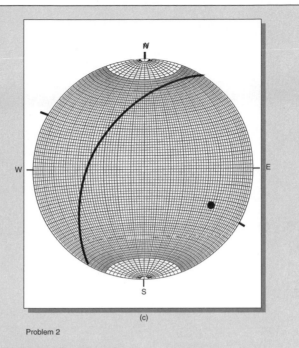

(c)

Problem 2

Fig. 7.11 Solution to Problem 2

dip direction of the pole is in this case $90 - 50 = 40°$ and $180 + 300 = 480°$, $480 - 360 = 120°$, respectively. The procedure for plotting poles is shown below.

Procedure for plotting poles
The important thing to remember when plotting poles is that the projection of the pole will always be on the opposite side of the stereonet to the projection of the plane.

Problem 3
Plot the pole representing a plane that is dipping at 70° towards 240° Calculate the dip and dip direction of the pole:

Dip of pole $= 90 - 70 = 20°$.

Dip direction of pole $= 180 + 240 = 420°$ which is equivalent to a whole circle bearing of $420 - 360 = 60°$.

Thus the pole to the plane is dipping at 20° towards 060° (i.e. 060/20).

The pole can then be plotted on the stereonet using the technique described in problem 1.

A more convenient way of plotting the projection of a pole to a plane which avoids calculating the dip and dip direction of the pole is illustrated below.

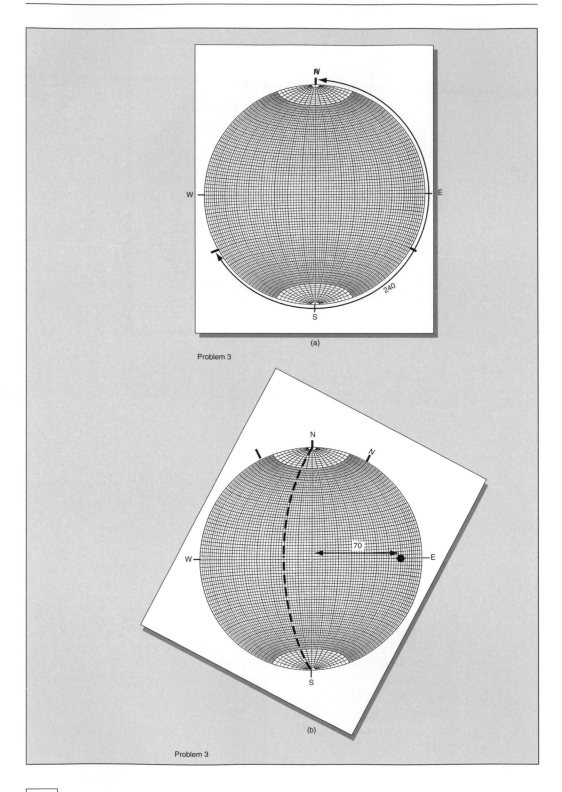

(a)

Problem 3

(b)

Problem 3

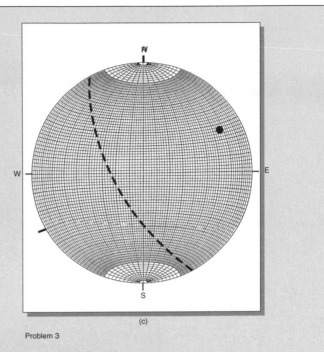

(c)

Problem 3

Fig. 7.12 Solution to Problem 3

Place the tracing paper over the stereonet and pierce it with the point of the drawing pin.

- Mark the North Point with a long tick and write 'N' above it.
- Count 240° clockwise from the North Point and mark the point with a small tick (Fig. 7.12(a)).
- Rotate the tracing paper such that the small tick is over the West Point. You could use the East Point but the West Point is closer (Fig. 7.12(b)).
- Now remember that you are plotting a pole. The plane will plot on the left side of the stereonet, so the pole will plot on the right side. Since the pole to a vertical plane will plot on the circumference and a pole to a horizontal plane will plot at the centre you will need to count 70° from the CENTRE towards the East Point and mark the point with a dot (Fig. 7.12(b)).
- Rotate the tracing paper such the large tick representing the North Point coincides with the North Point of the stereonet (Fig. 7.12(c)).

Some examples for you to practice on are given in problem 4.

Problem 4

Plot the great circles representing the following planes and the poles to the planes:

Plane	Orientation
A	324/36
B	103/30
C	210/75
D	247/87
E	065/05

The answers to Problem 4 are given in Fig. 7.13.

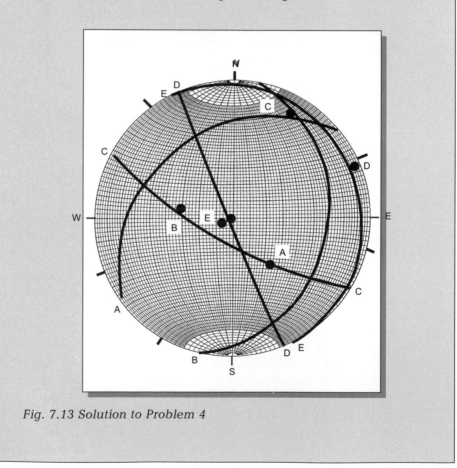

Fig. 7.13 Solution to Problem 4

Analysis of discontinuity orientation data

In order to identify the predominant sets in a rock mass and their average orientations it is necessary to plot the poles for the discontinuity orientation data and carry out a statistical analysis. This analysis makes use of simple graphical techniques and permits the identification of discontinuity clusters (i.e. sets) from a contoured diagram such as that shown in Fig. 7.14. There are a number of different techniques that can be carried out by hand. This can be a time consuming process, however, and is usually done using a computer software application.

A simple method for carrying out this process manually is to create a rectangular grid on a piece of paper such that the x and y grid increment

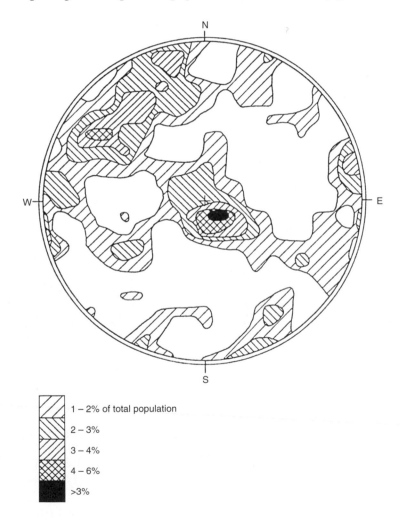

1 – 2% of total population

2 – 3%

3 – 4%

4 – 6%

>3%

Fig. 7.14 Lower hemispherical projection showing concentrations at discontinuities (i.e. discontinuity sets)

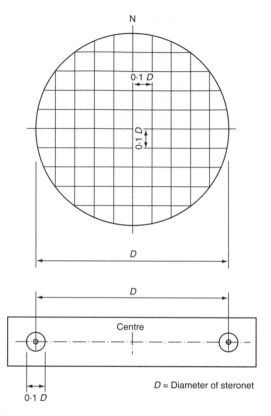

Fig. 7.15 Simple single cell system for counting poles

is one tenth of the diameter of the stereonet used to plot the poles as shown in Fig. 7.15(a). A counting device (see Fig. 7.15(b)) can be constructed from a piece of clear plastic film. The circles have a diameter which is one tenth of the diameter of the stereronet.

Counting is done by centring the tracing paper with the poles plotted on it over the grid and covering this with a clean piece of tracing paper. The counting device is moved such that one of the circles is positioned over the intersection of two grid lines and the number of poles within the circle is recorded on the top of tracing paper over the grid line intersection. For steeply dipping discontinuities (i.e. poles situated close to the circumference of the stereonet) a different procedure is used. In this case the counting device is centred over the grid and moved to a grid point on the circumference as shown in Fig. 7.16. The total number of poles in both half circles is recorded at both grid points, as shown in Fig. 7.16.

A more convenient method of carrying out this analysis is to use a graphical counting method. A wide variety of such methods have been devised (Kalsbech, 1965; Dimitrijevic and Petrovic, 1965; Stauffer, 1966; Denness, 1970, 1971). Of these, that devised by Dmitrijevic and Petrovic

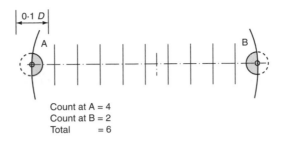

Count at A = 4
Count at B = 2
Total   = 6

Fig. 7.16 Procedure for counting poles close to the circumference of the stereonet

is the easiest to use. This counting net is constructed from a number of circles on a hemisphere each occupying 1% of the surface area. When these are projected on to the equal area net they become ellipsoidal to a degree consistent with their position on the hemisphere (see Fig. 7.17).

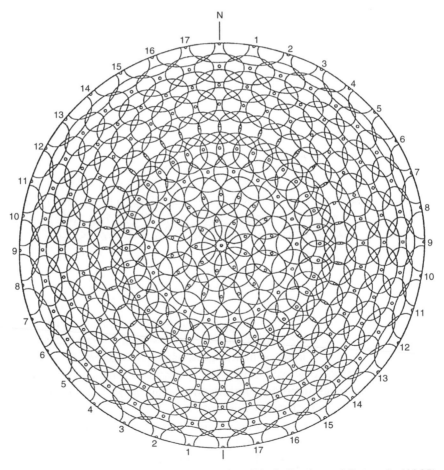

Fig. 7.17 Dimitrijevic counting net, after Dimitrijevic and Petrovic (1965)

235

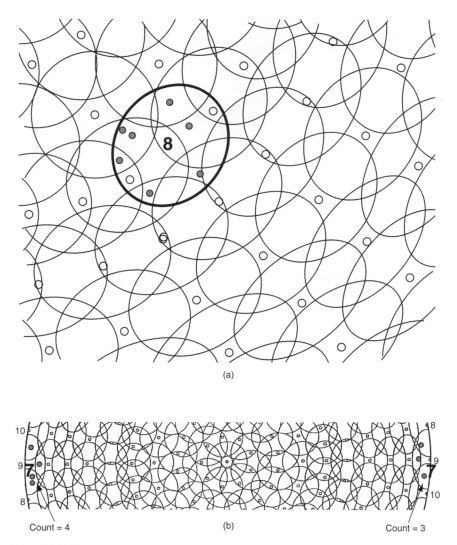

(a)

(b)

Count = 4

Count = 3

Fig. 7.18 Counting procedure using the Dimitrijevic counting net

A density distribution is obtained by counting the number of plotted points lying within each circle or ellipse as shown in Fig. 7.18. An even and overlapping distribution of ellipses allows the entire area of the net to be covered. No rotation is required during counting. The best results are obtained by plotting the poles on to tracing paper and overlaying this on to the counting net.

One of the disadvantages of using these counting nets to contour pole density is that the geometry of the net bears no relationship to the distribution of poles. When a cluster of poles falls across a boundary between two counting cells, a correct assessment of the pole concentration can only be obtained by allowing the counting device to 'float' from its original position and to centre it over the highest concentrations. This is achieved by rotating the counting net about its centre. The counting device shown in Fig. 7.15 has a distinct advantage here since it represents a single counting cell (when working away from the circumference of the stereonet) which can be moved to any position. The concept of using a rectangular grid as described above is to allow a coarse analysis of the data while ensuring that a systematic approach is maintained. Once this has been done finer detail may be obtained by using the single counting cell in a 'floating mode'. This can be time consuming and when checking contours of poles derived from a computer programme the Dmitrijevic net will be adequate. Contouring can be carried out visually interpolating between counts where necessary.

WORKED EXAMPLE

Problem 5

In this worked example the discontinuity data given in Appendix 3 have been plotted as poles using a stereonet and this is shown in Fig. 7.19(a).

Preparation
- Pierce the centre of the counting net with a drawing pin.
- Take the tracing paper with the poles plotted on it and align the drawing pin hole over the drawing pin in the counting net. Now rotate the tracing paper until the north point lines up with the north point on the counting net.
- Secure the tracing paper to avoid subsequent movement.
- Take a clean sheet of tracing paper and place this over the first sheet and secure the tracing paper to avoid subsequent movement. You can now remove the drawing pin.
- Trace the net circumference on the overlay, mark and identify the North reference position.

Counting
- For each whole circle/ellipse, count the number of points lying within or on the outline. Mark this value on the overlay at the centre of each ellipse/circle (see Fig. 7.18(a)).
- For each part-ellipse at the edge of the net, count the number of points lying within or on the outline. Combine this count with that from the part-ellipse on the opposite side of the net (numbers 1 to 17 assist this process) (see Fig. 7.18(b)). Mark the total value at the centre of both part-ellipses.

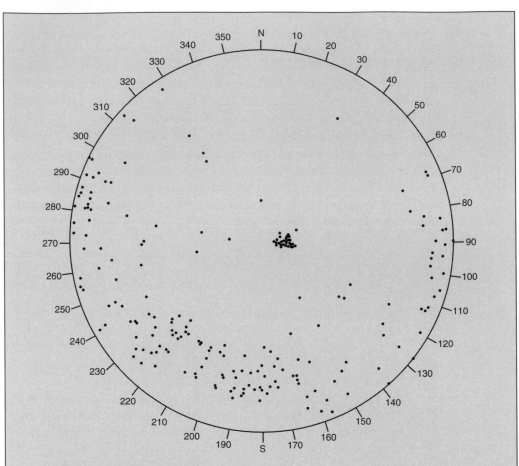

Fig. 7.19 Raw data for Problem 5

- Check that diagonally opposite part-ellipses have the same value.

The results of the counting procedure are shown in Fig. 7.20. The results of the analysis are best presented as a contoured diagram in which the pole count in each ellipse is expressed as a percentage of the whole population of poles. In Fig. 7.21 the results of the analysis carried out in the above worked example have been contoured.

It can be seen from Fig. 7.21 that a number of clusters of poles can be identified. These have been labelled A, B, C and D in rank order according to the density of poles. The maximum density of poles associated with each cluster is also shown in Fig. 7.21. The clusters represent sets of discontinuities. The relative importance of each set cannot be determined from the density of poles alone. The key attribute for importance, particularly where slope stability is concerned, is

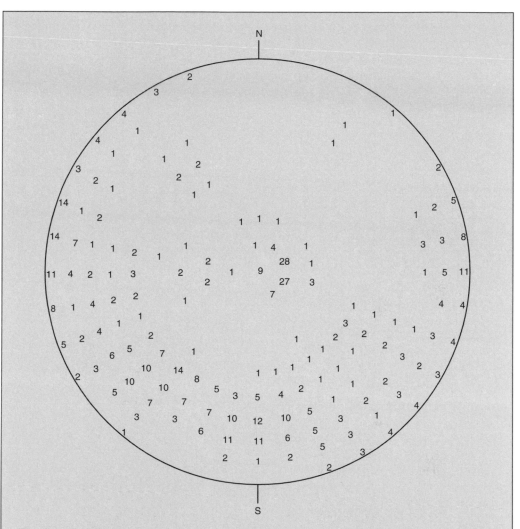

Fig. 7.20 Results of counting procedure showing density of poles per 1 × area of net

persistence. If, for example, set D has a very high persistence whereas set A had a very low persistence then in terms of importance set D will rank higher than set A even though more poles were found associated with set A. In some cases a single discontinuity with very high persistence can rank as the most important feature.

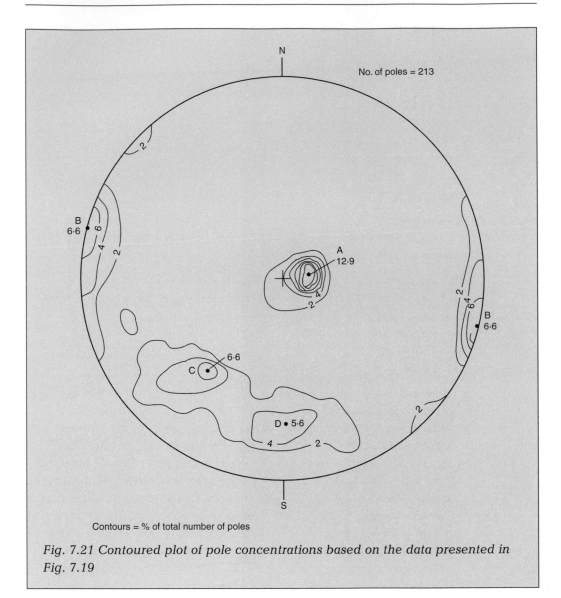

Fig. 7.21 Contoured plot of pole concentrations based on the data presented in Fig. 7.19

CHAPTER EIGHT

Types of rock slope failures: kinematic feasibility

Modes of failure

Overview

A rock mass may display one or more modes of failure depending on the following factors:

- presence or absence of discontinuity sets
- orientation of discontinuity sets in relation to that of the natural or excavated face
- discontinuity spacing in one and three dimensions
- shear strength of discontinuity walls
- persistence of discontinuities.

Table 8.1 describes the individual modes of failure that can occur in fractured rock masses.

Plane, wedge and toppling failure modes depend on the interaction of discontinuity orientation, face orientation and shear strength. This high degree of geometric control allows potential failure modes to be identified using kinematics. Kinematics is a branch of mechanics that deals with motion without reference to force or mass. Hence in a rock slope, blocks that have the freedom to move based on geometry alone may be regarded as kinematically feasible blocks. A block may only be regarded as unstable however, if it is capable of being removed from the rock mass without disturbing the adjacent rock and the disturbing forces are greater than the restoring forces. A particular mode of failure in a given face may be recognized by identifying the kinematically feasible block for that mode based on the relative orientation of the discontinuities and the face and the stability of the block by considering the shear strength characteristics of the discontinuities.

By constructing a series of overlays that are used with the contoured plot of pole concentrations, potentially unstable blocks relating to different modes of failure can be readily identified. In constructing the overlays a very simplistic approach is taken to define the shear strength of the discontinuities. The presence of water in the discontinuities and external loading (both static and dynamic) are ignored.

241

Table 8.1 Modes of failure in fractured rock masses

Failure mode	Sketch	Description	Comments
Plane failure		Plane failure occurs when a discontinuity dips in a direction close to that of the face and the magnitude of the dip is greater than the angle of friction for the discontinuity	This is one of the simplest modes of failure. For plane failure to occur in slopes there must be lateral release surfaces (Fig 8.1) which will allow a block of finite size to slide out of the face. A dangerous situation is created when the face is convex in plan. In such a case it is possible for the excavated face to act as the release surfaces
Wedge failure		Wedge failure occurs when the orientation of two discontinuities results in a line of intersection that dips in a direction close to that of the face and the dip of this line is significantly greater than the angle of friction for the discontinuities	This is the most dangerous mode of failure since no release surfaces are required
Circular failure		When the material is weak (as in soil slopes) or when the rock mass is heavily jointed or broken (as in a waste rock dump) the failure surface is likely to be circular	When the pattern of discontinuities is random (i.e. no sets) circular failure modes are likely. The treatment of circular failures is covered in Part 1 of this book

Block toppling	This type of failure occurs when long slender rock blocks (e.g. tabular or columnar blocks) dip into the face at relatively steep angles and rest on a basal discontinuity which dips out of the face at an angle less than the angle of friction for that discontinuity	This type of failure generally requires three sets of discontinuities. Two orientated such that their line of intersection dips into the face and one that dips in nearly the same direction as the face at a low angle
Flexural toppling	This commonly occurs when there is a set of closely spaced discontinuities which dip a relatively steep angle into the face	This type of failure often results in gradual movements behind the face at distance up to five times the height of the slope
Rockfalls	Rockfalls consist of free-falling blocks of different sizes which are detached from a steep rock face. The block movement includes bouncing, rolling, sliding and fragmentation. The detachment of relatively small fragments of rock from the face is known as ravelling	In the design of rock slopes the problem of rockfall is the prediction of the paths and the trajectories of the unstable blocks which detach from the rock slope so that suitable protection can be constructed

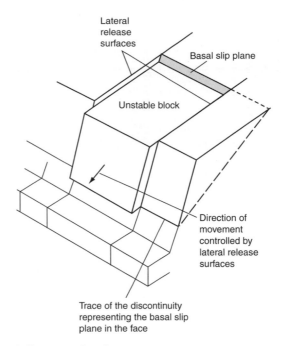

Lateral release surfaces

Basal slip plane

Unstable block

Direction of movement controlled by lateral release surfaces

Trace of the discontinuity representing the basal slip plane in the face

Fig. 8.1 Plane failure mechanism

Plane failure

In plane failure, movement of a rock block or mass occurs by sliding along a basal slip plane (Fig. 8.1). For movement to occur, the following three basic conditions must be satisfied (Matherson, 1983).

- The dip direction of the slip plane must lie within approximately 20° of the slope face dip direction.
- The slip plane must 'daylight' (outcrop) on the slope.
- The dip of the slip plane must exceed the friction angle on that plane.

Other factors such as the presence of water, the effect of water pressure, and the necessity for lateral release surfaces to exist, are ignored in this simple assessment.

The limit conditions described above can be portrayed on an overlay which is used with a contoured plot of pole density (e.g. see Fig. 7.21) to identify potential slip basal slip planes. The overlay is constructed in the following manner.

- Secure a piece of tracing paper over the equal area stereonet (Fig. 7.9). No rotation of the tracing paper is required.
- Mark the centre of the stereonet on the tracing paper by piercing it with a drawing pin.

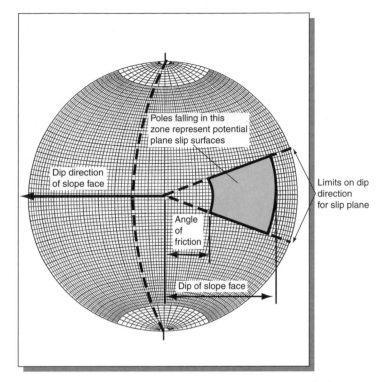

Fig. 8.2 Construction of the plane failure overlay (from Matherson, 1983)

- Mark the dip direction of the face with an arrow pointing towards the West along the EW line of the stereonet (Fig. 8.2).
- Draw a line from the centre of the stereonet to a point on the circumference 20° above the eastern end of the EW line. Draw a second line from the centre to a point 20° below the eastern end of the EW line (Fig. 8.2). These lines represent the limits on the dip direction of potential slip surfaces.
- Draw an arc with radius equal to the dip of the proposed slope face measured along the EW between the lines defining the limits on dip direction (Fig. 8.2).
- Draw an arc with radius equal to the angle of friction measured along the EW between the lines defining the limits on dip direction (Fig. 8.2).
- The resulting plane failure overlay has the appearance of a truncated wedge as shown in Fig. 8.2.

A qualitative assessment of potential stability can be made by super-imposing the overlay (correctly orientated according to the slope being assessed) on a pole plot of the same diameter derived from field measure-ments. At its simplest, a count of the number of poles within the defined region can give a semi-quantitative measure of failure potential, provided the sample is representative. When contouring has been carried out, the highest contour value within the defined region can be taken and used in a similar way.

In practice, uncertainty as to the value of friction and slope angle usually means that a first evaluation is carried out using the potential slip plane dip direction limits alone. This will lead to an appreciation of the importance of the data collected to the stability of the slope in question. The angle of friction of smooth rock surfaces is usually between 25° and 35° for a wide range of rock types. For preliminary assessments the angle of friction may be taken as 30°. Once the frictional characteristics are known a more accurate value can be used. Concentrations of poles in the region of potential instability defined by potential slip plane dip direction and friction angle limits will then give an indication of the optimum angle for the design slope.

WORKED EXAMPLE

Determine the optimum slope angle for a face with a dip direction of 360° in a rock mass with the discontinuity orientations shown in Fig. 7.21.

The objective here is to find the maximum slope angle which will minimize the likelihood of plane failure. We are not given any information about the angle of friction or the roughness of the discontinuities. In this case, therefore, a friction angle of 30° will be assumed.

The procedure for finding the optimum slope angle is as follows.

- Construct an overlay with a friction angle of 30° and a series of arcs representing slope face angles at 10° intervals as shown in Fig. 8.3.
- Centre the overlay over the contoured plot of pole density using a drawing pin.
- Rotate the overlay so that the arrow showing the dip direction of the slope face points towards 360° (i.e. North) as shown in Fig. 8.4.

Now the overlay has been orientated in the dip direction of the proposed slope the contours falling within the dip direction limits can be examined. A cluster of poles falls within these lines and these may result in plane failure if the slope face dips at angles greater than 50°. There is still a risk of plane failure at lower angles but since the density of poles is relatively low the risk is minimized. The likelihood of plane failure is greatest at slope angles greater than 60°. In this case release surfaces are provided by steeply dipping discontinuities represented by the clusters located on the circumference near the East and West points.

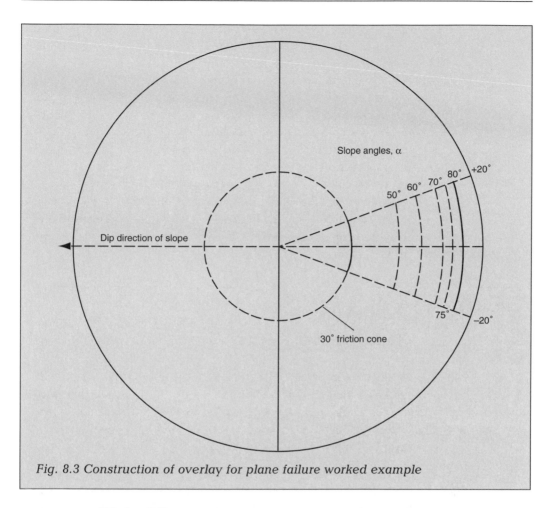

Fig. 8.3 Construction of overlay for plane failure worked example

Wedge failure

In wedge failure, two intersecting planes are present which dip towards each other. Movement takes place by sliding along both planes in the direction of the line of intersection as shown in Table 8.1.

For failure to occur, two basic conditions must be satisfied (Markland, 1972).

- The line of intersection must daylight on the slope.
- The dip of the line of intersection must exceed the friction angles of the planes.

Although this disregards the component of friction acting on both planes and possible water pressures, it is considered sufficiently accurate (and on the safe side) for preliminary assessments.

The limit conditions described above can be portrayed on an overlay which is used with a plot of principal discontinuity intersections to identify

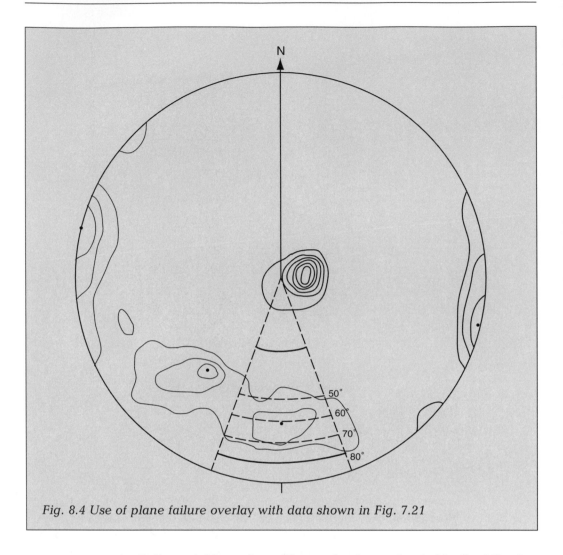

Fig. 8.4 Use of plane failure overlay with data shown in Fig. 7.21

potentially unstable wedges. The overlay is constructed in the following manner.

- Secure a piece of tracing paper over the equal area stereonet (Fig. 7.9). No rotation of the tracing paper is required.
- Mark the centre of the stereonet on the tracing paper by piercing it with a drawing pin.
- Mark the dip direction of the face with an arrow pointing towards the West along the EW line of the stereonet (Fig. 8.5).
- Trace the great circle representing the slope face (Fig. 8.5).
- Draw an arc beginning and ending at the great circle drawn in the preceding step with centre at the centre of the stereonet and radius equal to $(90 - \phi)$ where ϕ is the angle of friction (Fig. 8.5).

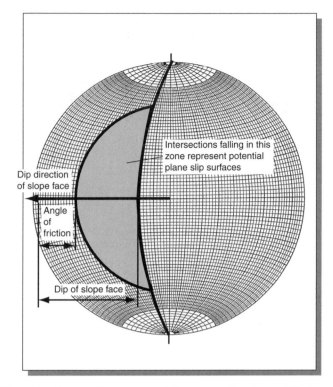

Fig. 8.5 Construction of the wedge failure overlay (from Markland, 1972)

- The finished overlay is a crescent. Intersections between pairs of discontinuities that fall into this crescent may represent potentially unstable wedges.

Superimposition of the overlay on an intersection plot of great circles (one for each cluster of poles) will allow a simple assessment of the possibility of wedge failure. The effect of variations in the friction angle or dip of the design slope can be judged. The objective in design is to select a parameter (like slope angle, face angle or dip, orientation of the face or dip direction) that allows exclusion of as many intersections as possible from the region of potential instability.

The large number of observations normally accompanying a stability assessment makes the task of graphically drawing each great circle and defining every intersection impractical. Instead, only the main pole concentrations found by contouring the pole plot are used. Great circles are drawn to these and the intersections identified. A qualitative assessment of their importance to stability is then made using the overlay described above. A semi-quantitative evaluation can be obtained by taking into consideration the relative importance of each pole concentration which may be assessed on the basis of persistence.

Plane failure may be regarded as a special form of wedge failure. A block undergoing plane failure has a single sliding surface and is bounded by lateral release surfaces. The intersection between the discontinuities representing the lateral release surfaces and the slip surface will appear within the crescent of the wedge failure overlay. It is necessary therefore to classify potentially unstable wedges into single plane sliding blocks (plane failure) and double plane sliding blocks (wedge failure). Hocking (1976) describes a simple tests based on the dip directions of the two planes, the line of intersection and the slope face. Consider two planes A and B with dip directions shown in Fig. 8.6. If dip direction of plane A

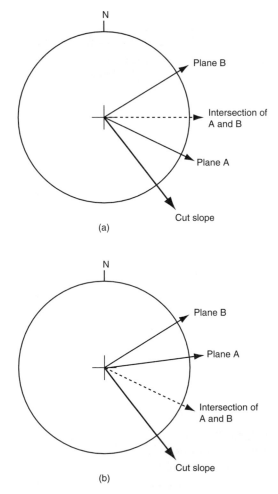

Fig. 8.6 A method for determining whether a potentially unstable wedge will slide on one or two planes (from Hocking, 1976). (a) Sliding on plane A. (b) Sliding on intersection of A and B

falls between the dip directions of the slope face and the line of inter-section between the planes, single plane sliding will occur on plane A (Fig. 8.6(a)). If neither plane has a dip direction between that of the slope face and the line of intersection between the planes sliding will take place along the line of intersection (i.e. double plane sliding). This test can be applied to all intersections which fall inside the crescent of the wedge failure overlay.

WORKED EXAMPLE

Identify the potentially unstable wedges in a slope face with dip 70° and dip direction 090° excavated in a rock mass with discontinuity orientations shown in Fig. 7.21.

We are not given any information about the angle of friction or the roughness of the discontinuities. In this case, therefore, a friction angle of 30° will be assumed.

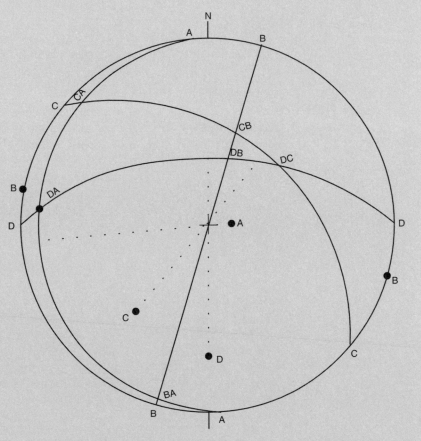

Fig. 8.7 Construction of intersection plot based on data given in Fig. 7.21

In order to identify the potentially unstable wedges in this face it is necessary to identify the highest concentrations of poles and to use these to determine the most likely lines of intersection that will outcrop in the proposed face. The peaks associated with each cluster of discontinuities shown in Fig. 7.21 have been labelled A to C in Fig. 8.7. The great circles associated with each of these poles have been drawn in Fig. 8.7 and the intersections labelled. The point of intersection between two great circles represents the orientation of the line of intersection between the planes they represent.

The overlay shown in Fig. 8.8 has been constructed using a slope dip of 70° and an angle of friction of 30°. This is superimposed onto the plot of intersections (Fig. 8.7). Alignment of centres is achieved using a drawing pin in a similar manner as that described in the worked example for plane failure. The overlay must be rotated so that the

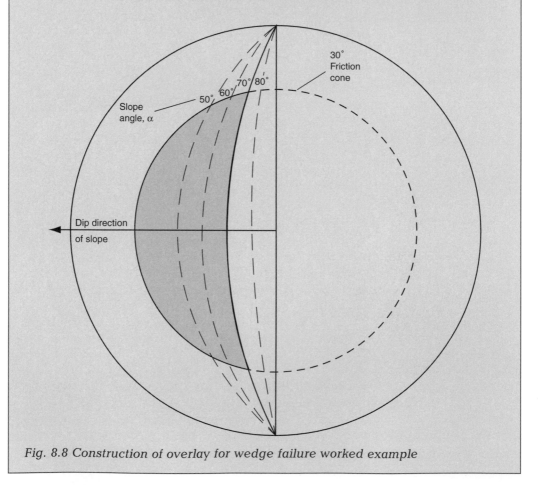

Fig. 8.8 Construction of overlay for wedge failure worked example

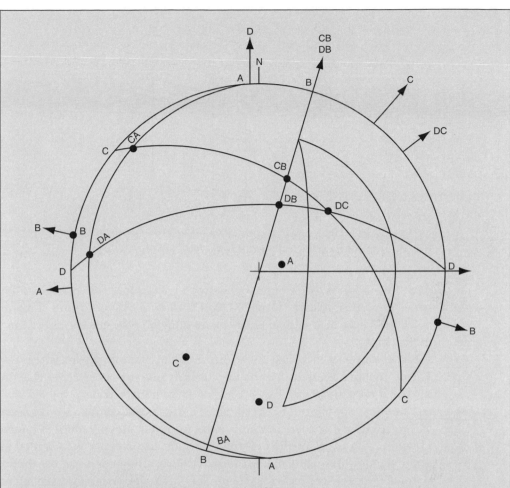

Fig. 8.9 Use of overlay with intersection plot

arrow representing the dip direction of the face is pointing towards 090°. The intersection DC falls within the zone of potential instability (Fig. 8.9). The Hocking test indicates that this wedge will slide along the line of intersection.

Toppling failure

Toppling failure involves either one or a combination of flexural toppling and block toppling (see Table 8.1).

Flexural toppling involves the overturning of rock layers like a series of cantilever beams. This is common in thinly bedded or foliated rocks such as shale, slate or schist or closely jointed rock where the fractures dip steeply into the face as shown in Table 8.1. Each layer tends to bend

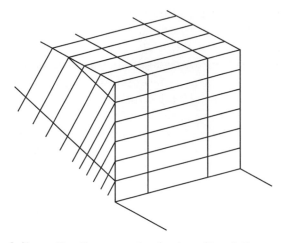

Fig. 8.10 Ideal discontinuity geometry for toppling failure

downhill under its own weight and hence transfers force downslope. If the toe of the slope is allowed to move either by sliding or overturning, flexural cracks will form in the upper layers to a significant distance behind to crest of the slope.

Block toppling involves the overturning of fracture-bounded blocks as rigid columns rather than having to fail in flexure (see Table 8.1). In the ideal case shown in Fig. 8.10 a set of relatively widely spaced discontinuities dipping out of the slope at a low angle combines with a set of more closely spaced discontinuities dipping steeply into the face together with a set of discontinuities which dip steeply in a direction approximately perpendicular to the face to form either columnar or tabular shaped blocks whose individual centre of gravity lies outside a basal pivot line or point located at the block edge. This combination of steeply dipping discontinuities rarely occurs in practice. The dip direction of these discontinuities need not be approximately parallel and perpendicular to the slope face to bring about block toppling failure. Any combination of dip directions can result in overturning blocks provided the spacing of the steeply dipping discontinuities is sufficiently close relative to the spacing of the low angle basal planes which dip out of the face. Such combinations can be identified by examining the orientation of the lines of intersection between discontinuities. If these dip steeply into the face in combination with a low dip set of basal planes, there is potential for toppling.

As with the other failure mechanisms an overlay may be constructed using the stereonet to identify the different toppling modes in a slope face from discontinuity orientation data. The criteria for the construction of the overlay are now described.

Criteria for flexural toppling

The mechanism of flexural toppling involves flexural slip between closely spaced layers (Fig. 8.11). The steeply dipping fractures must therefore be in a state of limiting equilibrium. The angle between the normal stress active in the slope face and the normal to the discontinuities which are undergoing flexural slip is γ in Fig. 8.11. If there are no external forces or water forces limiting equilibrium along the steeply dipping discontinuities will be achieved when $\gamma = \phi$, where ϕ is the angle of friction of these discontinuities. The angle γ can be expressed in terms of the dip of the slope face (θ) and the dip of the steeply dipping discontinuities (β) as

$$\gamma = \theta - (90 - \beta)$$

where $(90 - \beta)$ is the dip of the pole to the steeply dipping discontinuity.

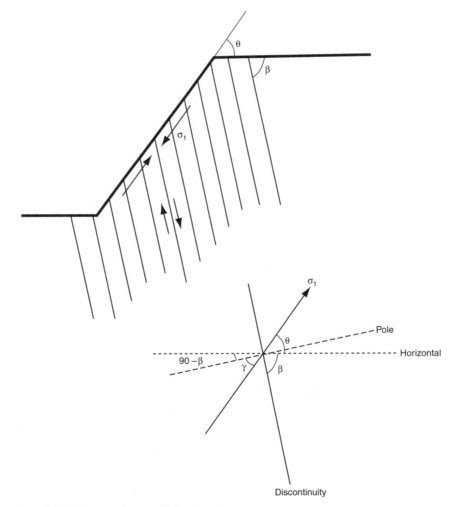

Fig. 8.11 Kinematic condition for flexural slip

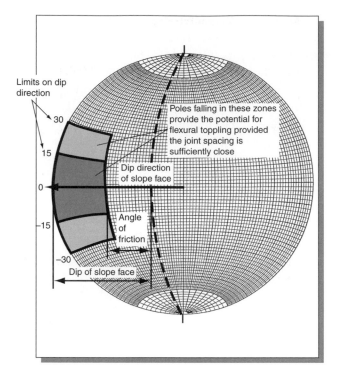

Fig. 8.12 Construction of overlay for flexural toppling

Limiting equilibrium will occur when $(90 - \beta) \leq \theta - \phi$. Flexural toppling can occur when the dip direction of the steeply dipping discontinuities is up to 30° away from that of the slope face, but is much more likely to occur when the difference in dip directions is less than 15°.

These criteria are combined to form a stability overlay in the following manner.

- Secure a piece of tracing paper over the equal area stereonet (Fig. 7.9). No rotation of the tracing paper is required.
- Mark the centre of the stereonet on the tracing paper by piercing it with a drawing pin.
- Mark the dip direction of the face with an arrow pointing towards the West along the EW line of the stereonet (Fig. 8.12).
- Trace the great circle representing the slope face (Fig. 8.12).
- Mark the point on the EW line which is a distance ϕ from the great circle representing the slope face towards the W point. Trace the great circle which passes through this point (Fig. 8.12). This great circle represents the limit of poles which dip at $(\theta - \phi)$.
- Trace two small circles starting at the circumference of the stereonet at points 30° above and below the W point and terminating at the great circle drawn in the preceding step (Fig. 8.12).

- Repeat the above step at points 15° above and below the W point (Fig. 8.12).

Criteria for block toppling (Matherson, 1983).

- The dip direction of the basal plane must lie within approximately 20° of the slope face dip direction.
- The dip of the basal plane must be less than the friction angle on that plane.
- The dip direction of lines of intersection between discontinuities must lie within approximately 20° of the slope face dip direction. For steep slopes this can be extended to 90°
- The dip of the lines of intersection must exceed $(90 - \phi)$, where ϕ is the friction angle on that plane.

The overlay based on the above criteria is constructed using the following procedure.

- Use the tracing paper with the flexural toppling overlay (Fig. 8.12).
- Draw a line from the centre of the stereonet to a point on the circumference 20° above the eastern end of the EW line. Draw a second

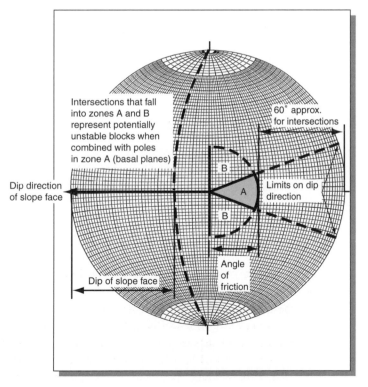

Fig. 8.13 Construction of overlay for block toppling

line from the centre to a point 20° below the eastern end of the EW line (Fig. 8.13). These lines represent the limits on the dip direction of basal planes and lines of intersection.

- Draw a semi-circle on the east side of the stereonet with radius equal to the angle of friction measured along the EW as shown in Fig. 8.13. This represents the upper limit on dip for basal planes and the lower limit on lines of intersection.

A stereonet does not have dimensions and so it is not possible to represent centre of gravity. Any potential for toppling (either flexural or block) identified using the overlays must be validated by examining the spacings of the discontinuity sets involved.

WORKED EXAMPLE

Using the discontinuity orientation data presented in Fig. 7.21, determine the face directions which are most likely to have a toppling failure hazard if the dip of the face is 70° and the angle of friction for all discontinuities is assumed to be 30°.

Solution
- Construct a toppling failure overlay as shown in Fig. 8.14. Note that since $\phi = 30°$ the edge of the wedge shaped zone defining the basal planes coincides with the edge of the semi-circular zone defining the critical discontinuity intersections for block toppling.
- Combine the contoured pole density plot (Fig. 7.21) with the intersection plot used in the wedge failure analysis (Fig. 8.7).
- Superimpose the overlay (Fig. 8.14) onto the combined pole density and intersection plot and centre using a drawing pin.
- Rotate the overlay to find the critical slope directions defined by combinations of poles falling into zones 1, 2 and 3 and intersection falling into zone 4.

Critical slope directions
Face direction 180°. The highest concentration of poles in set D fall inside zone 2 (see Fig. 8.15) suggesting that there is a significant hazard from flexural toppling if the spacing of set D is close. The intersection of sets B and D (Point DB in Fig. 8.15) is on the edge of the semi-circular shaped zone 4 used to identify block toppling. There is however only a small amount (2%) of low angle discontinuities belonging to set A which have poles falling into zone 3 (basal planes). This suggests that there is little potential for block toppling and the main hazard is from flexural toppling. The degree of hazard will

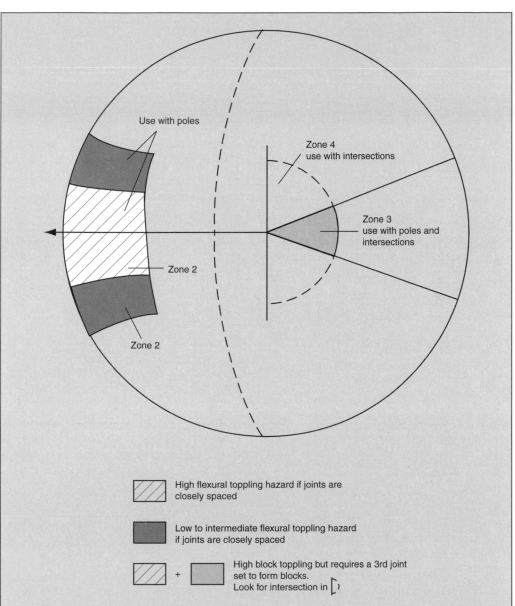

Fig. 8.14 Construction of combined flexural and block toppling overlay for worked example

depend principally on the spacing of set D. If set D is widely spaced then the hazard will be reduced.

Face direction 280°. Poles representing a steeply dipping set of discontinuities (set B in Fig. 8.16) fall with their highest concentration inside zone 2 indicating a high flexural toppling hazard. In addition,

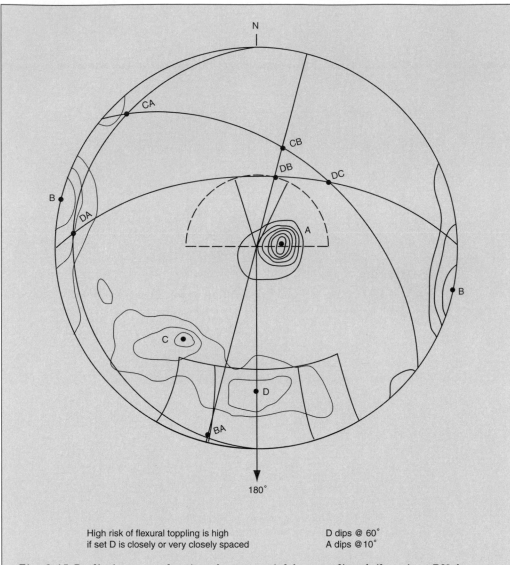

High risk of flexural toppling is high
if set D is closely or very closely spaced

D dips @ 60°
A dips @ 10°

Fig. 8.15 Preliminary evaluation the potential for toppling failure in a 70° face dipping towards 180° based on the data given in Fig. 7.21

the highest concentration of poles associated with discontinuity set A falls inside zone 3 indicating the presence of basal planes for block toppling. The line of intersection between sets B and D fall on the edge of zone 4 reinforcing the likelihood of block toppling failure. The hazard of block toppling is of course highest when the spacing of sets D and B are much less than that of set A. In this case the rock mass would be dominated by steeply dipping tabular or columnar blocks.

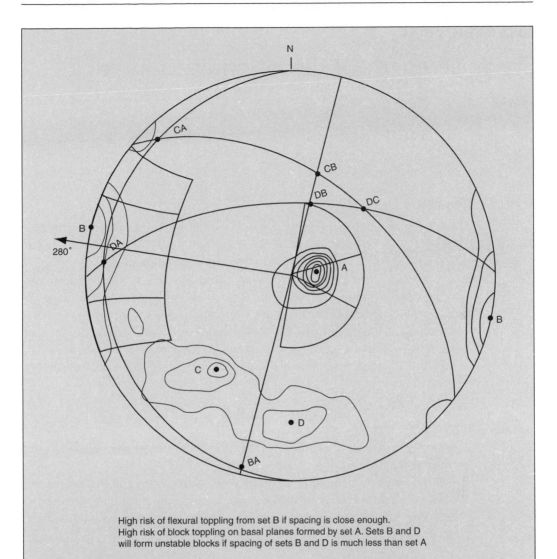

High risk of flexural toppling from set B if spacing is close enough.
High risk of block toppling on basal planes formed by set A. Sets B and D
will form unstable blocks if spacing of sets B and D is much less than set A

Fig. 8.16 Preliminary evaluation the potential for toppling failure in a 70° face dipping towards 280° based on the data given in Fig. 7.21

Face direction 100°. Poles representing set B fall inside zone 2 and only a small concentration of poles from set A fall inside zone 3. No intersections fall inside zone 4. This suggests that the principal hazard is associated with flexural toppling of set B. The spacing of set B will determine the how real this hazard is.

CHAPTER NINE

Shear strength of discontinuities in rock

Overview

Most failure modes in rock slopes involve sliding and/or toppling. Sliding generally occurs along discontinuities and hence the study of shear strength of discontinuities is critical to any rock slope design.

A simple way to estimate the shear strength of a single discontinuity is to take a sample of rock containing the discontinuity and then to incline the discontinuity from the horizontal until the upper half slides (Fig. 9.1). If the weight of the upper block is W and the gross area of contact is A, at an angle of inclination β the effective normal stress σ'_n and the shear stress τ down the line of maximum dip are respectively given by

$$\sigma'_n = \frac{W \cos \beta}{A} - u \qquad \text{and} \qquad \tau = \frac{W \sin \beta}{A} \tag{9.1}$$

where u is the water pressure within the discontinuity, generally zero in this particular test.

If the angle of inclination at the point of sliding is β_f, and the associated effective normal and shear stresses are σ'_n and τ_f, the ratio of shear stress to effective normal stress at the point of slip is, applying equations (9.1), given by

$$\frac{\tau_f}{\sigma'_n} = \frac{\sin \beta_f}{\cos \beta_f} = \tan \beta_f$$

This simple result forms the basis of the Mohr–Coulomb failure criterion used in soil mechanics:

$$\tau_f = \sigma'_n \tan \phi \tag{9.2}$$

The friction angle of a discontinuity can be defined using the following terms which refer to the displacements obtained during shear movements (Krahn and Morgenstern, 1979, 1980; Barton, 1980):

- Peak friction angle ϕ_p is evaluated on natural discontinuities, in correspondence with the maximum shear strength determined by roughness.

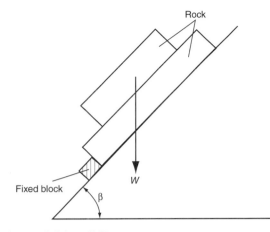

Fig. 9.1 Simple model for sliding

- Basic friction angle ϕ_b is evaluated on an artificially planar surface and is characteristic of the rock mineralogy (see Table 9.1 for typical values of ϕ_b).
- Residual friction angle ϕ_r is evaluated when the shear strength is stabilized on a minimum value. This is the lowest value. The value of the residual friction angle ϕ_r for most rocks is usually between $25°$ and $35°$. For unweathered rock $\phi_r = \phi_b$.

Discontinuities in rock, however, do not behave like soils (unless infilled with soil material). The rock on either side of the discontinuity will behave as rigid blocks. Effective stresses need only be considered when the discontinuity fills with water or other fluids (e.g. oil). In rock engineering, unlike soil engineering, normal stresses are normally expressed in terms of total stress and the effect of joint water pressure is treated separately in the stability computations.

The simplistic equation for shear strength presented in equation (9.2) is not really suitable for most natural discontinuities. The shear strength

Table 9.1 Basic friction angles for a range of rock materials (Barton and Choubey, 1977)

Rock type	ϕ_b dry: degrees	ϕ_b wet: degrees
Sandstone	26–35	25–34
Siltstone	31–33	27–31
Limestone	31–37	27–35
Basalt	35–38	31–36
Fine granite	31–35	29–31
Coarse granite	31–35	31–33
Gneiss	26–29	23–26

of natural discontinuities may be described as a function of several parameters that are also listed in Chapter 7 as key features requiring description and measurement:

- *Wall roughness.* This will result in dilatancy of discontinuities at low normal stresses. Rough discontinuities will have a higher shear strength at low normal stresses than smooth discontinuities. A smooth planar discontinuity will show purely friction resistance that will depend on the surface texture, mineralogy and the presence of moisture (Colback and Wiid, 1965).

- *Wall strength.* As the normal stress increases there will come a point at which shearing must involve considerable asperity breakage and a consequent reduction in the instantaneous frictional resistance. The wall strength will determine point at which asperity breakage will dominate the shearing process.

- *Wall coating.* Low friction minerals (e.g. chlorite, graphite and serpentenite) may coat the surface of discontinuities and significantly reduce the frictional resistance to sliding.

- *Infilling.* Where infilling thickness is greater than the asperity amplitude the mechanical properties of the infilling material will dominate the shear strength of the discontinuity.

- *Water* (or other incompressible fluids). When a discontinuity is full of incompressible fluid then the shear strength will be reduced by the joint fluid pressure.

- *Persistence.* Inpersistent discontinuities are characterized by rock bridges which contribute the cohesion component to the shear strength.

It is difficult both to evaluate these parameters and to analytically formulate a strength criterion equation which takes all these parameters into account. This has resulted in the development of empirical approaches that relate shear behaviour observations to a limited number of parameters which mainly govern the phenomenon.

Influence of surface roughness

At low normal stress a rough discontinuity surface will exhibit a higher apparent angle of friction than a smooth surface in the same rock material since shear displacements will be associated with dilatancy. As the normal stress increases there will be an increasing component of asperity breakage due to dilatancy being inhibited. Ultimately the instantaneous angle of friction will approach the residual angle ϕ_r. Patton (1966) studied this behaviour on model discontinuities of the type shown in Fig. 9.2(a) and proposed a bi-linear model to describe the shear strength behaviour of rough discontinuities as shown in Fig. 9.2(b).

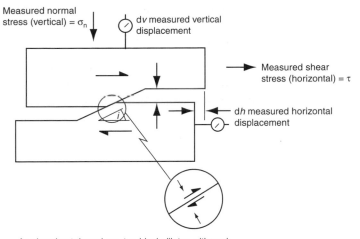

Measured normal stress (vertical) = σ_n

dv measured vertical displacement

Measured shear stress (horizontal) = τ

dh measured horizontal displacement

As shearing takes place, top block dilates with work being done against normal load through dv

To calculate stresses relative to actual plane of shearing

Then, $i = $ arc Tan $\dfrac{dv}{dh}$

$$\tau_i = (\tau \cos i - \sigma \sin i) \cos i$$
$$\sigma_i = (\sigma \cos i + \tau \sin i) \cos i$$

Where compression takes place then,

$$\tau_i = (\tau \cos i + \sigma \sin i) \cos i$$
$$\sigma_i = (\sigma \cos i - \tau \sin i) \cos i$$

(a)

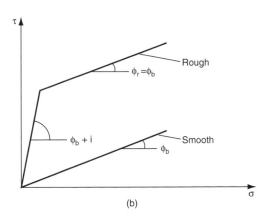

Rough

$\phi_r = \phi_b$

$\phi_b + i$

Smooth

ϕ_b

(b)

Fig. 9.2 (a) Idealized rough discontinuity. (b) Model for the shear strength of rough discontinuities (Patton, 1966)

265

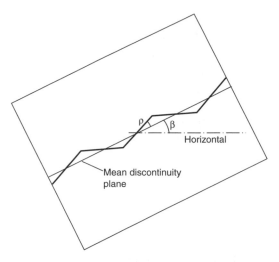

Fig. 9.3 Idealized saw-tooth model for discontinuity roughness

The behaviour at low normal stress may be incorporated into the shear strength criterion by regarding ϕ_p as being composed of two angular components (see Fig. 9.2(b)):

- ϕ_r the residual angle for an apparently smooth surface of the rock material, and
- a component i, called the effective roughness angle, due to visible roughness and other surface irregularities, such that

$$\phi_p = \phi_r + i \quad \text{(for unweathered surfaces } \phi_r = \phi_b) \qquad (9.3)$$

The significance of the roughness component i can be appreciated by considering the idealized saw-tooth model of Fig. 9.3 in which there is a uniform roughness inclined at an angle ρ to the mean discontinuity plane. If this model were tilted at an angle β to the horizontal, the saw-tooth surfaces would be inclined at angles $\beta - \rho$ and $\beta + \rho$. Only those surfaces inclined at $\beta - \rho$, however, will remain in contact and control sliding. Applying equations (9.1) to obtain the shear and normal stresses on the contact surfaces at the point of slip gives

$$\frac{\tau_f}{\sigma'_n} = \tan(\beta_f - \rho) = \tan\phi_b$$

Hence the apparent angle of friction, given by the overall angle of inclination at the point of slip, is

$$\phi_p = \beta_f = \phi_r + \rho \qquad (9.4)$$

The above result shows that the roughness component i for the saw-tooth model is theoretically equal to the saw-tooth angle ρ. Any shear displacement dh of this saw-tooth geometry must be associated with an opening, or dilation, dv of the discontinuity at a rate dv/dh = tan ρ.

Replacing ϕ in equation (9.2) by $\phi_b + i$ gives the well known result presented by Patton (1966).

$$\tau_f = \sigma_n' \tan(\phi_b + i) \qquad (9.5)$$

In a review of shear test results on rough discontinuities, Hoek and Bray (1981) concluded that effective roughness angles of between 40° and 50° could be applicable at effective normal stresses below about 0.7 MPa. Taking a basic friction angle of 30° in equation (9.5) implies that roughness features inclined at more than 60° would impart infinite shear strength to the discontinuity; this would indeed be the case if the rock adjacent to the discontinuity were infinitely strong. In practice, however, the shear stress rises until it is high enough to induce shearing through the rock material forming the roughness feature.

The shear behaviour discussed above is complicated by a number of additional geometrical and mechanical factors. In reality the surface topography of a discontinuity wall does not correspond to a series of regular 'saw-tooth' like ridges. In most cases the surface topography is irregular as shown by the profiles in Fig. 9.4.

Small-scale roughness (Table 7.1) is often associated with steep sided asperities which can induce a high rate of dilatancy. These asperities are generally characterized by short wavelengths and hence narrow bases which promote asperity breakage at relatively low stress. Intermediate and particularly large-scale roughness is commonly associated with much lower roughness angles and high wavelengths. Such asperities are not easily broken even at relatively high stress. If movement occurs on a rough discontinuity the resulting dilation can lead to an increase in normal stress if the adjacent rock mass is not free to move due to confinement as shown in Fig. 9.5(a)

Discontinuities adjacent to a free face, however, can dilate without inducing significant changes in normal stress, as shown in Fig. 9.5(b). A further complication arises where weathering effects cause a weakening and softening of the rock material adjacent to the discontinuity, or where an infill material has been deposited in a previously open discontinuity. The simple roughness angle model of equation (9.5) is, therefore, inadequate to describe the shear behaviour of real discontinuities. In order to address this problem it is necessary to conduct shear tests on real discontinuities under conditions that reflect those encountered in situ.

The interaction of roughness and wall strength

When natural or artificial rough discontinuities are sheared the relationship between shear stress and effective normal stress is curved, as

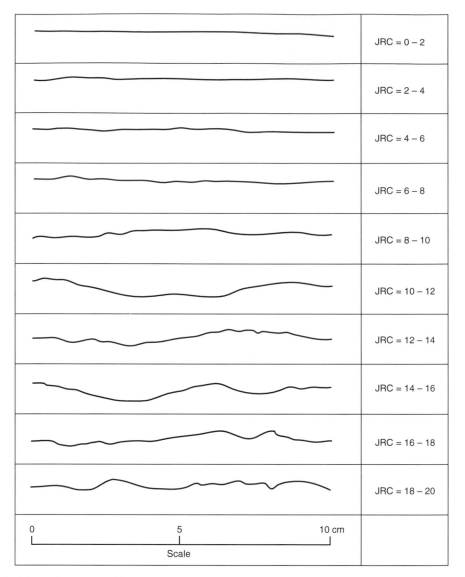

Fig. 9.4 Roughness profiles and corresponding JRC values, after Barton and Chouby (1977)

shown in Fig. 9.6, see Barton (1976), Brown *et al.* (1977), Krahn and Morgenstern (1979) and Bandis *et al.* (1981). The generally accepted explanation for this phenomenon is linked to the role of roughness features during the shearing process, and in particular to the degree to which a discontinuity rides over, or shears through, these asperities.

An important consequence of the non-linear shear stress–normal stress relation is observed when the linear models in equations (9.2) and (9.5) are

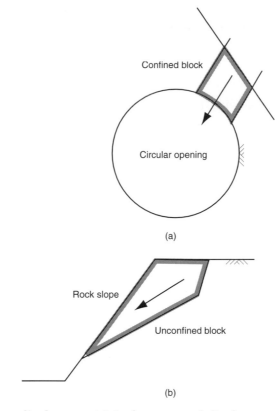

(a)

(b)

Fig. 9.5 Shear displacement (a) where normal displacement is restrained and (b) where normal displacement is permitted

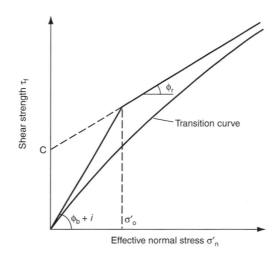

Fig. 9.6 Bi-linear shear strength model with empirical transition curve

269

applied to results obtained at even moderate normal stresses. Figure 9.6 indicates that extrapolation of the linear model would give a positive shear strength at zero normal stress. This component of shear strength is traditionally termed cohesion, c, in the following general form of the Coulomb shear strength criterion.

$$\tau_f = c + \sigma'_n \tan \phi \tag{9.6}$$

Equation (9.6) can provide an adequate prediction of the shear strength of a discontinuity over a specified limited range of effective normal stresses when the parameters c and ϕ have been determined by fitting a straight line to shear test results obtained over this same range of normal stresses. It should be made clear when using this model whether the cohesive strength is due to genuine cementing or whether it is due to roughness. Patton (1966) addressed this problem by formulating the bi-linear model, illustrated in Fig. 9.6, for explaining the behaviour of artificial discontinuities with regular roughness features. At effective normal stresses less than or equal to σ'_o the shear strength is given by equation (9.5). At normal stresses greater than or equal to σ'_o shear strength is given by

$$\tau_f = c_a + \sigma'_n \tan \phi_r \tag{9.7}$$

where c_a is the apparent cohesion derived from the asperities and ϕ_r is the residual angle of internal friction of the rock material forming the asperities.

The difficulty in applying Patton's bi-linear model is in the determination of σ'_o. The stress σ'_o is scale dependent. It will be low for small-scale roughness (typically measured in laboratory-scale shear tests) and much higher for large-scale roughness which is representative of field-scale slip surfaces. When dealing with relatively shallow slips that may occur in shallow surface excavations (e.g. a typical road cutting), the stresses are unlikely to be large enough to promote significant asperity breakage in the case of large-scale roughness. In such cases equation (9.5) can be used and the large-scale roughness angle estimated or measured from observations of the discontinuities in suitable exposures. Where intermediate and small-scale roughness is likely to play a significant role in the stability of rock slopes a model is required that takes account of asperity breakage without the requirement to determine σ'_o. Jaeger (1971) and Barton (1976) have proposed such models.

Jaeger (1971) proposed the following shear strength model to provide a curved transition between the straight lines of the Patton model

$$\tau_f = c_a(1 - e^{-d\sigma'_n}) + \sigma'_n \tan \phi_r \tag{9.8}$$

The parameters c_a and ϕ_r are, respectively, the intercept and slope of a straight line fitted to experimental shear strength data at high normal stresses, and d is an experimentally determined empirical parameter which controls the shape of the transition curve, as shown in Fig. 9.6.

A direct, practical approach to predicting the shear strength of discontinuities on the basis of relatively simple measurements has been developed by Barton and co-workers, and published in a sequence of papers throughout the 1970s and the early 1980s. Reviews of this empirical model, and also the shear strength of discontinuities in general, are presented by Barton (1976) and more recently by Barton and Bandis (1990). According to the Barton model, the shear strength τ_f of a discontinuity subjected to an effective normal stress σ'_n in a rock material with a basic friction angle ϕ_b is given by

$$\tau_f = \sigma'_n \tan \left(\phi_b + \text{JRC} \log_{10} \left(\frac{\text{JCS}}{\sigma'_n} \right) \right) \tag{9.9}$$

JCS is the uniaxial compressive strength of the rock material immediately adjacent to the discontinuity. The parameter JRC, the Joint Roughness Coefficient, provides an angular measure of the geometrical roughness of the discontinuity surface in the approximate range 0 (smooth) to 20 (very rough). It is important to appreciate, however, that JRC is not equal to the effective angle of roughness i used in the Patton model. If the discontinuity is unweathered, JCS is equal to the uniaxial compressive strength of the rock material σ_c, determined by point load index tests or compression tests on cylindrical specimens. If there has been softening or other forms of weathering along the discontinuity, then JCS will be less than σ_c and must be estimated in some way. The strength of discontinuity wall can be measured with a Schimdt rebound hammer as outlined in Chapter 7. The Schmidt hammer is one of the few methods available for estimating the strength of a surface coating of material.

The JRC, which is the key parameter in the Barton model, can be estimated in a number of ways. Barton and Choubey (1977) present a selection of scaled typical roughness profiles, reproduced in Fig. 9.4, which facilitate the estimation of JRC for real discontinuities by visual matching. Alternatively a simple tilt shear test can be conducted on a discontinuity specimen (Fig. 9.1) and the JRC can be back-figured from equation (9.9), using equations (9.1) to calculate σ'_n and τ_f.

Barton and Choubey (1977) suggest that the curves should be truncated such that the maximum allowable shear strength for design purposes is given by arctan $(\tau_f/\sigma'_n) = 70°$. Barton proposed that a high stress version of equation (9.9) could be obtained by replacing JCS by $(\sigma'_{1f} - \sigma'_3)$, where σ'_{1f} is the effective axial stress required to yield the rock material under an effective confining stress σ'_3. The failure stress σ'_{1f} can either be determined experimentally or can be estimated from an appropriate yield criterion such as the one proposed by Hoek and Brown (1980).

Scale effects

Bandis *et al.* (1981) examined the scale effects of the shear behaviour of discontinuities by means of experimental studies. Scale effects on the following parameters were studied:

- peak displacement
- dilatancy value
- JRC
- asperity failure
- size and distribution of the contact area
- limit size of specimens, on ultimate shear resistance and on a strongly jointed rock mass, for different values of normal stress.

Figure 9.7 shows the scale dependence of the laboratory specimen size on the three components of the shear strength of natural discontinuities. In particular, this illustrates how, by increasing the size of a specimen with a discontinuity, one obtains:

- a gradual increase in the displacement required to mobilize the maximum shear stress
- an apparent transition zone from a 'brittle' to 'plastic' mode of shear failure
- a reduction in the peak friction angle as a consequence of a decrease in peak dilatation and an increase in asperity failure
- a decrease of the residual shear strength.

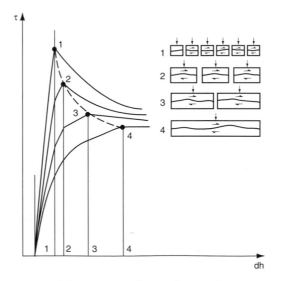

Fig. 9.7 Dependence of the test specimen size on the components of shear strength in natural discontinuities, after Bandis et al. *(1981)*

A heavily jointed rock mass exhibits a reduced stiffness by increasing the degree of freedom of the individual joint blocks which are able to rotate and mobilize all the roughness resistances at different scales. Consequently, as demonstrated by jointed model experiments (Bandis *et al.*, 1981), small blocks in a densely jointed mass may mobilize higher JRC values than larger blocks in a mass with wider-spaced joints.

The scale effect on peak shear strength implies that there should be a minimum size of test specimen that is considered as technically acceptable. Barton and Choubey (1977) suggested, as a first approximation, the natural block size of the rock mass or more specifically, the spacing of cross-joints.

The size of this scale effect for a tilt test can be calculated using the empirical formulas given by Barton and Bandis (1982) and Barton and Bakhtar (1983) and based on a large number of in situ and laboratory experiments. These are as follows:

$$JRC = JRC_0[L_n/L_0]^{-0.02JRC_0}$$
$$JCS = JCS_0[L_n/L_0]^{-0.03JCS_0}$$

(9.10)

where L_n and L_0 are the lengths of discontinuities at the in situ and laboratory scales respectively. JRC and JRC_0 are joint roughness coefficients at in situ and laboratory scales respectively. JCS and JCS_0 are joint wall compressive strengths at in situ and laboratory scales respectively.

Anisotropy induced by surface roughness

In some cases the surface roughness roughness may display a preferred orientation (e.g. undulations, slickensides). In these cases the shear strength mobilized by the discontinuity will be affected by the direction of sliding. Figure 9.8 shows shear test results parallel surface ridges (shear direction A) and normal to surface ridges (shear direction B) of a corrugated discontinuity. The resistance to sliding is very much greater across the corrugations than along them. These results emphasize the importance of recording the orientation of such roughness features pointed out in Chapter 7.

Influence of infilling

Papaliangas *et al.* (1990) and Toledo and DeFreitas (1993) present comprehensive investigations of the mechanisms controlling the shear strength of model rock joints infilled with a variety of soil materials. In general the shear strength of infilled discontinuities varies with filler thickness, from the strength of the soil filler alone to a maximum value that is smaller than that obtained from an unfilled discontinuity. When the filler thickness is smaller than the asperity height there are two phases of shear failure. During the first phase the mechanical properties of the discontinuity are controlled by the properties of the soil filler.

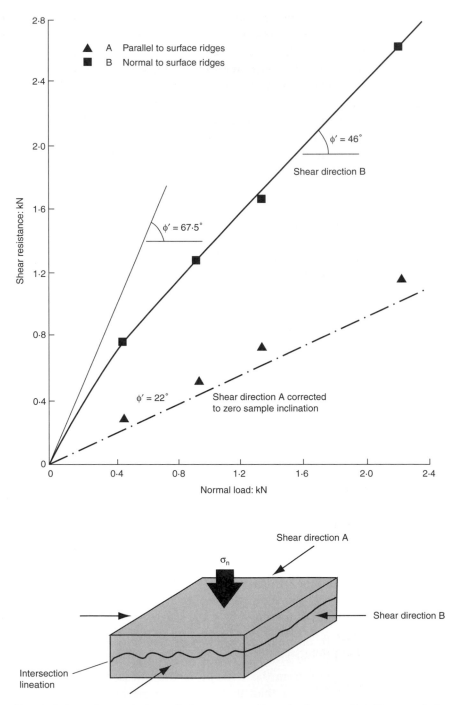

Fig. 9.8 Roughness induced shear strength anisotropy, after Brown et al. (1977)

After some displacement has taken place, the two rock surfaces come into contact and the strength of the discontinuity, from then on, is controlled mainly by the wall roughness and strength with associated dilatancy and asperity breakage depending on the normal stress. Toldeo and DeFreitas (1993) showed that rock asperities begin to interfere with the shear mechanisms at different filler thicknesses t_{crit}, depending on the grain size of the filler material. Discontinuities infilled with clay the ratio of t_{crit} to asperity height is unity or less. When the infill is a granular material this ratio is generally greater than unity and may in some cases be as high as two. This means that the minimum strength of discontinuities infilled with clay will occur at a lower infill thickness than in if the infill were granular for the same wall roughness. A guide to the strength of filled discontinuities and filling materials is given in Table 9.2.

Influence of surface texture and mineral coating on discontinuities

The basic friction angle measured on a saw cut surface may differ significantly from that determined from a shear test on a natural discontinuity in the same rock type where the dilatancy effects associated with roughness have been removed. Values may be higher due to increased textural interlocking or lower as a result of low friction minerals forming coating the discontinuity wall. Table 9.3 shows the differences in the basic angle of friction resulting from surface texture and mineral coating have been reported by Hencher and Richards (1989).

Table 9.3 highlights the need to determine the basic angle of friction from natural discontinuities instead of artificially smooth surfaces and to ensure samples tested are representative of the discontinuities likely to be encountered in the slope under consideration.

Influence of water pressure in discontinuities

In rock slopes, as in soil slopes, the influence of water pressure on stability cannot be overemphasized. In a soil under constant total stress an increase in pore water pressure will result in a reduction in effective stress and a lowering in shear strength. The same principle applies to discontinuities in rock. If the water pressure in a discontinuity increases the effective stress normal to the discontinuity is reduced and the shear strength along the discontinuity is lowered.

Cohesion

Here we must distinguish between true cohesion, which is a property of the intact rock, and apparent cohesion that is a result of expressing shear test data in a certain way.

Table 9.2 Shear strength of filled discontinuities and filling materials, after Barton (1974)

Rock	Description	Peak c': (MPa)	Peak ϕ'	Residual c'_r: MPa	Residual ϕ'_r
Basalt	Clayey basaltic breccia, wide variation from clay to basalt content	0.24	42°		
Bentonite	Bentonite seam in chalk	0.015	7.5°		
	Thin layers	0.09–0.12	12–17°		
	Triaxial tests	0.06–0.1	9–13°		
Bentonitic shale	Triaxial tests	0–0.27	8.5–2.9°		
	Direct shear tests			0.03	8.5°
Clays	Over-consolidated, slips, joints and minor shears	0–0.18	12–18.5°	0–0.003	10.5–16°
Clay shale	Triaxial tests	0.06	32°		
	Stratification surfaces			0	19–25°
Coal measure rocks	Clay mylonite seams 10 to 25 mm	0.012	16°	0	11–11.5°
Dolomite	Altered shale bed 150 mm thick	0.04	14.5°	0.02	17°
Diorite, granodiorite and porphyry	Clay gouge (2% clay, PI = 17%)	0	26.5°		
Granite	Clay filled faults	0–0.01	23.45°		
	Sandy loam fault filling	0.05	40°		
	Tectonic shear zone, schistose and broken granites, disintegrated rock and gouge	0.24	42°		
Greywacke	1–2 mm clay in bedding planes			0	21°
Limestone	6 mm clay layer			0	13°
	10–20 mm clay fillings	0.1	13–14°		
	<1 mm clay filling	0.05–0.2	17–21°		
Limestone, marl and lignites	Interbedded lignite layers	0.08	38°		
	Lignite/marls contact	0.1	10°		
Limestone	Marlaceous joints, 20 mm thick	0	25°	0	15–24°
Lignite	Layer between lignite and clay	0.014–0.3	15–17.5°		
Montmorillonite Bentonite clay	80 mm seams of bentonite (montorillonite) clay in chalk	0.36 0.016–0.2	14° 7.5–11.5°	0.08	11°
Schists, quartzites and siliceous schists	100–150 mm thick clay filling	0.03–0.08	320		
	Stratification with thin clay	0.61–0.74	41°		
	Stratification with thick clay	0.38	31°		
Slates	Finely laminated and altered	0.05	33°		
Quartz/kaolin/ pyolusite	Remoulded triaxial tests	0.042–0.9	36–38°		

Table 9.3 Examples of the differences in basic friction angle resulting from surface texture and mineral coating

Rock type	Difference in basic friction angle	Comments
Slightly decomposed granite	+6°	Difference between natural joint (corrected for dilatancy) saw cut surfaces. The increase in value due to surface texture associated with slight chemical decomposition
Dolomite	+10°	Difference between natural joint (corrected for dilatancy) saw-cut surfaces
Dolomite	−6°	Difference between coated and uncoated joints. Joints coated with minor amounts of bitumen and/or calcite
Monzonite	−19	Difference between joints with thin coating of chlorite and iron stained joints

In most cases, the cohesion component is ignored for rock discontinuities. It should only be considered when you are certain that cohesion will contribute to shear strength. This can occur when the critical set of discontinuities (i.e. the set on which sliding is most likely to occur) is inpersistent such that for sliding to occur there must be failure through intact rock. This is, however, very difficult to assess. Unless you have strong evidence to the contrary, take $c = 0$.

Apparent cohesion arises in Patton's equation for shear strength in order to describe strength using instantaneous values of friction angle and are purely an artifact of the mathematics (Fig. 9.6).

Methods for measuring the shear strength parameters for rock discontinuities

The shear strength parameters of discontinuities are best determined by direct shear testing. Ideally this should be carried out on representative samples in situ. Large-scale in situ tests can be conducted on isolated discontinuities by adopting a test set-up such as that described by Romero (1968) and illustrated in Fig. 9.9. Such tests are, however, expensive and can only be justified for major excavation projects or for research purposes. Apart from the cost the following factors often preclude in situ direct shear tests from being carried out:

- exposing the test discontinuity
- providing a suitable reaction for the application of the normal and shear stresses

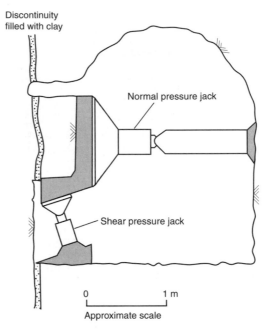

Discontinuity
filled with clay

Normal pressure jack

Shear pressure jack

0   1 m

Approximate scale

Fig. 9.9 In situ shear test on a clay-filled discontinuity, after Romero (1968)

- ensuring that the normal stress is maintained safely as shear displacement takes place.

The alternatives to in situ tests are laboratory direct shear tests. It is not possible, however, to test representative samples of discontinuities in the laboratory. As discussed in Chapter 7 the roughness features often occur on a scale of metres. These features will have a significant effect on the resistance to sliding of large blocks of rock. Samples of similar dimensions from the same joint will often have different roughnesses and individual samples will demonstrate different degrees of interlocking when sheared in different directions (Brown *et al.*, 1977). The only practical method for determining the shear strength of discontinuities is to accurately measure the basic angle of friction ϕ_b and estimate the roughness parameters (*i* or JRC) from field observations. The basic angle of friction is best measured on natural discontinuities using laboratory direct shear tests.

The general requirements for direct shear test apparatus is given in the relevant ISRM suggested method (Brown, 1981). The requirements set out by the ISRM may be achieved with minor modifications to the direct shear box apparatus used for testing soil specimens. Testing with these machines suffers from the following disadvantages:

- difficulty with mounting rock discontinuity specimens in the apparatus

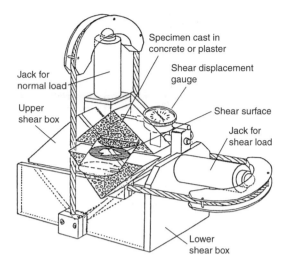

Fig. 9.10 Portable field shear box, after Hoek and Bray (1981)

- difficulty in maintaining the necessary clearances between the upper and lower halves of the box during shearing
- the load capacity of most machines designed for testing soils is likely to be inadequate for rock testing.

The most commonly used devices for direct shear testing of rock and rock discontinuities is the 'portable' shear box described by Ross-Brown and Walton (1975) and Hoek (1981) and shown in Fig. 9.10. Although very versatile the principal problems with using this device are as follows:

- The normal load is applied by means of a hydraulic jack on the upper box and acts against a cable loop attached to the lower box. This loading system results in the normal load increasing in response to dilation of rough discontinuities during shear. Continual adjustment of the normal load is required throughout the test.
- As the shear displacements increase the applied 'normal' load moves away from the vertical and corrections for this may be required.
- The constraints on horizontal and vertical movement during shearing are such that displacements need to be measured at a relatively large number of locations if accurate shear and normal displacements are required.
- The shear box is somewhat insensitive and difficult to use with the relatively low applied stresses in most civil engineering applications since it was designed to operate over a range of normal stresses from 0 to 154 MPa.

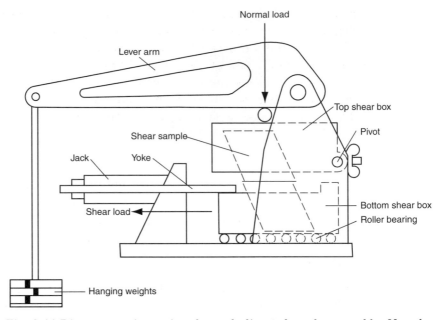

Fig. 9.11 Diagrammatic section through direct shear box used by Hencher and Richards (1982)

A more suitable device is described by Hencher and Richards (1982) and is shown in Fig. 9.11. The normal load is applied by means of a dead load system and therefore remains constant throughout the test. Tests can be carried out accurately at relatively low stresses. Vertical displacement is measured at a single point on the lever arm allowing a magnification of up to ten times providing a relatively high degree of sensitivity. The methodology for the measurement of shear strength of discontinuities using this apparatus is described by Hencher and Richards (1989).

The basic angle of friction represents the frictional component of shear strength for a smooth discontinuity. It will vary with surface texture, mineralogy and the presence of moisture. In most cases discontinuities within rock masses are not completely dry and hence measurements of frictional characteristics of discontinuities in the laboratory should be made with damp surfaces.

The simplest method for determining ϕ_b is by shearing artificially smooth rock surfaces. Suitable surfaces can be made by cutting the rock with a diamond saw. Saw-cut surfaces, however, may produce unreliable values of ϕ_b, particularly if the discontinuity walls are partially or completely weathered or have mineral coatings (Hencher and Richards, 1989). The worst case is where the discontinuity walls are coated with low friction minerals such as chlorite, graphite or serpentinite. In such cases the basic friction angle for the discontinuity may be $10°$ to $20°$ less than that which would be measured for a uncoated saw-cut surface.

Hencher and Richards (1989) recommend that the basic friction angle be measured on natural discontinuities in which corrections have been made for the incremental roughness angles. Direct shear tests are therefore carried out on rough discontinuities. The contribution to the dilation associated with surface roughness may be explained using equation (9.5).

$$\tau = \sigma \tan(\phi_b + i)$$

The dilational component, i, can be simply accounted for by the geometry of the sample. The incremental roughness angle can be calculated throughout the test by considering the incremental vertical (dv) and horizontal (dh) displacements as follows:

$$\mathrm{d}v/\mathrm{d}h = \tan i \qquad (9.11)$$

Stresses in the horizontal and vertical planes can then be resolved parallel and perpendicular to the inclined plane on which sliding is taking place (see Fig. 9.2(a)) using the expressions

$$\tau_i = (\tau \cos i - \sigma \sin i)\cos i$$
$$\sigma_i = (\sigma \cos i + \tau \sin i)\cos i \qquad (9.12)$$

where τ_i and σ_i are the shear and normal stresses on the actual plane of sliding (Fig. 9.2(a)) and i is the angle of roughness.

A worked example of a multi-stage direct shear test is given below.

WORKED EXAMPLE

A sample of a natural discontinuity in a quartzite is subjected to a multi-stage direct shear test at the following normal stresses: 500, 750, 1000 and 1500 kPa. During the tests horizontal and vertical displacements were measured. The results of the tests are shown in Fig. 9.12. The shear strength envelope based on the maximum shear stress is shown in Fig. 9.13. The failure envelope is slightly curved indicating some asperity failure is taking place at the higher normal stress levels. Table 9.4 shows the results of the direct shear tests and calculation of τ_i and σ_i at the maximum applied shear stresses.

The values of τ_i and σ_i are plotted in Fig. 9.13. The angle of friction derived from the gradient of the straight line produced from these data represents the basic angle of friction. The value of ϕ_b in this case is $30°$.

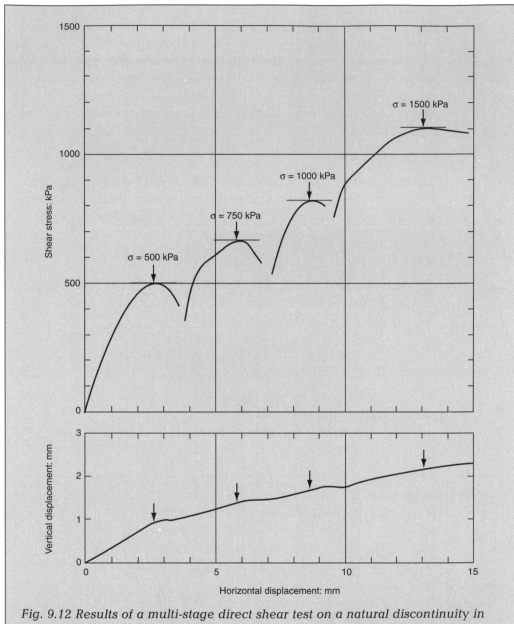

Fig. 9.12 Results of a multi-stage direct shear test on a natural discontinuity in quartzite

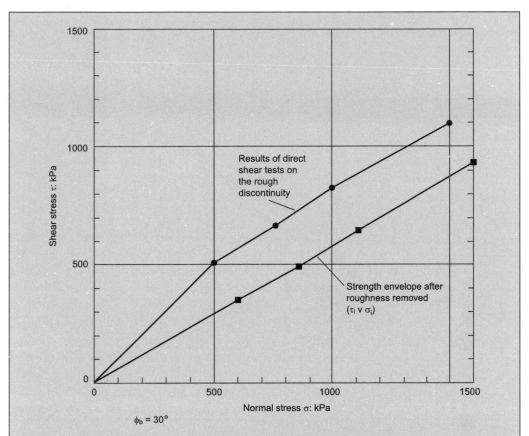

$\phi_b = 30°$

Fig. 9.13 Shear strength envelopes for a rough discontinuity in quartzite and for the same discontinuity with the roughness component removed

Table 9.4

Applied normal stress, σ_n: kPa	Max. shear stress, τ_f: kPa	Instantaneous dv/dh at max. shear stress	Angle of roughness i at maximum shear stress: degrees	σ_i: kPa	τ_i: kPa
500	506	0.268	15	593	347
750	667	0.199	11	849	497
1000	826	0.158	9	1103	651
1500	1107	0.105	6	1599	939

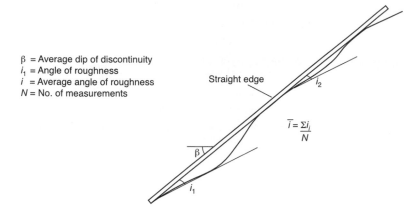

β = Average dip of discontinuity
i_1 = Angle of roughness
i = Average angle of roughness
N = No. of measurements

Straight edge

i_2

$\bar{i} = \dfrac{\Sigma i_i}{N}$

β

i_1

Fig. 9.14 Field measurement of roughness angle i

The angle of roughness *i* may be determined from field observations of discontinuities. Simple measurements of *i* can be made using a straight edge as shown in Fig. 9.14. The Joint Roughness Coefficient (JRC) is often determined in the field using a tilt test. The JRC value thus obtained may have to be corrected for scale using equation (9.10). The worked example below demonstrates how tilt test results are used to determine a JRC value.

WORKED EXAMPLE

A simple tilt test on a dry sample of siltstone, containing a specimen of a particular joint set, produced slip on the discontinuity at an inclination of 53°. The gross contact area during shearing was 89.3 cm^2, the volume of the upper block was 738 cm^3 and the mass of the upper block was 2.06 kg. The basic angle of friction measured using laboratory direct shear tests was 32°. A series of Schmidt hammer tests with the L-hammer gave an average rebound value of 19.5 for the natural discontinuity surface. Another discontinuity from the same joint set forms the sliding plane for a major rigid block failure mechanism in which the effective normal stress across the discontinuity is computed to be 85 kPa.

- Estimate the joint roughness coefficient JRC.
- Assuming that the sliding plane has the same JRC, estimate its shear strength at the computed effective normal stress.
- Use graphical or other methods to estimate the equivalent cohesion and angle of friction for this discontinuity over the effective normal stress range 80 to 90 kPa.

Solution

- The rock material has a basic friction angle ϕ_b of 32°. In the tilt test on the natural discontinuity, the upper block has a mass of 2.06 kg or a weight of 20.2 N. Hence, by equation (9.1), noting that there is no water pressure, at an inclination of 53° the effective normal stress σ'_n on the discontinuity is 1.36 kPa and the shear stress τ_f is 1.81 kPa. The upper block has a volume of 738 cm^3, indicating a unit weight of 0.0274 MN m^{-3}. Inputting this result, together with the average rebound value of 19.5, into equation (7.3) gives an estimated joint wall compressive strength σ_d of 30.2 MPa. Putting this result into equation (9.9) with $\sigma'_n = 1.36$ kPa, $\tau_f = 1.81$ kPa and $\phi_b = 32°$, then solving for JRC gives an estimated joint roughness coefficient of 4.83.

- Applying equation (9.9), to estimate the shear strength τ_f of the sliding plane at an effective normal stress of 85 kPa gives $\tau_f = 83.0$ kPa.

- Applying equation (9.9) again to estimate the shear strength τ_f of the sliding plane at effective normal stresses of 80 and 90 kPa gives $\tau_f = 78.5$ and 87.5 kPa, respectively. A straight line through these points, plotted in Fig. 9.15, has a slope and intercept indicating an angle of friction and cohesion of approximately 42° and 6.5 kPa, respectively. This figure also contains the complete shear strength envelope given by equation (9.9) for effective normal stresses in the range 0 to 100 kPa.

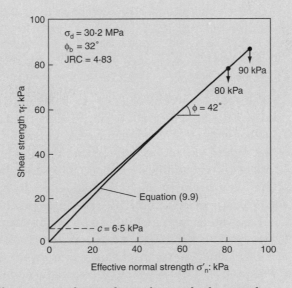

Fig. 9.15 Shear strength envelopes for worked example

CHAPTER TEN

Analysis of rock slopes

Limit equilibrium analysis

Overview: factor of safety

Rock slopes may fail in one or more different modes. These have been outlined in Table 8.1. In general, these modes involve either sliding, block
rotation or free-fall and bounce. Modes that involve sliding include
plane, wedge and circular failure, can be analysed in terms of determining
a factor of safety using limit equilibrium analysis. The other modes are
more complex and a factor of safety cannot be determined.

In the simple case where sliding occurs on a single plane, a factor of
safety may be calculated by resolving forces parallel and perpendicular
to the sliding surface. The components of forces perpendicular to the sliding surfaces are used in conjunction with the appropriate discontinuity
shear strength criterion. The sum of these restoring forces is compared
with the sum of the disturbing forces parallel to the sliding surface. This
gives the factor of safety.

Limiting equilibrium occurs when the ratio of the restoring forces to the
disturbing forces is unity, i.e.

$$\text{Factor of safety } F = \sum \text{restoring forces} / \sum \text{disturbing forces} = 1.$$

(10.1)

Clearly, for a stable slope $F > 1$. In reality the shear strength parameters
used may not be sufficiently accurate or constant along the failure surface.
Together with other uncertainties relating to the position and orientation
of the failure surface as well as the magnitude of external forces (e.g.
earthquake accelerations), the stability assessment will give a factor of
safety lying between limits of confidence. In practice, factors of safety
will be based on local knowledge based on case histories. In general, however, the values give in Table 10.1 may be taken as a guide. Priest and
Brown (1983) stipulated limits on the probabilities $P(F < 1.0)$ and
$P(F < 1.5)$ that the factor of safety of a given slope falls below 1.0 and
1.5 respectively. These values are reproduced in Table 10.1.

In calculating the factor of safety the following questions must be
answered.

Table 10.1 Acceptable values for factors of safety, after Priest and Brown (1983)

| Category of slope | Consequences of failure | Examples | Acceptable values of factor of safety F | | |
			Minimum Mean F	Maximum $P(F < 1.0)$	Maximum $P(F < 1.5)$
1	Not serious	Individual benches, small[*] temporary slopes not adjacent to haulage roads	1.3	0.1	0.2
2	Moderately serious	Any slope of a permanent or semipermanent nature	1.6	0.01	0.1
3	Very serious	Medium-sized and high slopes carrying major haulage roads or underlying permanent mine installations	2.0	0.003	0.05

[*] Small, height <50 m; medium, height 50 to 150 m; high, height >150 m.

- What is the critical sliding surface?
- What shear strength criterion should be used?
- What is the likely water pressure distribution along the discontinuities that form the boundaries of the sliding block?
- What external forces act on the sliding block?

What is the critical sliding surface?

In a rock mass containing different sets of discontinuities the set or sets that will form potential sliding surfaces may be identified using the approach described in Chapter 8. The actual discontinuity that will form the sliding surface within a set of discontinuities in most cases will be difficult to identify. In such cases the worst case is considered.

What shear strength criterion should be used?

In Chapter 9, three shear strength criteria are described. These are:

(a) $\tau_f = \sigma'_n \tan(\phi_b + i)$ at low normal stresses (Patton, 1966).

(b) $\tau_f = c + \sigma'_n \tan \phi_b$ at high normal stresses (Patton, 1966).

(c) $\tau_f = \sigma'_n \tan \left(\phi_b + \text{JRC} \log_{10} \left(\dfrac{\text{JCS}}{\sigma'_n} \right) \right)$ at any normal stress (Barton, 1971).

It is essential to identify clearly the bifurcation point between the two criteria (a) and (b) above (see Fig. 9.6). If $\tau_f = \sigma'_n \tan(\phi_b + i)$ is used for any normal stress the result is safe at low stresses and unsafe at high stresses. If $\tau_f = c + \sigma'_n \tan \phi_b$ is used for any normal stress the result is unsafe at low stresses and safe at high stresses (see Fig. 9.6). Because of

287

this difficulty, it is preferable to use the Barton criterion (c) above. Note that this requires the calculation of normal stress. This normal stress is the sum of the components of force perpendicular to the surface of sliding divided by the area of the surface of sliding.

Use of the Barton shear strength equation is straightforward for plane failure problems where typically a 2D section through the slope is analysed. For wedge failure a 3D analysis is required. For computer-based analysis the incorporation of the Barton equation is possible. The various solutions suitable for hand or spreadsheet calculation presented here are based on those proposed by Hoek and Bray (1981) and do not permit the use of this equation. The Patton equation is used with these solutions.

What is the likely water pressure distribution along the discontinuities that form the boundaries of the sliding block?

The current state of knowledge in rock engineering does not permit a precise definition of the ground water flow patterns in rock mass. Consequently, the only possibility open to the engineer is to consider a number of realistic extremes in an attempt to define the upper and lower bounds of the factor of safety and to assess the sensitivity of the slope to variations in ground water conditions. Four cases should be considered.

- *Dry slope.* The simplest case that can be considered is that in which the slope is assumed to be fully drained and no water pressures exist. Note that there may be moisture in the slope but, as long as no pressure is generated, it will not influence the stability of the slope.
- *Water in the tension crack at the back of sliding block.* A heavy rainstorm after a long dry spell of weather may result in the rapid build-up of water pressure in the tension crack which will offer little resistance to the entry of surface flood water unless effective surface drainage has been provided. Assuming that the remainder of the rock mass is relatively impermeable, the only water pressure that will be generated during and immediately after the rain, will be that due to the water in the discontinuity (Fig. 10.1(a)).
- *Water in tension crack and on the sliding surface.* The pressure distribution along the sliding surface is assumed to decrease linearly from the base of the tension crack to the intersection with the slope face (Fig. 10.1(b)). This is probably much simpler than that which occurs in practice but, since the actual pressure distribution is unknown, a linear distribution of water pressure is a reasonable assumption to make. It is possible that a more dangerous water pressure distribution could exist if the water was unable to drain freely out of the slope, e.g. if the face of the slope became frozen in the winter. In this case the full head of water in the slope acts at the base of the sliding surface

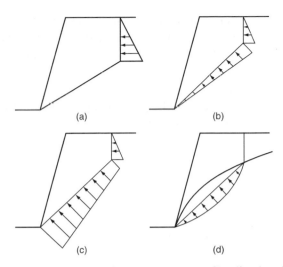

Fig. 10.1 Simplified schemes of water pressure distribution for slope stability analysis

(Fig. 10.1(c)). Such extreme conditions may occur from time to time and should be borne in mind by the engineer. In general, however, the use of this water pressure distribution would result in an excessively conservative slope and hence the distribution shown in Fig. 10.1(b) is normally used.

- *Saturated slope with heavy recharge.* If the rock mass is heavily fractured so that it becomes relatively permeable, a ground water flow pattern similar to that which would develop in a porous system would occur where water pressures are determined by the phreatic surface (Fig. 10.1(d)). In view of the uncertainties associated with the actual water pressure distributions which could occur within rock slopes when subjected to these conditions, there seems little point to refine this approach any further.

What external forces act on the sliding block?

It is possible for slopes to be subjected to static external forces such as foundation or surcharge loading. The resultants of the forces caused by these loadings should be resolved parallel and perpendicular to the sliding surface and combined with the disturbing and restoring forces in the calculation of the factor of safety.

In areas prone to seismic activity, the slope may be subjected to dynamic forces. An analysis can be carried out by using the limit equilibrium method described here by idealizing the dynamic loading occurrence with a time constant force that is proportional to the mass of the potentially unstable block according to a seismic coefficient that is expressed as a

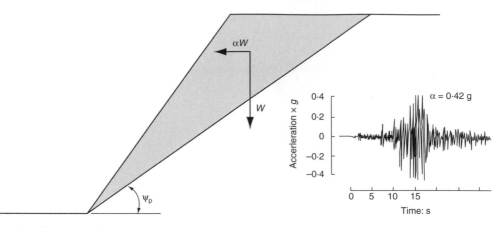

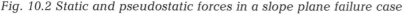

Fig. 10.2 Static and pseudostatic forces in a slope plane failure case

proportion of the acceleration due to gravity (e.g. 0.2 g). It is common for the seismic coefficient to be published in national codes of practice for structural engineering. The force representing the dynamic loading is assumed to be horizontal and equal to the product of the seismic coefficient and the mass of the block (Fig. 10.2).

Plane failure

Overview

Plane failure refers to the somewhat rare phenomenon of a block of rock sliding on a single plane. The comparative rarity of this type of failure stems from the geometrical conditions required for it to occur. These are summarized as follows.

(a) The plane on which sliding occurs must dip in a direction that is nearly parallel (i.e. within $\pm 20°$) of that of the slope face.

(b) The failure plane must 'daylight' (i.e. intersect) in the slope face. This means that the dip of the failure plane, ψ_p, must be less than the dip of the slope face, ψ_f (i.e. $\psi_p < \psi_f$).

(c) The dip of the failure plane must be greater than the angle of friction for this plane (i.e. $\psi_p > \phi$).

(d) Lateral release surfaces which provide negligible resistance to sliding must be present in the rock mass. These define the lateral boundaries of the slide.

Whereas conditions (a), (b) and (c) are often met in a slope face it is absence of suitably orientated lateral release surfaces that often prevent plane failure from taking place. For example, an ideal geometry for plane failure to occur would be a slope face dipping at say 70° towards 245° which is intersected by a set of smooth walled discontinuities

Faces act as release surfaces

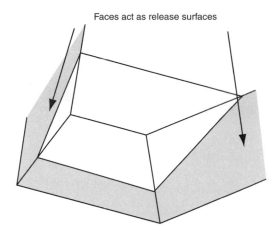

Fig. 10.3 Release surface created by slope geometry

($\phi = 30°$) dipping at 50° towards 245° and a set of vertical open discontinuities striking perpendicular to the slope face as shown in Fig. 8.1.

If the direction of dip of the sliding surface were just a few degrees different from that of the face, gravity would force the sliding block against the release surface on the down-dip side resulting in friction being mobilized along it. This would of course increase the resistance to sliding and thus have a stabilizing effect.

The greatest danger from plane failure arises when slopes are constructed with a convex 'nose' as shown in Fig. 10.3. In this case the side faces act as perfect lateral release surfaces. The opposite configuration of faces (i.e. a concave set of faces) will be very safe with respect to plane failure.

The wedge type of failure, which is discussed in the next section of this chapter, is a more general case for slope faces in fractured rock masses. Many rock engineers therefore treat plane failure as a special case of the more general 3D wedge failure analysis. Plane failure, which can be treated as a 2D case, should not be ignored completely since there are many valuable lessons to be learned from a consideration of the mechanics of this mode of failure. It is also useful for demonstrating the sensitivity of the slope to changes in shear strength, ground water conditions and external loads such as foundations and dynamic loading from earthquakes.

In analysing 2D slope problems in soils discussed in Part 1 of this book the approach taken was to consider a section through the slope of unit width at right angles to the slope face. The same approach is taken when analysing plane failure. In the 2D section under consideration the sliding surface is represented by a straight line with uniform shear strength properties and hence there no need to introduce slices. The

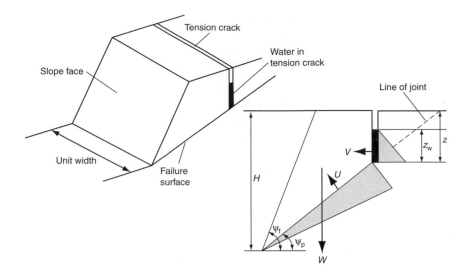

Fig. 10.4 2D section of slope used for the analysis of plane failure problems

area of the sliding surface can be represented as a straight line and the volume of the sliding block is represented by the area of the figure representing this block on the vertical section as shown in Fig. 10.4.

Plane failure analysis

The geometry of the slope considered in this analysis is defined in Fig. 10.4. The tension crack in Fig. 10.4 shown in the upper surface may be associated with another set of discontinuities suitably orientated or it may be a genuine tension-related feature. The tension crack can be positioned anywhere on the main slope or the upper slope. The analysis presented below can be used to investigate the sensitivity of the slope to the position of a tension crack. The following assumptions are made in this analysis:

(a) Both sliding surface and tension crack strike parallel to the sliding surface.

(b) The tension crack may be vertical and may be filled with water to a depth z_w.

(c) Water enters the sliding surface along the base of the tension crack and seeps along the sliding surface, escaping at atmospheric pressure where the sliding surface daylights in the slope face. The pressure distribution induced by the pressure of water in the tension crack and along the sliding surface is illustrated in Fig. 10.4.

(d) The forces W (the weight of the sliding block), U (uplift force due to water pressure on the sliding surface) and V (force due to water pressure in the tension crack) all act through the centroid of the sliding mass. This means that there are no moments which would tend to cause rotation of the block and hence failure is by sliding only. While this assumption may not be strictly true for actual slopes, the errors introduced by ignoring moments are small enough to neglect. The same assumption would be made when considering external forces such as foundation loading or rock anchor cable tension. In steep slopes with steeply dipping discontinuities, however, toppling failure may occur which will involve block rotation.

(e) The shear strength of the sliding surface is defined by cohesion (real or apparent) c, a basic friction angle ϕ_b and a roughness angle i when using the Patton shear strength model or by a basic friction angle ϕ_b, a Joint Roughness Coefficient (JRC) and a Joint Compressive Strength (JCS) when using Barton's shear strength model. The Barton shear strength model is preferred since it provides a reasonably accurate prediction of shear strength over a wide range of normal stresses. However, care should be taken in determining the magnitude of JRC in particular since it is scale dependent. The Patton model should really only be used in situations where there is reasonable certainty that either minimal or complete asperity breakage will occur during failure.

(f) A section perpendicular to the face of unit thickness is considered and it is assumed release surfaces are present so that there is no resistance to sliding at the lateral boundaries of the failure.

Using the shear strength parameters c and ϕ, the factor of safety given by the total force resisting sliding to the total force tending to induce sliding, is

$$F = \frac{cA + (W \cos \psi_p - U - V \sin \psi_p) \tan \phi}{W \sin \psi_p + V \cos \psi_p} \qquad (10.2)$$

Using the Barton shear strength model, the factor of safety is

$$F = \frac{A\sigma' \tan(\phi_b + JRC \log(JCS/\sigma'))}{W \sin \psi_p + V \cos \psi_p} \qquad (10.3)$$

$$\sigma' = \frac{W \cos \psi_p - U - V \sin \psi_p}{A} \qquad (10.4)$$

where, from Fig. 10.4,

$$A = (H - z) \cos ec\psi_p$$

$$U = \tfrac{1}{2}\gamma_w z_w (H - z) \cos ec\psi_p \qquad (10.5)$$

$$V = \tfrac{1}{2}\gamma_w z_w^2$$

For a vertical tension crack in the upper slope surface (as illustrated in Fig. 10.4):

$$W = \tfrac{1}{2}\gamma H^2[(1 - (z/H)^2)\cot\psi_p - \cot\psi_f] \qquad (10.6)$$

and for a vertical tension crack in the main slope face

$$W = \tfrac{1}{2}\gamma H^2[(1 - z/H)^2 \cot\psi_p(\cot\psi_p - 1)] \qquad (10.7)$$

When the tension crack is not vertical the above equations cannot be used and it is often easier to determine A, W, z, z_w using a scale drawing on graph paper.

When the geometry of the slope and the depth of water in the tension crack are known, the calculation of a factor of safety is straightforward. In some cases, however, it may be necessary to compare a range of slope geometries, water depths and the influence of different shear strengths. In such cases a spreadsheet may be used to carry out the analysis and to create graphs showing the sensitivity of the factor of safety to these changes.

The worked example below illustrates a more complex problem involving planar failure.

WORKED EXAMPLE

A bridge is to be constructed across a steep sided gorge cut into a massive sandstone. In order to minimize the span a site has been chosen where one face of the gorge forms a convex 'nose'. An examination of the rock mass at this location has revealed two highly persistent discontinuities dipping in approximately the same direction as the main gorge face. Details of these discontinuities are given below. Although the rock mass appears to stable there is some concern over the possibility of plane failure occurring on discontinuity A with the side faces of the 'nose' acting as lateral release surfaces.

Discontinuity	Dip	JRC	JCS: MPa
A	35°	10	5
B	80°	5	5

The geometry of the slope is shown in Fig. 10.5.

The foundations for the bridge are to be placed 2 m behind the crest of the slope. It will impose a 0.2 MN/m vertical load and a 0.1 MN/m horizontal load (direction same as dip direction of the slope face) on the rock mass.

The site is in an earthquake zone and the maximum likely ground acceleration determined from previous earthquake events is $0.2g$.

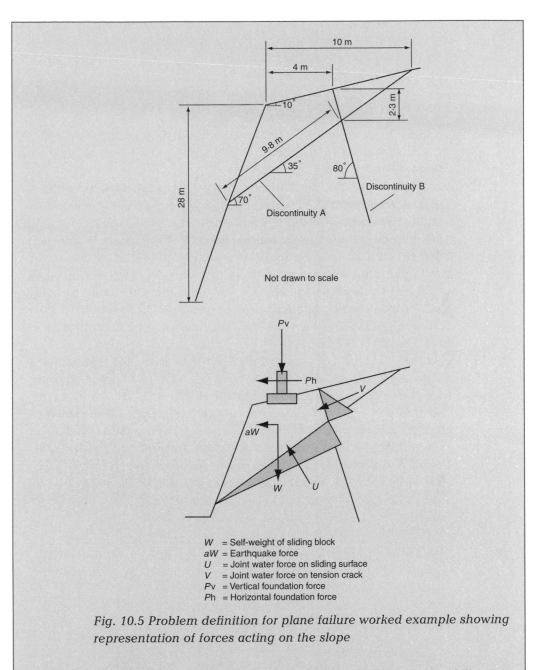

Fig. 10.5 Problem definition for plane failure worked example showing representation of forces acting on the slope

Determine the factor of safety against sliding for the following conditions.

(a) The natural slope with no foundation loading, no earthquake loading and no water in the discontinuities.

(b)   As (a) but after a period of prolonged rainfall.

(c)   As (b) but with the earthquake loading.

(d)   As (b) but with the foundation loading.

(e)   As (b) but with the foundation loading and the earthquake loading.

Assume reasonable values for any parameters not given.

Solution

The forces acting on the slope are also shown in Fig. 10.5. It has been assumed that during the period of prolonged and heavy rainfall discontinuity B is maintained full of water and that water drains freely along discontinuity A. A value of 30° is assumed for the basic angle of friction and 25 kN/m^3 is assumed for the unit weight of the rock mass.

A section perpendicular to the face and of unit width has been taken for the analysis (Fig. 10.5). The cross-sectional area of the sliding block is 27.5 m^2. This was determined by drawing the cross-section to scale on graph paper and counting squares. The self-weight of the sliding block is $25 \times 1 \times 27.5 = 687$ kN/m.

Since a number of different force combinations are to be examined it is good practice to resolve all the forces parallel and perpendicular to the sliding surface (discontinuity A) and place the results in a table such as shown in Table 10.2. The calculations are shown in Fig. 10.6.

The parallel and perpendicular components can now be summed for the different combinations of forces under consideration (see Table 10.3). The normal stress acting on the failure surface is required by the Barton shear strength model. This may be calculated for the 1 m

Table 10.2 Resolution of forces

Force	Magnitude: kN/m	Parallel component: kN/m	Perpendicular component: kN/m
Self-weight of sliding block, W	687	+394	+563
Pseudo static force representing the earthquake loading, aW	137	+112	−79
Force from water in discontinuity B, V	27	+25	−11
Uplift force from water in discontinuity A, U	113	0	−113
Vertical foundation loading, P_v	200	+115	+164
Horizontal foundation loading, P_h	100	+82	−57

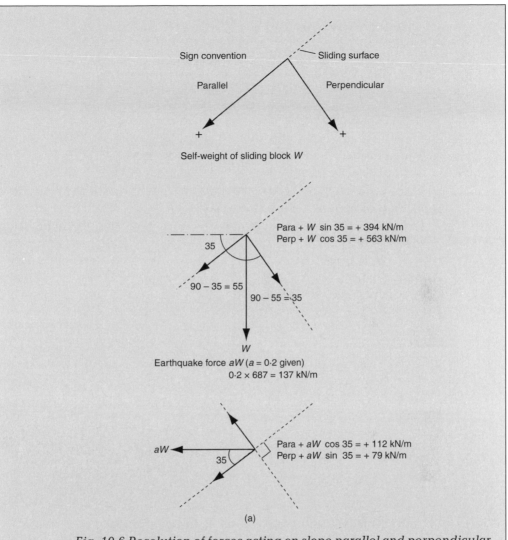

Fig. 10.6 Resolution of forces acting on slope parallel and perpendicular
to the sliding surface

wide section of slope under consideration by simply dividing the sum of
the perpendicular components $(\sum Z)$ by the length of the sliding
surface (L) which is 9.8 m in this example. The values of normal stress
for each of the five cases considered are shown in Table 10.3.

The factor of safety in each case has been calculated using

$$F = \frac{\tan(\phi_b + \mathrm{JRC}\log(\mathrm{JCS}/\sigma'))\sum Z}{\sum X} \qquad (10.8)$$

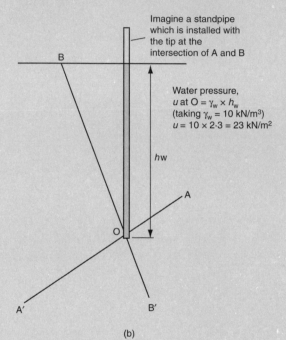

Joint water forces V and U

Assume that joint B fills with water up to the ground surface and is maintained full of water by the heavy rainfall. Assume that joint A is free draining (i.e. there is a spring where the joint outcrops in the face).

Calculate the water pressure at the intersection of joints A and B

Imagine a standpipe which is installed with the tip at the intersection of A and B

Water pressure,
u at $O = \gamma_w \times h_w$
(taking $\gamma_w = 10$ kN/m³)
$u = 10 \times 2 \cdot 3 = 23$ kN/m²

hw

Fig. 10.6 Continued

The sensitivity of the factor of safety to water pressure in the discontinuities and external loads is clearly seen in Table 10.3. The dry slope with no external loads has a factor of safety of 1.91. The effect of filling discontinuity B with water due to prolonged and heavy rainfall has resulted in the factor of safety dropping to 1.47. The slope is still safe, however. The combination of an earthquake event and heavy rainfall, however, results in slope failure. The construction of bridge results in a dangerously low factor of safety during heavy and prolonged rainfall and failure if an earthquake occurs at the same time. This result would justify the need to carry out a more rigorous analysis of the slope using a boundary element or finite element approach.

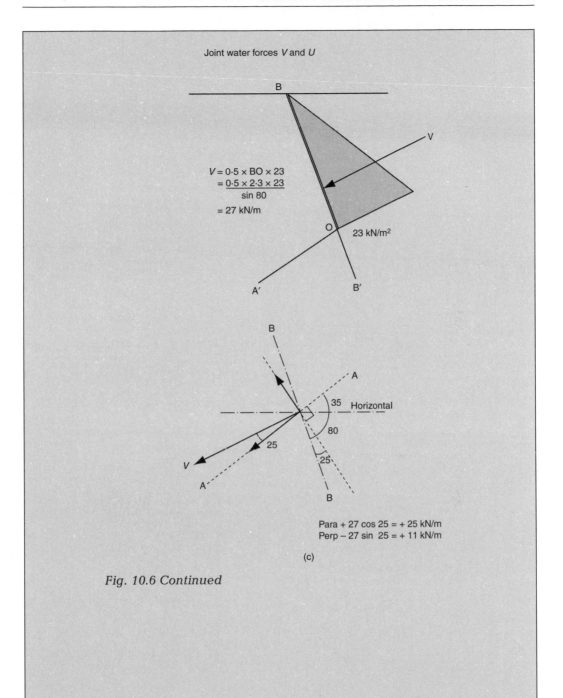

Joint water forces V and U

$V = 0.5 \times BO \times 23$
$= \dfrac{0.5 \times 2.3 \times 23}{\sin 80}$
$= 27$ kN/m

23 kN/m²

Para $+ 27 \cos 25 = + 25$ kN/m
Perp $- 27 \sin 25 = + 11$ kN/m

(c)

Fig. 10.6 Continued

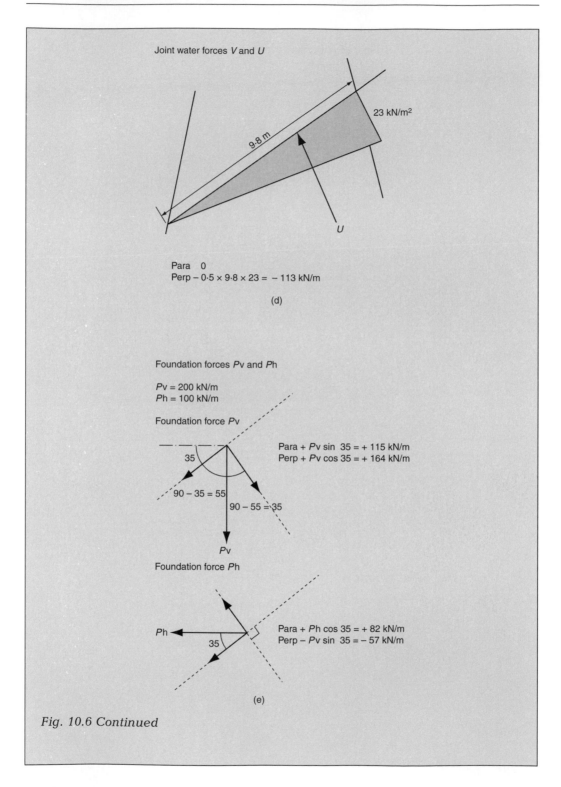

Joint water forces V and U

9.8 m

23 kN/m²

U

Para   0
Perp $-0.5 \times 9.8 \times 23 = -113$ kN/m

(d)

Foundation forces Pv and Ph

$Pv = 200$ kN/m
$Ph = 100$ kN/m

Foundation force Pv

35

$90 - 35 = 55$

$90 - 55 = 35$

Para $+ Pv \sin 35 = +115$ kN/m
Perp $+ Pv \cos 35 = +164$ kN/m

Pv

Foundation force Ph

Ph

35

Para $+ Ph \cos 35 = +82$ kN/m
Perp $- Pv \sin 35 = -57$ kN/m

(e)

Fig. 10.6 Continued

Table 10.3 Calculation of factor of safety for cases (a) to (e)

Case	Sum of parallel components $\sum X$: kN/m	Sum of perpendicular components $\sum Z$: kN/m	Normal stress on 1 m wide section $\sigma' = (\sum Z)/L$: kPa $L = 9.8\,\text{m}$	Factor of safety
(a) Dry slope	394	563	57	1.67
(b) Saturated slope	$(394 + 25 + 0) = 419$	$(563 - 11 - 113) = 439$	45	1.27
(c) Saturated slope + earthquake	$(394 + 112 + 25 + 0) = 531$	$(563 - 79 - 11 - 113) = 360$	37	0.85
(d) Saturated slope + foundation	$(394 + 25 + 0 + 115 + 82) = 616$	$(563 - 11 - 113 + 164 - 57) = 546$	56	1.04
(e) Saturated slope + earthquake + foundation	$(394 + 112 + 25 + 0 + 115 + 82) = 728$	$(563 - 79 - 11 - 113 + 164 - 57) = 467$	48	0.77

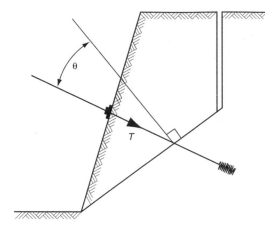

Fig. 10.7 Reinforcement of a slope

Stabilization by rock bolts or anchors

When it is established that a particular slope is unstable, it becomes necessary to consider whether it is possible to stabilize the slope by applying restoring external loads using rock bolts or anchors (Fig. 10.7). The design involves optimizing the capacity and orientation of the bolts/anchor to achieve the desired factor of safety. This is illustrated in the worked example below.

WORKED EXAMPLE

Determine the minimum rock anchor cable tension to give a factor of safety of 1.5 for the conditions given in part (e) of the previous worked example.

Solution

In order to maintain stability during a heavy storm and an earthquake rock anchors may be used to increase the forces resisting sliding. The cable tension required to achieve a given factor of safety will vary with orientation of the anchor. It is necessary to find the orientation that gives the minimum cable tension. In calculating the minimum cable tension it is convenient to consider the angle the anchor makes with the normal to the failure plane. The geometry of the problem is shown in Fig. 10.8.

The determination of the relationship between cable tension and orientation is made difficult when using the Barton shear strength model because of the asperity breakage term (JCS/σ'). The force combinations considered in part (a) resulted in normal effective stresses

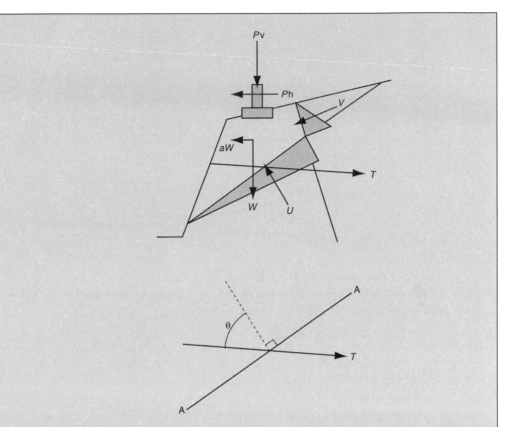

Fig. 10.8 Forces associated with slope reinforcement in worked example

ranging from 37 to 57 kPa and range of the equivalent roughness angles (derived from $JRC\log(JCS/\sigma')$) 19° to 21°. The average equivalent roughness angle is approximately 20°. A reasonable estimate of the minimum cable tension may be determined by using this average equivalent roughness angle in the following expression:

$$F = \frac{(467 + T\cos\theta)\tan(30 + 20°)}{728 - T\sin\theta} = 1.5$$

$$(467 + T\cos\theta)\tan 50° = 1092 - 1.5T\sin\theta$$

$$560 + 1.2T\cos\theta = 1092 - 1.5T\sin\theta$$

$$T(1.2\cos\theta + 1.5\sin\theta) = 532$$

The resulting relationship is shown in Fig. 10.9. The minimum cable tension occurs when $\theta = 50°$. The minimum cable tension required to achieve a factor of safety of 1.5 is 277 kN/m run of slope.

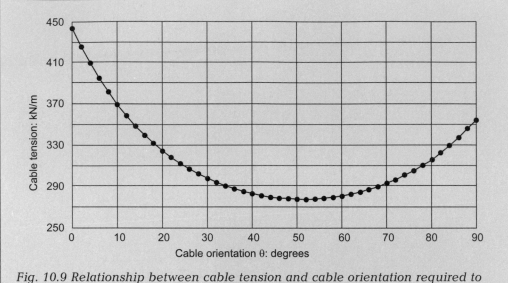

Fig. 10.9 Relationship between cable tension and cable orientation required to achieve a factor of safety of 1.5 in the worked example

Wedge failure

Overview

The combination of discontinuity orientations and face orientations commonly give rise to wedge (actually tetrahedral) shaped blocks which can slide out of the face along the line of intersection between two of the discontinuities as shown in Fig. 10.10. The rock face shown in Fig. 10.11 illustrates the surfaces on which a wedge-shaped block of rock has slipped out. This is a particularly dangerous failure mechanism since slip can occur without any topographic or structural release features. It was a wedge failure that resulted in the complete destruction of the Malpasset Arch Dam in France in 1959.

As wedge failure generally involves sliding on more than one surface the two-dimensional analysis adopted for plane failure will not be suitable. A three-dimensional analysis is required. The basic mechanics of the wedge failure are very simple. Because of the large number of geometric variables involved, the mathematical treatment is very complex. In reality, this means that complete analysis of wedge failure can only be carried out by computer. Full stability analyses of wedge-shaped blocks are described by Bray and Brown (1976), Hoek and Bray (1981) and Priest (1985).

The analysis of the stability of wedge-shaped blocks follows the same principles as plane failure, except that it is necessary to resolve forces

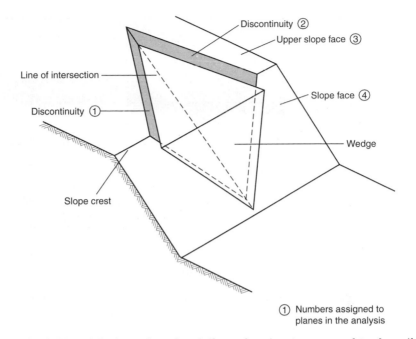

Fig. 10.10 *Pictorial view of wedge failure showing terms used to describe the various components*

on both the sliding surfaces. The procedure is to calculate the weight of the wedge and the area of each face. The self-weight of the wedge together with any external forces (e.g. water, earthquake loading, foundation loading and support forces) are then resolved into their normal and shear components acting on each of the two sliding surfaces.

Hoek and Bray (1981) have proposed various methods for wedge analysis. In order of advancing complexity (and hence accuracy) these are summarized in Table 10.4. With the exception of the comprehensive method, these methods are explained below.

Hoek and Bray Simplified Method

A simplified analysis for the stability of wedge-shaped blocks described by Hoek and Bray (1981) is outlined below. We have called this the 'Hoek and Bray Simplified Method'. It is assumed that sliding is resisted by friction alone and external forces are ignored. The geometry of the wedge is shown in Fig. 10.12. The factor of safety for the wedge defined in Fig. 10.12 is given by

$$F = \frac{(R_1 + R_2)\tan\phi}{W\sin\psi_i} \tag{10.9}$$

where R_1 and R_2 are the normal reactions provided by the planes 1 and 2 (note the steeper plane is always plane 2) in Fig. 10.12.

305

Fig. 10.11 Rock face in Dolerite showing evidence of wedge failure in upper part of slope (note the steeply dipping intersections which are orientated obliquely to the main slope face), Porth-Gaih, Wales (photograph: Marcus Matthews)

In order to find R_1 and R_2, resolve horizontally and vertically in the view along the line of intersection (see Fig. 10.12):

$$R_1 \sin(\beta - \tfrac{1}{2}\xi) = R_2 \sin(\beta + \tfrac{1}{2}\xi) \tag{10.10}$$

$$R_1 \cos(\beta - \tfrac{1}{2}\xi) - R_2 \cos(\beta + \tfrac{1}{2}\xi) = W \cos \psi_i \tag{10.11}$$

Solving for R_1 and R_2 and adding gives

$$R_1 + R_2 = \frac{W \cos \psi_i \sin \beta}{\sin \tfrac{1}{2}\xi} \tag{10.12}$$

Table 10.4 Hoek and Bray methods of wedge analysis

Method	Requirements	Friction angle $(\phi_1 = \phi_2)$	Friction angle $(\phi_1 \neq \phi_2)$	Cohesion	Joint water pressure	Other external forces (e.g. foundation loading, reinforcement)	Tension crack	Inclined upper slope
Simplified	Stereonet and Hoek and Bray chart	Yes	No	No	No	No	No	No
Improved	Stereronet and scientific calculator or spreadsheet	Yes	Yes	Yes	Yes	No	No	Yes
Short	Spreadsheet	Yes	Yes	Yes	Yes	No	No	No
Comprehensive	Spreadsheet/computer program	Yes	Yes	Yes	Yes	Yes	Yes	Yes

The terms 'Short' and 'Comprehensive' are names used by Hoek and Bray (1981). The terms 'Simplified' and 'Improved' are names we have given these methods which were not specifically named by Hoek and Bray.

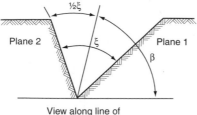

View along line of
intersection

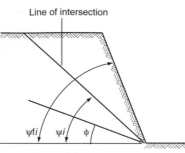

View at right angles to
line of intersection

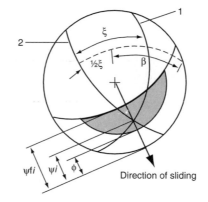

Steroplot of wedge failure geometry

Fig. 10.12 Wedge failure geometry

Hence

$$F = \frac{\sin \beta}{\sin \frac{1}{2}\xi} \frac{\tan \phi}{\tan \psi_i} \qquad (10.13)$$

Noting that $\tan \phi / \tan \psi_i$ is the factor of safety, F_p, for a planar failure, this equation may be expressed by:

$$F_\mathrm{w} = kF_\mathrm{p} \qquad (10.14)$$

where F_w is the factor of safety of a wedge supported by friction only. F_p is the factor of safety of a plane failure in which the slope face is inclined at ψ_fi and the failure plane in inclined at ψ_i. The factor k depends upon the included angle of the wedge, ξ, and upon the angle of tilt, β. Values of k are plotted for a range of values of β and ξ in Fig. 10.13.

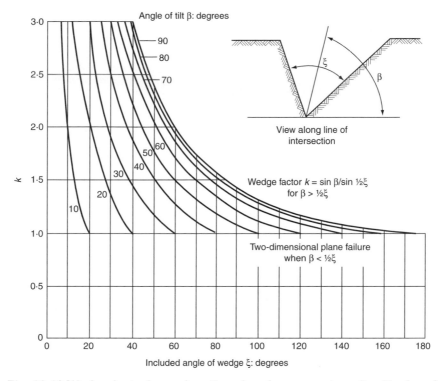

Fig. 10.13 Wedge factor k *as a function of wedge geometry, after Hoek and Bray (1981)*

WORKED EXAMPLE

It is proposed to construct a slope in a very strong greywacke. The orientation of the main slope face in terms of dip direction and dip is 208/65. The ground above the main slope is inclined with an orientation 218/10. Two highly persistent discontinuities with orientations 128/55 and 267/70 are located such that their line of intersection will intersect the main slope face 20 m below the crest. A third discontinuity with orientation 198/85 intersects the upper slope at a point 5 m behind the crest measured along the outcrop of the discontinuity which is dipping at 55°. This situation is summarized in Table 10.5.

The shear strength parameters of the two discontinuities forming the sliding surfaces of the wedge are shown in Table 10.6. The 3D visualization of the geometry of the problem described above is shown in Fig. 10.14. Determine the kinematic feasibility of wedge failure occurring and the factor of safety for the wedge using the Hoek and Bray simplified method.

Table 10.5 Summary of plane orientations

Plane No.	Description	Dip direction: degrees	Dip: degrees
1	Discontinuity forming one of the sliding surfaces	128	55
2	Discontinuity forming the steepest sliding surface	267	70
3	Upper slope face	218	10
4	Main slope face	208	65
5	Tension crack	198	85

Table 10.6 Summary of plane properties

	Shear strength parameters	
Plane No.	ϕ_b: degrees	Roughness angle i: degrees
1	30	15
2	30	20

Solution

Kinematic feasibility

The overlay for the kinematic feasibility of wedge failure has been drawn in Fig. 10.15. It will be seen from Fig. 10.15 that the line of intersection of discontinuities 1 and 2 fall inside the unstable zone when a basic angle of friction (ϕ_b) of 30° is used. When the roughness angle is added the unstable zone reduces in size and no longer includes the line of intersection. We can conclude that for a dry slope with no external forces the wedge will be stable. We cannot guarantee however that it remain stable if the discontinuities fill with water or external loads are applied.

It will be seen from Fig. 10.15 that the dip directions of discontinuity 1 or 2 do not fall in between that of the line of intersection and the main slope face. Based on Hocking (1976) this suggests that if the wedge slides out of the face it will maintain contact with both discontinuities. This assumes that there are no water pressures developed in the discontinuities.

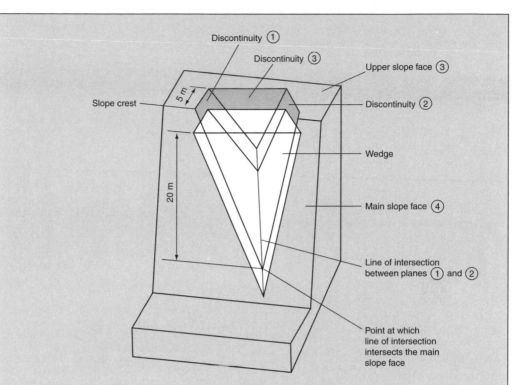

Fig. 10.14 3D visualization of wedge in worked example

Factor of safety using the Hoek and Bray simplified method
From the stereoplot shown in Fig. 10.16:

> Included angle $= 68°$
> Angle of tilt $\beta = 80°$
> Dip of line of intersection $\psi_i = 33°$

$$k = \frac{\sin \beta}{\sin \frac{1}{2}\xi} = \frac{\sin 80}{\sin 34} = 1.76$$

($k = 1.75$ from chart).
   Using shear strength parameters for discontinuity 1

$$\phi_b + i = 30 + 15 = 45°$$

$$F = k\frac{\tan(\phi_b + i)}{\tan \psi_i} = 1.76\frac{\tan 45}{\tan 27} = 3.45$$

Using shear strength parameters for discontinuity 2

$$\phi_b + i = 30 + 20 = 50°$$

$$F = k\frac{\tan(\phi_b + i)}{\tan \psi_i} = 1.76\frac{\tan 50}{\tan 27} = 4.12$$

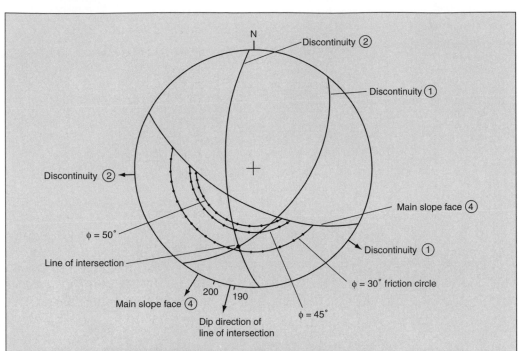

Fig. 10.15 Kinematic feasibility analysis for the wedge and slope described in the worked example

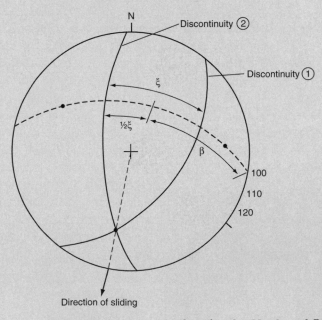

Fig. 10.16 Stereoplot used to provide data for the Hoek and Bray simplified method in the worked example

The actual factor of safety fall between these two values. The average of these two factors of safety is 3.79. As expected from the kinematic feasibility analysis both factors of safety are greater than 1 for the case of the dry slope with no external forces.

The analysis presented above ignores the presence of external forces that often serve to reduce the factor of safety. Thus the factor of safety derived from this analysis may be regarded as greater than the true factor of safety. Nevertheless, the Hoek and Bray simplified method can still be used to check the output of computer analyses of wedge failures by adopting the following reasoning.

(a) Accept that the full three-dimensional analysis of a wedge failure is too complex for hand calculation and so use the Hoek and Bray simplified approach.

(b) Because this method ignores factors such as water pressure, earthquake forces, foundation loadings, etc., the resulting factor of safety will be higher than that obtained by full analysis by computer which takes into account those external forces reducing factor of safety.

(c) If, however, your computer generated factor of safety is *not lower but is higher* than that generated by the Hoek and Bray simplified method then we have a paradox that requires further investigation.

(d) Try entering the input data to the computer program again — there may have been a data entry error. If the paradox remains, verify the computer program by inputting data for the Hoek and Bray simplified version of the real problem, i.e. without external forces, with a horizontal upper face, and with the same angle of friction for both sliding surfaces. Now the computer generated factor of safety and the hand calculated factor of safety should be nearly the same. If they are not, contact the supplier of the computer program!

Note that the form of the Hoek and Bray simplified analysis makes Barton's shear strength equation difficult to use. Accordingly, the Patton shear strength equation should be used with the lowest roughness angle for the two planes under consideration.

Hoek and Bray Improved Method

The inclusion of shear strength parameters for each sliding surface, water pressure and an upper slope face that may be obliquely inclined with respect to the main slope face requires a more complex analysis. We call

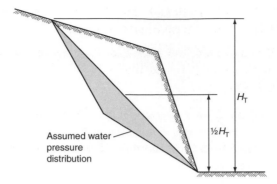

Fig. 10.17 View normal to the line of intersection E showing the total wedge height, H$_T$ and the water pressure distribution

this the 'Hoek and Bray Improved Method'. The water pressure distribution assumed for this analysis is based on the assumption that the sliding surfaces are free draining. The resulting pressure distribution is shown in Fig. 10.17 with the maximum pressure occurring along the line of intersection. Hoek and Bray (1981) suggest that this distribution is representative of extreme conditions resulting from prolonged heavy rainfall.

The geometry of the wedge used in this analysis is shown in Fig. 10.18. The lines of intersection of the various planes in this problem are identified by the letters A, B, C, D and E as follows.

A   Intersection of plane 1 with the slope face (plane 4).
B   Intersection of plane 2 (the steepest of the two sliding surfaces) with the slope face (plane 4).
C   Intersection of plane 1 with upper slope surface (plane 3).

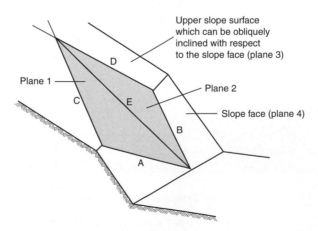

Fig. 10.18 Pictorial view of wedge showing system used to identify intersection lines and planes

D   Intersection of plane 2 with upper slope surface (plane 3).
E   Intersection of planes 1 and 2.

It is assumed that sliding of the wedge always occurs along the line of intersection between plane 1 and plane 2. The factor of safety of this slope is given by:

$$F = \frac{3}{\gamma H_T}(c_1 X + c_2 Y) + \left(A - \frac{\gamma_w}{2\gamma}X\right)\tan\phi_1 + \left(B - \frac{\gamma_w}{2\gamma}Y\right)\tan\phi_2 \quad (10.15)$$

derived from the a more detailed analysis described by Hoek *et al.* (1973) where

c_1 and c_2 are the cohesion values for discontinuities 1 and 2 respectively
ϕ_1 and ϕ_2 are the angles of shearing resistance for discontinuities 1 and 2 respectively
γ is the unit weight of rock
γ_w is the unit weight of water
H_T is the total height of the wedge (see Fig. 10.17)
X, Y, A, B are dimensionless parameters which depend upon the geometry of the wedge:

$$X = \frac{\sin\theta_{BD}}{\sin\theta_{DE}\cos\theta_{Bn1}} \quad (10.16)$$

$$Y = \frac{\sin\theta_{AC}}{\sin\theta_{CE}\cos\theta_{An2}} \quad (10.17)$$

$$A = \frac{\cos\psi_2 - \cos\psi_1\cos\theta_{n1.n2}}{\sin\psi_E\sin^2\theta_{1A.n2}} \quad (10.18)$$

$$B = \frac{\cos\psi_2 - \cos\psi_1\cos\theta_{n1.n2}}{\sin\psi_E\sin^2\theta_{n1.n2}} \quad (10.19)$$

where ψ_1 and ψ_2 are the dips of planes 1 and 2 respectively and ψ_E (ψ_i in the Hoek and Bray simplified method) is the dip of the line of intersection (line E in Fig. 10.18). The remaining angles required in the above equation may be measured from the stereoplot as shown in the worked example below.

WORKED EXAMPLE

Problem

Using the data given in the preceding worked example, determine the factor of safety for the wedge using the Hoek and Bray improved method:

(a)   with $c_1 = c_2 = 0$ and no joint water pressure
(b)   with $c_1 = c_2 = 0$ and joint water pressure
(c)   with $c_1 = c_2 = 20\,\text{kPa}$ and joint water pressure.

Solution

This method requires a number of angular measurements to be made from a stereoplot of the main slope face the upper slope face and the two discontinuities forming the wedge. Unlike the previous method the different shear strength parameters can be entered for each discontinuity and the effect of water pressure can be considered.

It should be remembered that for this method the height term H_T refers to the total height of the wedge (i.e. the vertical distance from the intersection of the line of intersection with the main slope face and the upper slope face as shown in Fig. 10.19). H_T may be determined using:

$$H_T = H\left(1 + \left(\frac{1}{\tan\psi_E} - \frac{1}{\tan\psi_{4E}}\right)\frac{\sin\psi_{3E}\sin\psi_E}{\sin(\psi_E - \psi_{3E})}\right) \tag{10.21}$$

where

H = height of slope crest above the intersection of the line of intersection with the slope face

ψ_E = dip of the line of intersection

ψ_{4E} = apparent dip of main slope face in the direction of the line of intersection

ψ_{3E} = apparent dip of upper slope face in the direction of the line of intersection

The above angles can be measured from the stereoplot shown in Fig. 10.20.

$\psi_E = 33°$

$\psi_{4E} = 64°$

$\psi_{3E} = 9°$

$H = 20\,\text{m}$, given in question.

$$H_T = 20\left(1 + \left(\frac{1}{\tan 33} - \frac{1}{\tan 64}\right)\frac{\sin 9 \sin 33}{\sin(33 - 9)}\right) = 24.4\,\text{m}$$

The following angular measurements were made using the stereoplot shown in Fig. 10.20.

$\theta_{n1.n2} = 113°$

$\theta_{B.n1} = 54°$

$\theta_{A.n2} = 61°$

$\theta_{BD} = 66°$

$\theta_{AC} = 61°$

$\theta_{CE} = 29°$

$\theta_{DE} = 28°$

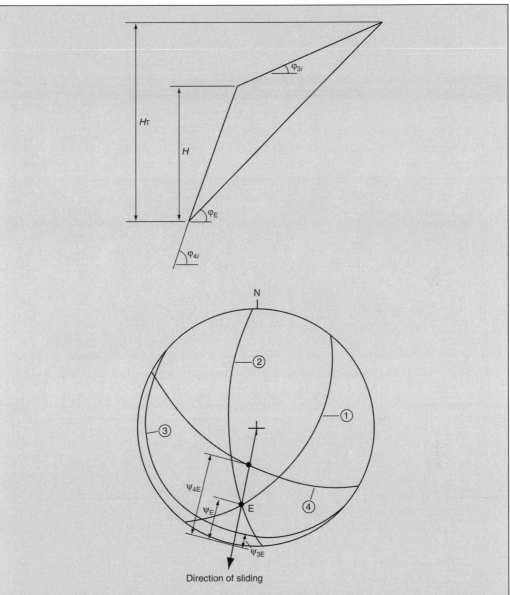

Fig. 10.19 Diagrammatic representation of the parameters required to determine the relationship between the wedge crest height H and the total wedge height H_T

The factor of safety is calculated using equations (10.15) to (10.19).

Case (a)
The unit weight of water γ_w and the cohesion parameters c_1 and c_2 are set to zero (see Table 10.7).

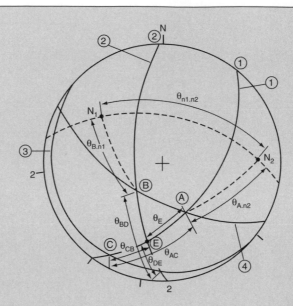

Fig. 10.20 Stereoplot used to provide data for the Hoek and Bray improved method in the worked example

Note that the factor of safety is lower than the average determined using the Hoek and Bray simplified method.

Case (b)

For case (b) the effect of water pressure developed along the sliding surface is considered by setting the unit weight of water γ_w to $9.81\,\text{kN/m}^3$ (see Table 10.8).

Note, we are taking $\gamma_w = 9.81\,\text{kN/m}^3$ and not the simplification used for convenience in hand calculations of $\gamma_w = 10\,\text{kN/m}^3$. In a spreadsheet it is just as easy to be accurate by putting $\gamma_w = 9.81\,\text{kN/m}^3$ as slightly inaccurate by putting $\gamma_w = 10\,\text{kN/m}^3$. The changed input and output are highlighted in Table 10.8. The addition of water pressure in the two discontinuities has resulted in a significant drop in the factor of safety.

Case (c)

For case (c) both water pressure and cohesion are considered. The cohesion parameters c_1 and c_2 have both been set to $20\,\text{kPa}$ (see Table 10.9).

Note that the changed input and output are highlighted in Table 10.9. The addition of cohesion has resulted in an increase in the

Table 10.7 Spreadsheet layout for Hoek and Bray Improved Method

Wedge Stability Calculation Sheet (Hoek and Bray improved method)

Input data		Function value		Parameter value	
ψ_1	55 degrees	$\cos \psi_1$	0.5736	A	1.5325
ψ_2	70 degrees	$\cos \psi_2$	0.3420	B	1.2268
ψ_i	33 degrees	$\sin \psi_i$	0.5446		
$\theta_{n1.n2}$	113 degrees	$\cos \theta_{n1.n2}$	−0.3907		
		$\sin \theta_{n1.n2}$	0.9205		
θ_{BD}	66 degrees	$\sin \theta_{BD}$	0.9135	X	3.3106
θ_{DE}	28 degrees	$\sin \theta_{DE}$	0.4695		
$\theta_{B.n1}$	54 degrees	$\cos \theta_{B.n1}$	0.5878		
θ_{AC}	61 degrees	$\sin \theta_{AC}$	0.8746	Y	3.7211
θ_{CE}	29 degrees	$\sin \theta_{CE}$	0.4848		
$\theta_{A.n2}$	61 degrees	$\cos \theta_{A.n2}$	0.4848		
ϕ_1	45 degrees	$\tan \phi_1$	1.0000		
ϕ_2	50 degrees	$\tan \phi_2$	1.1918		
γ	$25 \,\text{kN/m}^3$	$\gamma_w / 2\gamma$	0.0000		
γ_w	$0 \,\text{kN/m}^3$	$3c_A / \gamma H_T$	0.0000		
c_1	0 kPa	$3c_B \gamma H_T$	0.0000		
c_2	0 kPa				
H_T	24.4 m				

Factor of safety $= 2.99$

factor of safety of about 47% when compared with case (b). The fact that cohesion has this marked effect on the factor of safety means that it should only be included when there is reasonable certainty that it exists for the discontinuities in question.

Summary

The results of the above cases are summarized in the Table 10.10.

Table 10.8 Spreadsheet layout for Hoek and Bray Improved Method with water pressure

Wedge Stability Calculation Sheet (Hoek and Bray improved method)		
Input data	**Function value**	**Parameter value**
ψ_1 — 55 degrees	$\cos\psi_1$ — 0.5736	A — 1.5325
ψ_2 — 70 degrees	$\cos\psi_2$ — 0.3420	B — 1.2268
ψ_i — 33 degrees	$\sin\psi_i$ — 0.5446	
$\theta_{n1.n2}$ — 113 degrees	$\cos\theta_{n1.n2}$ — −0.3907	
	$\sin\theta_{n1.n2}$ — 0.9205	
θ_{BD} — 66 degrees	$\sin\theta_{BD}$ — 0.9135	X — 3.3106
θ_{DE} — 28 degrees	$\sin\theta_{DE}$ — 0.4695	
$\theta_{B.n1}$ — 54 degrees	$\cos\theta_{B.n1}$ — 0.5878	
θ_{AC} — 61 degrees	$\sin\theta_{AC}$ — 0.8746	Y — 3.7211
θ_{CE} — 29 degrees	$\sin\theta_{CE}$ — 0.4848	
$\theta_{A.n2}$ — 61 degrees	$\cos\theta_{A.n2}$ — 0.4848	
ϕ_1 — 45 degrees	$\tan\phi_1$ — 1.0000	
ϕ_2 — 50 degrees	$\tan\phi_2$ — 1.1918	
γ — 25 kN/m^3	$\gamma_w/2\gamma$ — 0.1962	
γ_w — 9.81 kN/m^3	$3c_A/\gamma H_T$ — 0.0000	
c_1 — 0 kPa	$3c_B\gamma H_T$ — 0.0000	
c_2 — 0 kPa		
H_T — 24.4 m		

Factor of safety $= 1.47$

Table 10.9 Spreadsheet layout for Hoek and Bray Improved Method with cohesion and water pressure

Wedge Stability Calculation Sheet (Hoek and Bray improved method)						
Input data		Function value			Parameter value	
ψ_1	55 degrees	$\cos \psi_1$	0.5736		A	1.5325
ψ_2	70 degrees	$\cos \psi_2$	0.3420		B	1.2268
ψ_i	33 degrees	$\sin \psi_i$	0.5446			
$\theta_{n1.n2}$	113 degrees	$\cos \theta_{n1.n2}$	−0.3907			
		$\sin \theta_{n1.n2}$	0.9205			
θ_{BD}	66 degrees	$\sin \theta_{BD}$	0.9135		X	3.3106
θ_{DE}	28 degrees	$\sin \theta_{DE}$	0.4695			
$\theta_{B.n1}$	54 degrees	$\cos \theta_{B.n1}$	0.5878			
θ_{AC}	61 degrees	$\sin \theta_{AC}$	0.8746		Y	3.7211
θ_{CE}	29 degrees	$\sin \theta_{CE}$	0.4848			
$\theta_{A.n2}$	61 degrees	$\cos \theta_{A.n2}$	0.4848			
ϕ_1	45 degrees	$\tan \phi_1$	1.0000			
ϕ_2	50 degrees	$\tan \phi_2$	1.1918			
γ	25 kN/m^3	$\gamma_w/2\gamma$	0.1962			
γ_w	9.81 kN/m^3	$3c_A/\gamma H_T$	0.0984			
c_1	20 kPa	$3c_B\gamma H_T$	0.0984			
c_2	20 kPa					
H_T	24.4 m					

Factor of safety = 2.17

Table 10.10 Summary of factors of safety determined using the Hoek and Bray Improved Method showing the influence of cohesion and water pressure

Case	Description	Factor of safety
(a)	Dry slope $c_1 = c_2 = 0$	2.99
(b)	Wet slope $c_1 = c_2 = 0$	1.47
(c)	Wet slope $c_1 = c_2 = 20$ kPa	2.17

Hoek and Bray Short Method

Hoek and Bray (1981) present the following two methods, which have been developed for maximum speed and efficiency, for solving the wedge problem.

- A short method for a wedge with a horizontal slope crest and with no tension crack. Each plane may have a different friction angle and cohesive strength and the influence of water pressure on each plane is included in the solution. The influence of an external force is not included in this solution.
- A comprehensive method which included the effects of a super-imposed load, a tension crack and an external force such as that applied by a rock bolt/anchor.

The short method can be easily incorporated into a spreadsheet. The comprehensive method is more complex, but more flexible than the short method. In most cases this solution would be programmed on a PC using a suitable programming language such as C++ or Visual Basic. A detailed description of the development of this method is beyond the scope of this book and only the short solution is presented here.

The short method presented is for the computation of the factor of safety for translational slip of a tetrahedral wedge formed in a rock slope by two intersecting discontinuities, the slope face and the upper ground surface. The influence of a tension crack (or some other discontinuity that changes the geometry from a tetrahedron to a trapezoid) is not considered in this method. The method allows for different strength parameters and water pressures on the two planes of weakness. It is assumed that the slope crest is horizontal, i.e. the upper ground surface (plane 3) is either horizontal or dips in the same direction as the slope face (plane 4) or at 180° to this direction. When a pair of discontinuities is selected at random from a set of field data, it is not known whether:

- the planes could form a wedge (the line of intersection E may plunge too steeply to daylight in the slope face or it may be too flat to intersect the upper ground surface)
- one of the planes overlies the other (this affects the calculation of the normal reactions on the planes)
- one of the planes lies to the right or the left of the other plane when viewed from the bottom of the slope.

In order to resolve these uncertainties, the short method has been developed in such a way that either of the planes may be labelled 1 (or 2) and allowance has been made for one plane overlying the other. In addition, a check on whether the two planes do form a wedge is included in the

Table 10.11 Input data for the Hoek and Bray Short Method

Parameter	Description	Dip ψ: degrees	Dip direction α: degrees	Friction angle ϕ: degrees	Cohesion c: kPa	Water pressure u: kPa
Plane 1	Sliding surface	M	M	M	O	O
Plane 2	Sliding surface	M	M	M	O	O
Plane 3	Upper slope	O	O	N/A	N/A	N/A
Plane 4	Main slope face	M	M	N/A	N/A	N/A
H	Height of the crest of the slope above the intersection (see Fig. 10.19) in m [M]					
γ	Unit weight of rock in kN/m^3 [M]					
η	Overhang index. If slope face overhangs the toe, $\eta = -1$, if the slope does not overhang, $\eta = +1$ [M]					

M or [M] = mandatory data, O = optional data, N/A = not applicable.
If it is assumed that the discontinuities are completely filled with water and that the water pressure varies from zero at the free faces to a maximum at some point on the line of intersection, then $u_1 = u_2 = \gamma_w H_T/6$ where H_T is the overall height of the wedge.

solution at an early stage. Depending upon the geometry of the wedge and the magnitude of the water pressure acting on each plane, contact may be lost on either plane and this contingency is provided for in the solution.

The geometry of the problem is illustrated in Fig. 10.10. Remember the various planes are denoted in the following manner:

1   discontinuity forming one of the sliding surfaces
2   discontinuity forming the other sliding surface
3   upper ground surface
4   the main slope face.

The input data required for the solution of the problem are given in Table 10.11.

Other terms used in the solution are:

F = factor of safety against wedge sliding calculated as the ratio of the resisting to the mobilized shear forces
A = area of a face of the wedge (e.g. A_1 = area of plane 1)
W = weight of the wedge
N = effective normal reaction on a plane (e.g. N_1 = effective normal reaction on plane 1)

323

S = shear force mobilized on a plane (e.g. S_1 = shear force mobilized on plane 1)

x, y, z = co-ordinate axes with origin at 0. The z axis is directed vertically upwards, the y axis is in the dip direction of plane 2

A = unit vector in the direction of the normal to plane 1 with components (a_x, a_y, a_z)

B = unit vector in the direction of the normal to plane 2 with components (b_x, b_y, b_z)

F = unit vector in the direction of the normal to plane 4 with components (f_x, f_y, f_z)

G = vector in the direction of the line of intersection of planes 1 and 4 with components (g_x, g_y, g_z)

I = vector in the direction of the line of intersection of planes 1 and 2 with components (i_x, i_y, i_z)

$i = -i_z$

q = component of **G** in the direction of **B**

r = component of **A** in the direction of **B**

$$k = |\mathbf{I}|^2 = i_x^2 + i_y^2 + i_z^2 \tag{10.21}$$

$$z = W/A_2 \tag{10.22}$$

$$p = A_1/A_2 \tag{10.23}$$

Assuming contact on both planes:

$$n_1 = N_1/A_2 \tag{10.24}$$

$$n_2 = N_2/A_2 \tag{10.25}$$

$$|zi|/\sqrt{k} = SA_2 \tag{10.26}$$

Assuming contact on plane 1 only:

$$m_1 = N_1/A_1 \tag{10.27}$$

and the denominator of F is the mobilized shear stress S_1/A_1.

Assuming contact on plane 2 only:

$$m_2 = N_2/A_2 \tag{10.28}$$

and the denominator of F is the mobilized shear stress S_2/A_2.

The sequence of calculations is as follows.

The factor of safety of a tetrahedral wedge against sliding along a line of intersection may be calculated as

$$(a_x, a_y, a_z) = \{\sin \psi_1 \sin(\alpha_1 - \alpha_2), \sin \psi_1 \cos(\alpha_1 - \alpha_2), \cos \psi_1\} \tag{10.29}$$

$$(f_x, f_y, f_z) = \{\sin \psi_4 \sin(\alpha_4 - \alpha_2), \sin \psi_4 \cos(\alpha_4 - \alpha_2), \cos \psi_4\} \tag{10.30}$$

$$b_y = \sin \psi_2 \tag{10.31}$$

$$b_z = \cos \psi_2 \tag{10.32}$$

$$i = a_x b_y \tag{10.33}$$

$$g_z = f_x a_y - f_y a_x \tag{10.34}$$

$$q = b_y (f_z a_x - f_x a_z) + b_z g_z \tag{10.35}$$

If $nq/I > 0$, or if $\eta(f_z - q/i)\tan\psi_3 > \sqrt{(1 - f_z^2)}$ and $\sigma_3 = \alpha_4 \pm (1 - \eta)\pi/2$, no wedge is formed and the calculations should be terminated.

$$r = a_y b_y + a_z b_z \tag{10.36}$$

$$k = 1 - r^2 \tag{10.37}$$

$$z = (\gamma H q)/(3g_z) \tag{10.38}$$

$$p = -b_y f_x/g_z \tag{10.39}$$

$$n_1 = \{(z/k)(a_z - rb_z) - pu_1\}p/|p| \tag{10.40}$$

$$n_2 = \{(z/k)(b_z - ra_z) - u_2\} \tag{10.41}$$

$$m_1 = (za_z - ru_2 - pu_1)p/|p| \tag{10.42}$$

$$m_2 = (zb_z - rpu_1 - u_2) \tag{10.43}$$

If $n_1 > 0$ AND $n_2 > 0$, there is contact on both planes, so

$$\text{Factor of safety } F = \frac{(n_1 \tan\phi_1 + n_2 \tan\phi_2 + |p|c_1 + c_2)\sqrt{k}}{|zi|} \tag{10.44}$$

If $n_2 < 0$ AND $m_1 > 0$, there is contact on plane 1 only, so

$$\text{Factor of safety } F = \frac{m_1 \tan\phi_1 + |p|c_1}{[z^2(1 - a_z^2) + ku_2^2 + 2(ra_z - b_z)zu_2]^{1/2}} \tag{10.45}$$

If $n_1 < 0$ AND $m_2 > 0$, there is contact on plane 2 only, so

$$\text{Factor of safety } F = \frac{m_2 \tan\phi_2 + c_2}{[z^2 b_y^2 + kp^2 u_1^2 + 2(rb_z - a_z)pzu_1]^{1/2}} \tag{10.46}$$

If $m_1 < 0$ AND $m_2 < 0$, contact is lost on both planes and the wedge floats as a result of water pressure acting on planes 1 and 2. In this case the factor of safety fall to zero.

WORKED EXAMPLE

Problem

Using the data given in the first worked example in wedge failure analysis, determine the factor of safety for the wedge using the Hoek and Bray short method:

(a) with $c_1 = c_2 = 0$ and no joint water pressure
(b) with $c_1 = c_2 = 0$ and joint water pressure
(c) with $c_1 = c_2 = 20\,\text{kPa}$ and joint water pressure.

faces to a maximum at some point on the line of intersection. The water pressure on the discontinuities is given by:

$$u_1 = u_2 = \frac{\gamma_w H_T}{6} \qquad (10.47)$$

where H_T is the overall height of the wedge. In the Hoek and Bray spreadsheet rapid method the upper slope is assumed to be horizontal

Table 10.14 Spreadsheet layout for Hoek and Bray Short Method with cohesion and water pressure

		Input data				
Plane	Description	Dip ψ: degrees	Dip Dir. α: degrees	Friction angle ϕ: degrees	Cohesion: kPa	Water pressure: kPa
1	Sliding surface	55	128	45	20	32.7
2	Sliding surface	70	267	50	20	32.7
3	Upper slope	10	218	Assumed horizontal		
4	Main slope	65	208			

Unit weight of rock	γ		$25\,\text{kN/m}^3$
Height of slope crest above intersection	H		$20\,\text{m}$
Overhang index	n		1

Calculations				Tests and output	
a_x	−0.5374	q	0.4554	−0.90 wedge formed	
a_y	−0.6182	r	−0.3848	0.23	
a_z	0.5736	k	0.8520	0.91 wedge formed	
f_x	−0.7769	z	103.8015	Factor of safety	
f_y	0.4668	p	0.9985	Contact on both planes	2.39
f_z	0.4226	n_1	53.2676		
b_y	0.9397	n_2	35.8602		
b_z	0.3420	m_1	39.4699		
I	−0.5050	m_2	15.3647		
g_z	0.7311				

$H_T = H = 20\,\text{m}$ (see Table 10.13) and hence:

$$u_1 = u_2 = \frac{9.81 \times 20}{6} = 32.7\,\text{kPa}$$

Note that the changed input and output are highlighted in Table 10.13. As in the preceding worked example, we see the water pressure reduces the factor of safety significantly.

Case (c)

In case (c) the influence of cohesion is considered (see Table 10.14).

Note that the changed input and output are highlighted in Table 10.14. As with the analysis in the preceding worked example in which cohesion was considered, the factor of safety has increased by about 40%.

Summary

The results of the above cases are summarized in Table 10.15.

Table 10.15 Summary of factors of safety determined using the Hoek and Bray Short Method showing the influence of cohesion and water pressure

Case	Description	Factor of safety
(a)	Dry slope $c_1 = c_2 = 0$	2.95
(b)	Wet slope $c_1 = c_2 = 0$	1.69
(c)	Wet slope $c_1 = c_2 = 20\,\text{kPa}$	2.39

Hoek and Bray Comprehensive Method

Overview

This method allows for different strength parameters and water pressures on the two sliding surfaces and for water pressure in the tension crack. There is no restriction on the inclination of the upper slope and the influence of external forces such as pseudo-static earthquake loading, foundation loading and rock anchor cable tension are included. This analysis is described in full in Appendix 2 of Hoek and Bray (1981). A number of computer programs that are commercially available are based on this approach.

The method is not described here because it is of such complexity that it can only be managed in a computer program or spreadsheet. Nevertheless, the following shows the output of this computer analysis based on the same problem addressed in the preceding worked examples. This is so that the Hoek and Bray comprehensive method can be compared with the very much simpler (and therefore less accurate) hand calculations and spreadsheet-aided hand calculations of the Hoek and Bray simplified, improved and short methods described above.

We have developed a spreadsheet for the Hoek and Bray comprehensive method called the 'Surrey-wedge'. This is available for download for educational purposes only from the University of Surrey web site (http://www.surrey.ac.uk/CivEng/research/geotech/index.htm). This spreadsheet, 'Surrey-wedge', has been used in the following worked example.

WORKED EXAMPLE

Problem

Using the data given in the first worked example in wedge failure analysis, determine the factor of safety for the wedge using the Hoek and Bray comprehensive method:

(a)   with $c_1 = c_2 = 0$ and no joint water pressure with and without the tension crack

(b)   with $c_1 = c_2 = 0$ and joint water pressure with and without the tension crack

(c)   with $c_1 = c_2 = 20\,kPa$ and joint water pressure with and without the tension crack

(d)   with $c_1 = c_2 = 0$, joint water pressure, tension crack and a vertical foundation loading applied to the upper slope of $1000\,kN$

(e)   with $c_1 = c_2 = 0$, joint water pressure, tension crack, a vertical foundation loading applied to the upper slope of $1000\,kN$ and dynamic loading from an earthquake event such that $a = 0.2g$

(f)   repeat (d) and (e) with $c_1 = c_2 = 20\,kPa$.

Solution

The comprehensive analysis does not require any measurements to be made using the stereoplot. The input data are shown in Table 10.16. The dry slope case requires zero to be entered for the unit weight of water. Zero is also entered for the cohesion of the two sliding surfaces and for the external forces T and E.

The comprehensive analysis permits the tension crack to be considered. The following analyses have been done with and without the tension crack in order to provide a comparison with the previous analyses. The results of the comprehensive method are shown in Table 10.17.

The variation in the factor of safety with the addition of water pressure and cohesion is similar to that seen for the previous analyses. The tension crack breaks the sliding wedge into a smaller unit and introduces an extra force associated with water filling the tension

---

Table 10.16 Spreadsheet layout for Hoek and Bray Comprehensive Method, dry slope

Input data

Plane	Description	Dip ψ: degrees	Dip Dir. α: degrees	Friction angle ϕ: degrees	Cohesion: kPa
1	Sliding surface	55	128	45	0
2	Sliding surface	70	267	50	0
3	Upper slope	10	218		
4	Main slope	65	208		
5	Tension Crack	70	165		

				Force: kN	
External force T		0	0	0	
External force E		0	0	0	

Unit weight of rock	γ		$25\,\mathrm{kN/m}^3$
Unit weight of water			$0\ \mathrm{kN/m}^3$
Height of slope crest above intersection	H		$20\,\mathrm{m}$
Distance of tension crack from crest, measured along the trace of plane 1	L		$5\,\mathrm{m}$
Overhang index	n		$1 - 1 =$ overhanging slope

crack. This extra force is largely responsible for the reduction in the factor of safety in cases (b) and (c).

In cases (d), (e) and (f) external forces from foundation loading and an earthquake. The foundation loading is vertical and hence the dip is entered as 90° in the input data table. The dynamic loading from the earthquake is considered by the inclusion of a pseudo-static force of

Table 10.17 Summary of factors of safety determined using the Hoek and Bray Comprehensive Method showing the influence of cohesion and water pressure

Section	(a)	(b)	(c)
Water pressure considered	No	Yes	Yes
Cohesion considered	No	No	Yes ($c_1 = c_2 = 20$ kPa)
F (without tension crack)	2.95	1.96	2.65
F (with tension crack)	2.95	1.53	2.06

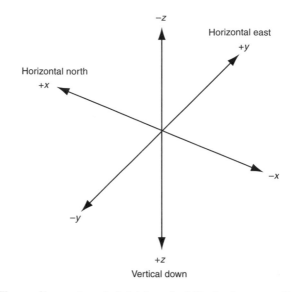

Fig. 10.21 Three-dimensional right-handed Cartesian coordinate system, viewed from above

ψ_A and dip direction α_A (i.e. plunge and trend) its Cartesian components can be determined from

$$A_x = |\mathbf{A}| \cos \alpha_A \cos \psi_A$$

$$A_y = |\mathbf{A}| \sin \alpha_A \cos \psi_A$$

$$A_z = |\mathbf{A}| \sin \psi_A \qquad (10.48)$$

The worked example below shows how the Cartesian components of a number of forces are determined using equations (10.48).

WORKED EXAMPLE

Problem

Find the resultant of the forces shown in Table 10.20.

Solution

Forces A and C have a downward sense and hence the dips are shown as positive numbers. Force B has an upward sense which is the dip is shown as a negative number. The Cartesian components of these forces have been calculated using equations (10.48) and the results are shown in Table 10.21.

The Cartesian components of the resultant force R are found by summing the x, y and z components of forces A, B and C. The magnitude of

Table 10.20

Force	Dip ψ: degrees	Dip direction α: degrees	Magnitude: kN
A	25	170	571
B	−50	060	250
C	90	—	1000

Table 10.21 Calculation of resultant forces (all units kN)

Force	X	Y	Z
A	−509.6	89.9	241.3
B	80.4	139.2	−191.5
C	0	0	1000.0
Resultant R	−429.3 (R_x)	229.1 (R_y)	1049.8 (R_z)

the resultant force R is given by

$$|\mathbf{R}| = \sqrt{R_x^2 + R_y^2 + R_z^2} \tag{10.49}$$

Hence the magnitude of the resultant of forces A, B and C in the above example is 1157.1 kN.

The dip ψ_R and dip direction α_R of the resultant force with Cartesian components R_x, R_y and R_z are given by

$$\psi_R = \arctan\left(\frac{R_z}{\sqrt{R_x^2 + R_y^2}}\right) \tag{10.50}$$

$$\alpha_R = \arctan\left(\frac{R_y}{R_x}\right) + Q \tag{10.51}$$

The dip of the resultant force in the above example is 65.1°. The arctan function used in the determination of dip direction will only return angles between ±90°. Equation (10.51) includes a parameter Q which ensures that α_R lies in the correct quadrant and in the range 0 to 360°. This parameter depends upon the signs of R_x and R_y as listed in Table 10.22.

In the above example $R_x < 0$ and $R_y > 0$, therefore $Q = 180°$. The dip direction of R is 151.9°.

Table 10.22 Values of the parameter Q

R_x	R_y	Q
≥ 0	≥ 0	0
< 0	≥ 0	180°
< 0	< 0	180°
≥ 0	< 0	360°

Circular failure

Circular failure is most likely to occur in rock masses that display random discontinuity orientations. Statistical analysis of discontinuity data as described in Chapter 7 will enable the state of random orientations to be identified. The critical failure surface and the associated factor of safety may be calculated using the methods described in Part 1 of this book.

Circular failure charts

Hoek and Bray (1981) produced a series of charts to provide a rapid determination of the factor of safety for a circular failure in randomly jointed rock or waste dumps for a variety of slope geometries and ground water conditions. The charts are numbered 1 to 5 to correspond with the ground water conditions shown in Appendix 2 (Fig. A2.1). The charts are provided in Appendix 2 (Figs. A2.2 to A2.6).

Use of the circular failure charts

The procedure for the determination of a factor of safety using the circular failure charts is outlined in Fig. 10.22 and Fig. 10.23.

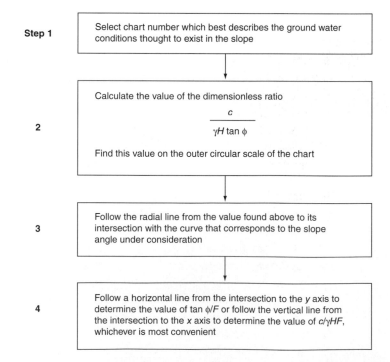

Fig. 10.22 Determination of factor of safety using circular failure charts

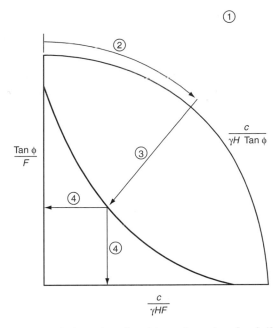

Fig. 10.23 Sequence of steps involved in using circular failure charts to find the factor of safety of a slope

WORKED EXAMPLE

A 25 m high slope with a face angle of 50° is to be excavated in a highly fractured sandstone with a unit weight of 25 kN/m³. Analysis of discontinuity orientations in this rock mass shows no evidence of any discontinuity sets. The shear strength parameters for the rock mass have been estimated as $c = 60$ kPa, $\phi = 35°$. The ground water table was found at a depth of 18 m.

Solution

The ground-water conditions indicate the use of chart No. 2 (Fig. A2.3 in Appendix 2). The value of $c/\gamma H \tan \phi = 60/(25 \times 25 \times \tan 35°) = 0.14$. The corresponding value of $\tan \phi/F$, for a 50° slope, is 0.56. Hence, the factor of safety of the slope is 2.13.

Because of the speed and simplicity of using these charts, they are ideal for checking the sensitivity of the factor of safety of a slope to a wide range of conditions and the authors suggest that this should be their main use. These charts are useful for the construction of drawings of potential slides and also for estimating the friction angle when back-analysing existing circular slides. They also provide a start for a more sophisticated circular failure analysis in which the location of the circular failure surface having the lowest factor of safety is found by iterative methods.

Toppling failure

Toppling failure (unlike those modes discussed above) involves rotation of columns or blocks of rock about some fixed base. The basic geometrical conditions governing the toppling and sliding of a single block on an inclined surface are shown in Fig. 10.24. For toppling to occur the vector representing the weight W of the block must fall outside the base b. This is only likely when the block is tall and slender ($h > b$). In the field the toppling mechanism is often more complicated resulting from the geometrical characteristics of the discontinuities and the slope face. Goodman and Bray (1976) described a number of different types of toppling which have been observed in the field. Some of these are outlined in Table 8.1. The principal types of toppling failure are as follows:

- *Flexural toppling (Fig. 10.25)*. This occurs in rock masses which display continuous columns (or sheets) of rock, which are separated by steeply dipping discontinuities. The columns break in flexure as they bend forward towards the slope face. It is most commonly found in rock masses with a high degree of structural anisotropy such as shale and slate. Sliding, undermining or erosion of the toe of the slope allows the toppling process to start. It retrogresses backwards into the slope, producing deep tension cracks which occur in a zone that can extend up to ten times the slope height back from the slope crest. The lower part of the is often covered with disorientated and disordered blocks which can make the toppling mechanism difficult to identify.
- *Block toppling (Fig. 10.26)*. This type of toppling failure is common in masses which display columnar or tabular shaped blocks of rock resulting from relatively widely spaced discontinuities dipping out of the face at a relatively low angle combined with two sets of more closely spaced discontinuities which combine to give lines of intersection that dip steeply into the face. The columns at the base of the slope are too short to topple. These columns, however, undergo sliding because the toppling blocks in the upper part of the slope are pushing them forward. This sliding action promotes further toppling higher up in the slope and thus the failure is both complex (mixed modes – sliding and toppling) and progressive. The base of the failure is

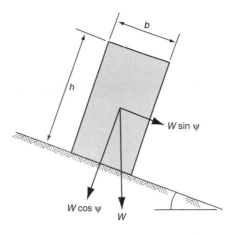

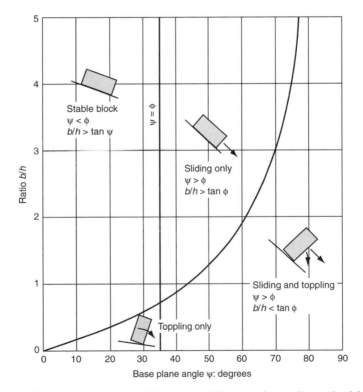

Fig. 10.24 Geometry and conditions for sliding and toppling of a block on an inclined plane

Fig. 10.25 Diagrammatic representation of flexural toppling

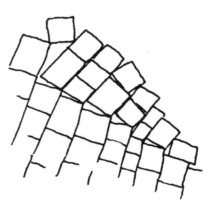

Fig. 10.26 Diagrammatic representation of block toppling

better defined than in a flexural topple and has the appearance of a stairway rising from one cross-joint to the next.

- *Block-flexure toppling (Fig. 10.27)*. This type of toppling is developed in rock masses which display mixed spacing of discontinuities dipping into the face. This is usually associated with interbedded rocks such as sandstone and shale.

Toppling failure may be induced in slopes by external means such as sliding or other factors not associated with the pattern of discontinuities. Some of these secondary toppling modes are illustrated in Fig. 10.28.

Analysis of Toppling Failure

Unlike the other modes of failure discussed early in this chapter the limiting equilibrium analysis of toppling cannot be expressed in terms of a

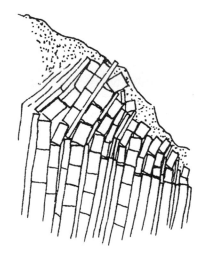

Fig. 10.27 Diagrammatic representation of block flexure toppling

factor of safety. A simple limit equilibrium analysis suitable for block toppling involving a regular system of blocks is presented by Goodman and Bray (1976). This should not be regarded as a design tool but serves to provide a check for a few special cases.

The analysis considers the equilibrium of the forces of each block starting from the uppermost block and determining the interaction force with the adjacent block down to the slope toe. When the lower force obtained on the toe block is positive (downward directed) the slope is unstable, when it is zero, the limiting equilibrium condition is reached.

By comparing the inter-block force just sufficient to prevent toppling with that just sufficient to prevent sliding, the slope may be divided into the following groups:

- a set of sliding blocks in the toe region
- a set of stable blocks at the top
- an intermediate set of toppling blocks.

This approach also allows the cable forces required to stabilize the slope to be estimated. The analysis is relatively simple and can be carried out using a calculator or spreadsheet. More accurate and comprehensive analyses consider the equilibrium of every block in the face. These make use of numerical methods such as finite elements or boundary elements and require a powerful computer.

The analysis is based on an idealized model in which a slope angle θ is excavated in a rock mass with layers dipping at $90 - \alpha$ (α is the dip of the basal plane) towards the face (see Fig. 10.29). The base of the moving

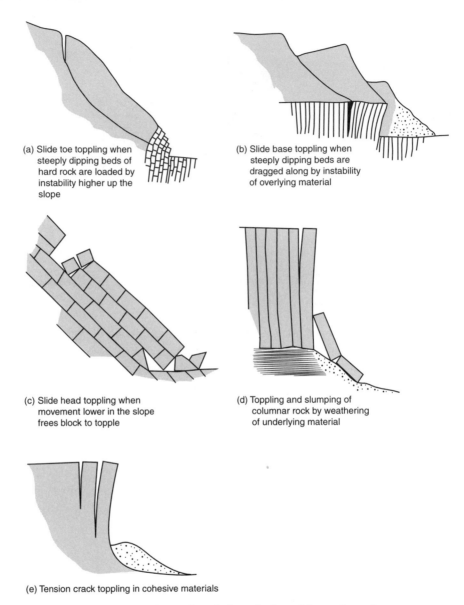

(a) Slide toe toppling when steeply dipping beds of hard rock are loaded by instability higher up the slope

(b) Slide base toppling when steeply dipping beds are dragged along by instability of overlying material

(c) Slide head toppling when movement lower in the slope frees block to topple

(d) Toppling and slumping of columnar rock by weathering of underlying material

(e) Tension crack toppling in cohesive materials

Fig. 10.28 Examples of toppling failure induced by external means

blocks is stepped upwards with an overall inclination β. The upper slope has inclination θ_u and the width of each block is Δx.

Fig. 10.30(a) shows a typical block (n) with the forces developed on the base (R_n, S_n) and on the interfaces with adjacent blocks (P_n, Q_n, P_{n-1}, Q_{n-1}). When the block is one of the toppling set, the points of application of all the forces are known, as shown in Fig. 10.30(b).

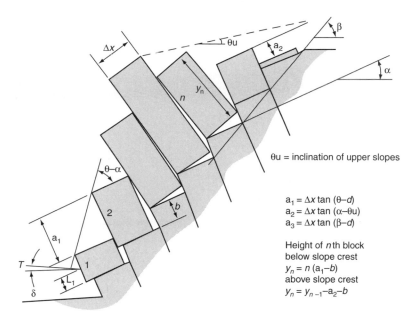

θu = inclination of upper slopes

$a_1 = \Delta x \tan(\theta - d)$
$a_2 = \Delta x \tan(\alpha - \theta u)$
$a_3 = \Delta x \tan(\beta - d)$

Height of nth block
below slope crest
$y_n = n(a_1 - b)$
above slope crest
$y_n = y_{n-1} - a_2 - b$

Fig. 10.29 *Model for limiting equilibrium analysis of toppling on a stepped base, after Hoek and Bray (1981)*

If the nth block is below the slope crest:

$$M_n = y_n \tag{10.52}$$

$$L_n = y_n - a_1 \tag{10.53}$$

If the nth block is the crest block:

$$M_n = y_n - a_2 \tag{10.54}$$

$$L_n = y_n - a_1 \tag{10.55}$$

If the nth block is above the slope crest:

$$M_n = y_n - a_2 \tag{10.56}$$

$$L_n = y_n \tag{10.57}$$

In all cases $K_n = 0$.

For an irregular array of blocks, y_n, L_n and M_n can be determined graphically.

Assuming limiting equilibrium for the sides of the block:

$$Q_n = P_n \tan \phi \tag{10.58}$$

$$Q_{n-1} = P_{n-1} \tan \phi \tag{10.59}$$

where ϕ is the angle of friction for the discontinuities dipping into the slope and this is assumed to be equal to the angle of friction for the basal discontinuities.

343

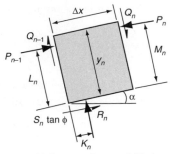

(a) Forces acting on nth block

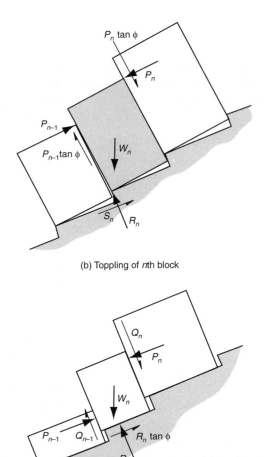

(b) Toppling of nth block

(c) Sliding of nth block

Fig. 10.30 Limiting equilibrium conditions for toppling and for sliding of the nth block

By resolving perpendicular and parallel to the base,

$$R_n = W_n \cos \alpha + (P_n - P_{n-1}) \tan \phi \qquad (10.60)$$

$$S_n = W_n \sin \alpha + (P_n - P_{n-1}) \qquad (10.61)$$

Considering rotational equilibrium, it is found that the force P_{n-1} which is just sufficient to prevent toppling has the value:

$$P_{n-1,t} = \frac{P_n(M_n - \Delta x \tan \phi) + (W_n/2)(y_n \sin \alpha - \Delta x \cos \alpha)}{L_n} \qquad (10.62)$$

when the block under consideration is one of the sliding set (see Fig. 10.30(c)).

$$S_n = R_n \tan \phi \qquad (10.63)$$

In this case the magnitudes and points of application of all the forces applied to the sides and base of the block are unknown. The procedure suggested by Hoek and Bray (1981) is to assume that conditions of limiting equilibrium are established on the side faces so that

$$R_n = W_n \cos \alpha + (P_n - P_{n-1}) \tan \phi \qquad (10.64)$$

$$S_n = W_n \sin \alpha + (P_n - P_{n-1}) \qquad (10.65)$$

Thus the force P_{n-1}, which is just sufficient to prevent sliding, has the value:

$$P_{n-1,s} = P_n - \frac{W_n(\tan \phi \cos \alpha - \sin \alpha)}{1 - \tan^2 \phi} \qquad (10.66)$$

The assumption introduced here is arbitrary, but has no effect on the calculations of the overall stability of the slope.

Numerical methods

Numerical methods can be applied in rock slope analysis to model the various mechanisms of primary and secondary toppling failures. The limit equilibrium method is based on a restrictive hypothesis of rock slope behaviour that can be overcome using appropriate numerical methods. The limit equilibrium method considers only potentially unstable masses and failure surfaces and solves only the equilibrium of forces. Joint stiffness, rock block elasticity, crack propagation and progressive slope failure are not taken into account. Furthermore, the limit equilibrium method does not consider the influence of the stress state evolution on the equilibrium conditions of potentially unstable rock masses.

The numerical methods that are currently used to model toppling and other modes of failure are: the finite element method (FEM), the distinct element method (DEM) and the boundary element method (BEM).

FEM application in slope analysis allows one to model:

- the history of stress state (it is possible to carry out the simulation of constructions and excavations)
- the different deformation behaviour of rock and soil formations
- the constitutive laws of the geological media (it is possible to consider the elasto-plasticity or the creeping of the materials)
- the principal faults and the joints.

The principal limitations of FEM application to slope stability problems carried out up to now are:

- only small deformation fields have been considered
- the discontinuities are considered as completely persistent and crack propagation phenomena are not simulated
- the slope stability problems are examined in static fields and only the incipience of toppling movement or shearing movement have been assessed by examining tension and shear zones in the modelled area.

The FEM has been used for investigating the toppling failure by several authors. Kalkani and Piteau (1976) examined the stress states induced by the water table uplift in an elastic field using a 2D FEM model. FEM analyses were carried out to investigate the toppling failure at Hell's Gate Bluffs in Canada. The toppling instability phenomena were predicted on the basis of the extention of the tensile stress zones.

Brown *et al.* (1980) used FEM to study the Nevis Bluff rock slope failure in New Zealand. The study involved the back-analysis of a failed slope and joint elements were introduced into the FEM model for flexural toppling analysis.

Evans *et al.* (1981) used the FEM to model rock slope mechanical behaviour in the Grose Valley in New South Wales. FEM analyses were carried out to investigate the mechanism of secondary toppling failure in this rock slope. The FEM model set up for the analysis included joint elements, whilst geological phenomena such as rock weathering and rock creep behaviour were taken into account.

The DEM application in slope analysis allows one to model:

- the rock mass as a system of blocks
- the different phases of the slope formation
- the rock block movements
- the interactive behaviour of a discontinuous medium with a continuous medium when the DEM method is coupled with the BEM or with the finite difference method.

The principal limitations of DEM application to slope stability problems carried out up to now are:

- the rock slopes are modelled as a system of blocks. The block contacts are perfectly persistent
- cracking propagation phenomena in the rock are not predicted.

The DEM has been used to investigate the toppling failure by several authors, e.g. Cundall *et al.* (1975), Hocking (1978), Voegele (1978), Ishida *et al.* (1987).

Rockfalls

Overview

Rock blocks which become detached from a face and roll or bounce down it are a major hazard to highways and railways. Although such rockfalls generally do not pose the same degree of economic risk as some of the failure mechanisms discussed earlier they do cause major disruption to transport routes and can cause the same number of fatalities as other forms of slope failure.

Rockfalls are generally initiated by some event that causes a change in the forces acting on the rock. These may be associated with climatic or biological events or with construction activity. The types of events that bring about rockfalls include:

- joint water pressure increases due to heavy rainfall
- freeze–thaw processes in temperate and cold climates
- chemical decomposition of the rock in humid tropical climates
- root growth or leverage by roots moving in high winds
- vibrations due to earthquakes or construction activities

Construction activities are likely to increase the potential for rockfall by one or two orders of magnitude in comparison to climatic and biological events (Hoek, 1998).

The most important factors controlling the trajectory of a falling block of rock is the slope geometry and type of slope surface. Slopes that act as 'ski jumps' are the most dangerous as a high horizontal velocity is imparted to the falling block causing it to bounce a long way out from the toe of the slope. Clean faces of hard rock are also dangerous because they have a high coefficient of restitution and hence do not retard the movement of the falling block to any significant degree. In contrast, talus material, screes or gravels have a low coefficient of restitution and thus absorb a considerable amount of the energy of the falling block and in many cases stop it completely. Other factors such as block size and shape, frictional characteristics of the rock surfaces and whether or not the block breaks on impact are all of lesser importance.

Since slope geometry and the coefficient of restitution can be considered controlling parameters for the analysis of rockfall, relatively crude

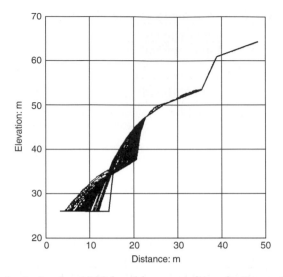

Fig. 10.31 Trajectories for 1000 boulders weighing between 200 and 20 000 kg, after Hoek (1998)

simulation models (such as that developed by Hoek (1986)) are capable of producing reasonably accurate results. More refined models such as those developed by Bozzolo *et al.* (1988), Hungr and Evans (1988), Spang and Rautenstrauch (1988) and Azzoni *et al.* (1995), produce more accurate results provided representative input data are available.

Most rockfall models make use of the Monte Carlo simulation technique to vary the parameters included in the analysis. This technique is similar to the random process of throwing a die. Thus each parameter such as slope angle and coefficient of restitution can be varied in a random manner within preset limits. The results of such an analysis are shown in Fig. 10.31. The purpose of the analysis shown in Fig. 10.31 was to determine the spread of first impacts at the base of a proposed rock cutting so that an effective catch ditch and barrier fence could be designed.

Rockfall Hazard Rating System

In order to identify slopes which are particularly hazardous and which require remedial work or further investigation simple classification schemes have been developed, particularly in the USA and Canada where there are hundreds of kilometres of highways and railways in mountainous terrain. These schemes are based on visual inspection and simple calculations so that surveys can be carried out rapidly. One of the most widely accepted is the Rockfall Hazard Rating Scheme (RHRS) developed by the Oregon State Highway Division (Pierson *et al.*, 1990). Table 10.23 gives a summary of the scores for different categories included in the classification scheme.

Table 10.23 Rockfall Hazard Rating System (RHRS), after Pierson et al. (1990)

Category	Rating criteria and score			
	Points 3	Points 9	Points 27	Points 81
Slope height	7.6 m	15.2 m	22.9 m	30.5 m
Ditch effectiveness	Good catchment: all or nearly all of falling rocks are retained in the catch ditch	Moderate catchment: falling blocks occasionally reach the roadway	Limited catchment: Falling rocks frequently reach the roadway	No catchment: no ditch or ditch totally ineffective
Average vehicle risk	25%	50%	75%	100%
% of decision sight distance	Adequate site distance, 100% of low design value	Moderate sight distance, 80% of low design value	Limited site distance, 60% of low design value	Very limited sight distance, 40% of low design value
Roadway width including paved shoulders	13.4 m	11.0 m	8.5 m	6.1 m
Geological character:	CASE 1: for slopes where discontinuities are the dominant structural feature			
Structural condition	Discontinuous joints, favourable orientation	Discontinuous joints, random orientation	Discontinuous joints, adverse orientation	Continuous joints, adverse orientation
Rock friction	Rough, irregular	Undulating	Planar	Clay infilling, slickensided or low friction mineral coating
Geological character	CASE 2: for slopes where differential erosion or oversteepened slopes is the dominant condition that controls rockfall. Common slopes that are susceptible to this condition are: layered units containing easily weathered rock that erodes undermining more durable rock			
Structural condition	Few differential erosion features	Occasional erosion features	Many erosion features	Major erosion features
Difference in erosion rates	Small difference; erosion features develop over many years	Moderate difference; erosion features develop over a few years	Many erosion features; erosion features develop annually	Major erosion features; erosion features develop rapidly
Block size	300 mm	610 mm	914 mm	1219 mm
Quantity of rockfall/event	0.8 m^3	1.5 m^3	2.3 m^3	3.1 m^3
Climate and presence of water on slope	Low to moderate precipitation; no freezing, no water on slope	Moderate precipitation or short freezing periods or intermittent water on slope	High precipitation or long freezing periods or continual water on slope	High precipitation and long freezing periods or continual water on slope and long freezing periods
Rockfall history	Few falls; rockfall only occurs a few times a year or less	Occasional falls; rockfall can be expected several times a year	Many falls; frequent rockfalls during a certain season, e.g. winter freeze–thaw	Constant rockfalls; rockfalls occur frequently throughout the year

Table 10.24 Decision site distances for different posted speed limits

Posted Speed Limit: km/hr	DSD: m
48	137
64	183
80	229
97	305
113	335

In Table 10.23, the average vehicle risk (AVR) refers to the percentage of time that a vehicle will be present in the rockfall hazard zone. The AVR is determined using the following expression:

$$AVR = \frac{\text{Average Daily Traffic (cars/hr)} \times \text{Slope Length (km)}}{\text{Posted Speed Limit (km/h)}} \times 100\%$$

(10.67)

Per cent decision sight distance (DSD) refers to the length of roadway a driver must have to make a complex or instantaneous decision. The DSD is critical when obstacles on the road are difficult to perceive, or when unexpected or unusual manoeuvres are required. Throughout the rockfall section the sight distance can change significantly. Horizontal and vertical highway curves along with obstructions such as rock outcrops can severely limit a driver's ability to notice a rockfall. The percentage DSD is determined from the expression:

$$\%DSD = \frac{\text{Measured Site Distance (m)}}{\text{Decision Site Distance (m)}} \times 100\%$$

(10.68)

The DSD is determined using Table 10.24.

Table 10.25

Parameter	Value	Rating
Slope height	23 m	27
Ditch effectiveness	Moderate	9
AVR	30	7 by interpolation
%DSD	Very limited	81
Road width	8.5 m	27
Geological structure	Random	9 (CASE 1)
Rock friction	Rough	3 (CASE 1)
Block size	914 mm	27
Climate and water	High precip. & freezing	27
Rockfall history	Many falls	27
Total score		244

An example of how the RHRS presented in Table 10.23 is used in a rock mass dominated by discontinuities is shown in Table 10.25.

Based on the data given in Table 10.23 the range of scores is 30 to 810. The RHRS does not include recommendations for actions to be taken for different ratings. This is because such decisions will depend on many factors such as budget, importance of route etc which cannot be built into a general scheme. Local highway authorities using such schemes will be able to define suitable actions based on the rating. For example a rating less than 300 may be assigned a low priority whereas slopes with ratings greater than 500 are identified for urgent remedial action.

CHAPTER ELEVEN

Stabilization, maintenance and protection of rock slopes

Overview

Rock slopes can be become unstable or dangerous by the following means:

- blocks sliding out (i.e. planar and wedge failure)
- blocks toppling or flexing
- blocks becoming detached and rolling or falling.

By good design based on rational analysis, we can minimize the risk of these events. For example, we can sometimes alter height, angle and orientation of a slope to optimize the design. Above all we can utilize good construction methods that do not fracture and weaken the rock mass. Where we are not free to alter the design parameters of height, angle and orientation of the slope, then we will be required to consider other stabilization measures as discussed below. Normally, we will also need to take steps to ensure that any deterioration of the slope is slowed down sufficiently to ensure that a reasonable design life can be attained. We will thus be concerned with the stabilization of the slope, the protection of the slope, and the protection of property and people from rockfalls, as described below. Note that benching is key to all of these considerations of best practice.

Construction

By adopting the most appropriate construction methods within the overall design the long term stability and hence the whole-life costs of the slope can be optimized. Construction may be considered under the following headings:

- scaling and trimming
- blast design
- benching.

Scaling and trimming

Scaling refers to the removal of loose, overhanging or protruding blocks using hand pry bars, hydraulic splitters or jacks, and explosives. Trimming involves drilling, blasting and scaling to remove small ragged or protruding rock in overhang areas where repetitive scaling would otherwise be required. These operations require specialized experience and form an important precursor to the excavation of a rock slope. They would be carried out in the region above the excavation with the objective of minimizing rockfall during and after construction.

On steep or precipitous slopes scaling is usually carried out by workers on ropes. The work is generally slow and often made difficult because of severe access problems. Before work begins, an experienced engineering geologist should thoroughly inspect each location and make decisions on the rocks to be removed.

Blasting procedures

The amount the rock mass is broken and loosened during excavation is dependent upon the method used. The ultimate in damage control is machine excavation. This is used in weak rocks and highly fractured hard rock. Where the combination of fracture spacing and rock strength preclude machine excavation, explosives must be used to break the rock. The objective here should be to break up the rock within the excavation and preserve that beyond it in the soundest possible condition. Uncontrolled blasting results in rough uneven contours, overbreak, overhangs, excessive shattering, and extensive tension cracks in the crest of the slope. Blasting damage, therefore, can lead to significantly higher scaling, excavation, remedial treatment and maintenance costs. The results of blast shock waves and gases along faults, joints, bedding and discontinuities, although not readily apparent on the blasted face, can lead to loosening of the rock. This sometimes occurs well behind the face, allowing easier infiltration of surface water, which may lead to unfavourable ground water pressures and unnecessary frost action.

Blast-hole patterns and powder loads must be properly balanced so that advantage is taken of the energy released by the explosive and the desired blast effects are obtained with minimum damage to the rock. Control of the degree of fragmentation can also facilitate handling the muck and ensure that the blasted rock is suitable for use as fill. Although guidelines can be specified, the blasting design should be based on practical experience with the rock in question and can best be determined in the field.

Controlled blasting involves drilling closely spaced, carefully aligned drill holes, which are loaded with a light explosive charge, and detonated in a specified sequence with respect to the main blast. An example of controlled blasting is illustrated in Fig. 11.1. The explosive load is designed to generate a shock wave and gas pressure just sufficient to break the rock between the drill holes, but not to cause radial fractures

353

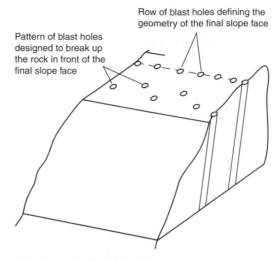

Fig. 11.1 Example of controlled blasting

in the rock behind the proposed slope face. This may be achieved by one of the following methods:

- use of an explosive with a relatively low detonation velocity
- ensuring that there is an air gap between the explosive and the wall of the drill hole (the pressure level generated by the explosive can be reduced by an order of magnitude if the diameter of the explosive is half that of the drill hole).

Three methods of controlled blasting are commonly used. These methods, which involve the simultaneous detonation of a row of closely spaced, lightly charged holes, are designed to create a clean separation surface between the rock to be blasted and the rock which is to remain. In each case a row of blast holes are drilled along the line of the face. Production blast holes are drilled in front of these. The methods differ in the blast sequence and hole spacing. Details of each method are given in Table 11.1.

A common misconception is that the only step required to control blasting damage is to introduce pre-splitting or smooth blasting techniques. When correctly performed, these blasts can produce very clean faces with a minimum of overbreak and disturbance. However, controlling blasting damage starts long before the introduction of pre-splitting or smooth blasting. A poorly designed blast can induce cracks several metres behind the last row of blast holes. Clearly, if such damage has already been inflicted on the rock, it is far too late to attempt to remedy the situation by using smooth blasting to trim the last few metres of excavation. On the other hand, if the entire blast has been correctly designed and executed, smooth blasting can be very beneficial in

Table 11.1 Methods of controlled blasting

Blasting method	Blast sequence	Typical hole spacing	Explosive load w_c: kg/m	Burden	Use
Pre-split blasting	Pre-split line is detonated first then production blast	10–12 blast hole diameters	$w_c = d_h^2/12\,200$ d_h = blast hole diameter. Adjustments should be made to take account of rock strength and fracture state	2–3 times blast hole depth	General use where smooth face and minimal blast damage is required
Trim blasting	Trim line holes are detonated last	16–20 blast hole diameters	As for pre-split blasting	Can be less than 2–3 times blast hole depth	As above but in closely fractured rock and in cases where vibration levels must be minimized
Line blasting	As for pre-split blasting	<10 blast hole diameters	As for pre-split blasting	As for pre-split blasting	Used for the excavation of tight corners and where stable faces cut to close dimensional tolerances are required

trimming the final excavation face. The key is in optimizing the production blast for the strength and fracture state of the rock mass.

Excavation lifts and related procedures

For quality control, each lift of rock excavation generally should not exceed about 10 m in height. Benches this high or lower generally prove to be the best for achieving effective scaling and rock bolting. Also, the accuracy of drilling and, hence, the quality of controlled blasting tend to decrease with increased height of excavation lifts.

Rock bolting, anchoring, scaling and similar work, if possible, should be carried out as each successive lift is excavated and completed. This will ensure that the slope above the working area is safe.

When heavy equipment, such as hydraulic backhoes and tractors with rippers, is used in areas of soft, weathered, or highly broken rock, care should be taken to avoid loosening the final cut face of the slope by equipment operations.

Benches

Figure 11.2 shows the form of a slope incorporating benches. The geometry of the benches is governed by the physical and mechanical characteristics of the rock mass. Bench height should provide a safe and efficient slope and an optimum overall slope angle. Bench height can be greater in stronger rock, and the bench face can be terminated at the

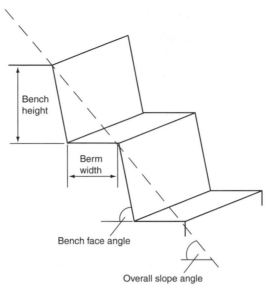

Fig. 11.2 Form of excavated slope incorporating benches

base of weaker horizons and water-bearing zones. Without affecting the overall slope angle, higher benches generally will allow for wider berms, giving better protection and more reliable and easier access for the regular cleaning of debris. The width of berms should be governed by the size of the equipment working on the bench and by the nature of the slope-forming material, but should generally be no less than 7 m.

Design of the bench face angle is governed to a large extent by the attitude of unfavorable structures in the slope to prevent excessive rockfalls onto the berms. If the bench faces are nearly vertical, high tensile stresses are likely to develop near bench crests causing tension cracks and overhangs to form. Ideally bench faces should be inclined.

Berms should be equipped with drainage ditches to intercept surface run-off and water from drain holes and other drainage facilities and divert it off the slope and away from problem areas. These ditches should be kept open and free of all debris and ice to ensure adequate performance. Ditch linings (e.g. clay, sprayed concrete, grout, asphalt, polyethylene sheeting) may be required if ditch leakages are anticipated or develop afterward. Berm surfaces should be graded to assist the collection of water in ditches and also to facilitate general drainage in a direction away from potential areas of instability. Care must be taken so that ditches do not create problems by channelling water from one area to another.

Benched slopes are considered an expensive option because of increased land-take and construction time. The advantages of a benched slope, however, more than justify the expense. Benched slopes:

- minimize rockfalls during and after construction
- provide rockfall catchments (in general, the faces of benches can be considerably steeper than the overall slope angle; hence, any rocks that do fall remain on the benches)
- reduce excessive weathering in weak rocks such as shale and mudstone
- reduce erosion due to run-off since energy of surface flow is dissipated and controlled
- reduce whole life maintenance costs
- provide access to all parts of the face for maintenance.

Benches appear to have no effect on the slope with respect to deep-seated failure. Although shear stresses increase with slope height, the direction of the maximum shear stress is supposedly independent of slope height and bench geometry. Benches may require the top of the slope to be moved back and, therefore, considerable additional excavation.

Stepped benches may be used on slopes cut in highly weathered rock material to control erosion and to establish vegetation. Figure 11.3 shows stepped cut slopes that consist of 0.6 to 1.2 m high benches with approximately similar berm width and an overall slope angle based on stability analysis. The design objective is that the material weathering from each rise will fill up the step of the bench and finally create a practically uniform overall slope. The steps are constructed horizontally to avoid the longitudinal movement of water, which could cause considerable erosion. Seeding and mulching or other suitable methods of slope stabilization can be readily applied; however, for rapidly ravelling slopes, about half of the bench or step width should be filled before seeding is done to prevent smothering of the seed.

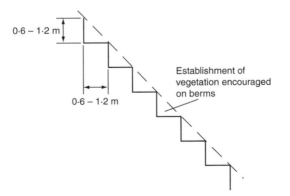

Fig. 11.3 Stepped benches

Stabilization

The stabilization of rock faces can be achieved by a combination of the following methods:

- by removing water by provision of drainage channels and by preventing water from entering the rock mass
- by preventing movement by use of retaining walls, anchors, dowels, gabions, etc.
- by providing benching to give access to all sections of the slope for maintenance, protection from rockfalls, and provision of drainage channels.

Stabilization methods give a positive solution to the problem in that either the driving forces are reduced or the resisting forces are increased. Because of the complex nature of the rock mass, the effectiveness of these methods is often difficult to assess quantitatively. For example, in attempting to prevent water from entering the rock mass behind the slope face, how can the designer be certain that the measures are 100% effective. It is generally assumed that such measures may only be 50 to 75% effective.

Stabilization measures reduce the likelihood of rocks moving out of place and generally should be considered first in the remedial treatment of rock slopes. Stabilization also reduces the rate of deterioration, a process that leads ultimately to failure.

Surface and sub-surface drainage

Methods that can be applied to improve either the surface or sub-surface drainage conditions and, hence, increase the stability of the slope should be given high priority in the design of any slope. Drainage measures, as compared with other possible measures, frequently result in substantial benefits at significantly lower cost. Often large failures, involving several thousand cubic metres of material, cannot be controlled within practical limits by any means other than some form of drainage.

Surface drainage control

Adequate surface drainage facilities, particularly if the rocks are relatively soft or susceptible to erosion, can substantially improve the stability of a slope where unfavourable ground-water conditions exist. Areas behind the upper portions of unstable slopes should be thoroughly inspected to determine whether surface water is flowing toward unstable areas or into the ground or both. The following methods have been used successfully to control surface drainage:

- drain water-filled depressions that occur above the working area from which water could seep into unstable zones

- reshape the surface of the area to provide controlled flow and surface run-off
- above the crest of the slope, use concrete, grout, asphalt, or rubber/plastic liners to temporarily or permanently seal or plug tension cracks and other obviously highly permeable areas that appear to provide avenues for excessive water infiltration (sealing cracks will also prevent frost action in the cracks)
- provide lined (e.g. paved, grouted) or unlined surface ditches, culverts, surface drains, flumes, or conduits to divert undesirable surface flows into non-problem areas
- minimize removal of vegetative cover and establish vegetative growth.

Drain holes

The purpose of sub-surface drainage facilities is, ideally, to lower the water table and, hence, the water pressure to a level below that of potential failure surfaces. A practical approach that appears to be best suited for most rock slope problems encountered in civil engineering is to incorporate a system of drain holes to depress the water level below the zones in which failure would theoretically take place in the slope. The drain holes should be designed to extend behind the critical failure zone. The direction of the drain holes may depend to a large degree on the orientation of the critical discontinuities. The optimum drain hole design is to intersect the maximum number of discontinuities for each metre of hole drilled.

The effectiveness of drains depends on the size, permeability and orientation of the discontinuities. A drain does not have to produce any noticeable flow of water to be effective; it may have flow only under extreme conditions. Furthermore, the absence of damp spots on the rock face does not necessarily mean that unfavourable ground water conditions do not exist. Groundwater may evaporate before it becomes readily apparent on the face, particularly in dry climates.

Drain holes usually are inclined upward from the horizontal about 5°. In weak, soft or weakly cemented rocks, however, the holes may have to be inclined slightly downward to prevent erosion at the drain hole outlet. In this case, a small pipe can be left in the mouth of the drain hole to retard erosion. Spacing of drain holes can range from 7 to 30 m, but 10 to 15 m spacing generally is used. For high rock cuts, installing drain holes at different levels on the slope may be advantageous to increase the effectiveness of the overall drainage system. For certain conditions on a slope, a series of drain holes may best be installed in a fan pattern so that drill machine setup and moving time is minimized.

Drain holes should be thoroughly cleaned of drill cuttings, mud, clay, and other materials; drain holes not properly cleaned may have their effectiveness reduced by 75%. High-pressure air, water, and in some instances a detergent should be used to clean drain holes. In highly fractured ground, care should be taken to ensure that caving does not block drain holes. If caving is significant, perforated linings should be installed so that drain holes remain open. If freezing conditions exist, drain hole outlets should be protected from ice build-up that could cause blockage. Insulating materials, such as straw, sawdust, gravel, or crushed rock, have been used for this purpose.

Other methods for sub-surface drainage of slopes include pumped wells, drainage galleries, shafts, and trenches. Noteworthy discussions concerning this subject have been presented by Záruba and Mencl (1969), Baker and Marshall (1958), Hoek and Bray (1981) and Cedergren (1967). Only under special circumstances are sub-surface drainage methods other than subhorizontal drain holes used for most cut slopes.

Sprayed concrete

Slopes prone to spalling, rockfalls and sliding of small volumes of rock can be stabilized effectively by spraying the face with concrete. Shotcrete is the term used for a sprayed concrete comprising mortar with aggregate as large as 20 mm in size whereas the term gunite is used to describe a similar material that contains smaller aggregate. The shotcrete or gunite is normally applied in layers between 80 and 100 mm thick. When these materials are applied to an irregular rock surface, the resulting surface configuration is smoother (Fig. 11.4). The sprayed concrete helps to maintain the adjacent rock blocks in place by means of its bond to the rock and its initial shear and tensile strength acting as a membrane. The

Fig. 11.4 Sprayed concrete applied to a rock face (photograph reproduced with permission of the John Grimes Partnership)

result is that a composite rock–concrete structure is developed on the surface of the rock. There is no transfer of load from the rock mass to the sprayed concrete lining. In that the interlocking quality of the surface blocks is improved, the sprayed concrete acts as reinforcement and not as support. The more quickly the concrete is applied after excavation, the more effective the results are.

Deterioration of sprayed concrete can result from frost action, ground water seepage, or rock spalling due to lack of bond with the underlying rock. It is therefore important that, prior to spraying with concrete, the face should be thoroughly scaled and trimmed and any unfavourable ground water flows should be drained to provide long term stability. Weep holes should be drilled or installed through the hardened sprayed concrete lining and into the rock to prevent development of water pressure behind the face. Short flexible plastic pipes are placed in cracks or holes drilled into water-bearing broken rock and protection against freezing may be provided using rock wool, geotextiles, or even heating cables.

Initially dry rock surfaces are preferred in the sprayed concrete process, although the use of admixtures and careful control of nozzle water can give successful applications on wet surfaces. Where alteration products, such as clay or mud, exist on discontinuities, care should be taken to clean such surfaces by air or water jet to ensure a good bond between the concrete and rock. As a rule of thumb, weak material should be removed to a depth at least equal to the width of the weak zone before the sprayed concrete is applied. These areas are where the sprayed concrete will be most effective and hence extra attention is required in its application.

Sprayed concrete can be used in combination with steel wire mesh and rock bolts to give structural support and also to form buttresses for small loads. Where it is applied to mesh, all loose material should be removed from the rock surface and the mesh fabric tightened.

Support and reinforcement systems

Support systems offer only passive resistance to the loads imposed by the slope as the rock mass undergoes deformation caused by slope failure or stress relief. Support systems include the following.

- *Buttresses*. These are generally used to support overhanging blocks that are too large or too dangerous to remove.
- *Retaining walls*. These are used to prevent large blocks in the slope from failing and to control or correct failures by increasing the resistance to slope movement. Retaining walls have the advantage of lessening weathering of the rock slope thus offering permanent

Fig. 11.5 Example of a retaining wall incorporated into a rock slope (photograph reproduced with permission of the John Grimes Partnership)

protection. The space along railways and highways is often too narrow for normal gravity or cantilever types of walls, but anchors or rock bolt tiebacks may be used to overcome this problem. Tied-back walls need only have the strength required for bending and shear resistance between rock bolts. An example of such a retaining wall is shown in Fig. 11.5.

- *Dowels*. These comprise steel reinforcing bars that are cemented into boreholes; they are not subjected to any post-tensioning. Dowels basically increase the shearing resistance across potential failure surfaces. A typical application of dowels is shown in Fig. 11.6. They can also anchor restraining nets and cables, catch nets, catch fences, cable catch walls and cantilever rock sheds. The maximum available anchor force is determined from the tensile and shear resistance of the steel cross section at the assumed failure surface crossed by the anchor.

In contrast to support systems, reinforcement systems add strength to the rock mass by increasing the tensile strength and by increasing the resistance to shear along discontinuities. There is a wide variety of reinforcement systems. The more commonly used systems include:

- rock bolts
- rock anchors
- anchored cable nets
- cable lashing.

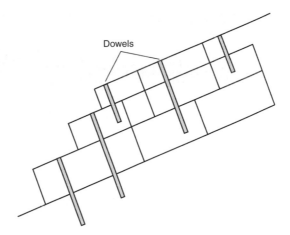

Fig. 11.6 Example of the use of dowels

Rock bolts and rock anchors

Rock bolts are generally regarded as a short, low-capacity reinforcement comprising a bar or tube fixed into the rock and tensioned to a predetermined load (Fig. 11.7). For most rock bolts the free length is fully bonded using grout or resin soon after the bolt has been stressed. Rock bolts are used to:

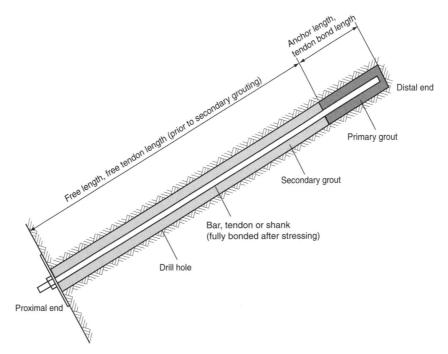

Fig. 11.7 Typical configuration for rock bolts

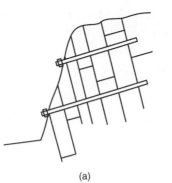

(a)

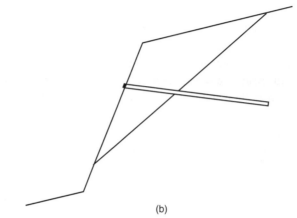

(b)

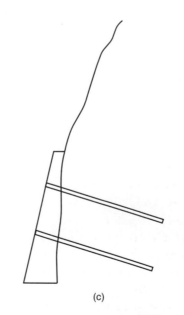

(c)

Fig. 11.8 Typical applications of rock bolts and anchors

- tie together blocks of rock to increase the effective base width in order to prevent toppling (Fig. 11.8(a)) or to increase the resistance to sliding on discontinuity surfaces (Fig. 11.8(b))
- anchor structures such as retaining walls and catch nets (Fig. 11.8(c)).

Anchors are used for supporting large masses of unstable rock and hence are generally longer and of higher capacity than rock bolts. They generally comprise tensioned cables which are fixed into the rock. Details of design and construction considerations for rock bolts and rock anchors are given by Hoek and Londe (1974), BS 8081: 1989 (1989) and BA80/99 (1999).

The main advantage of using tensioned bars or cables is that no movement has to take place before the anchor develops its full capacity; thus, deformation and possible tension cracking of the slope are minimized. A further advantage of this system is that a known anchor force is applied, and proof loading can be accomplished during installation of each anchor. The tension in the bar or cable can be re-checked, if necessary, at any subsequent time to determine whether the load is being maintained and re-tensioned if necessary. A greater degree of confidence in the anchor design is thus provided.

The reinforcement is usually designed by using the limit equilibrium method of analysis as described in Chapter 10. For the reinforcement to be effective the bolt or anchor must be fixed beyond the assumed failure surface. The anchor force required for stability is applied partly as an increased compressive normal stress on the failure plane, thus increasing frictional resistance, and partly as a force resisting the driving force that is causing instability of the slope. The value of the component forces will depend on both the design geometry and orientation of the anchors and the characteristics of the assumed failure surface as illustrated in Fig. 11.9.

Rock bolts and anchors are often installed in arrays as shown in Fig. 11.10. Varying rock mass conditions within a rock slope may require rock bolts of different capacities to be used. Figure 11.10 shows two rock bolt systems in use. In weak or soft rocks beams may be required to spread the load at each anchor/bolt head as shown in Fig. 11.11. This type of construction is often referred to as an anchored beam.

Adequate corrosion protection for rock reinforcement systems should be provided. The degree and type of protection depends on the design life of the reinforcement system, the corrosivity of the rock mass and the consequences of failure in terms of personal injury and damage to property and infrastructure. Corrosion protection involves encapsulation of the bar or cable in grout or resin, coating with materials such as zinc, a combination of grout and a plastic sheath (double corrosion protection) or using corrosion resistant materials. These protection systems are

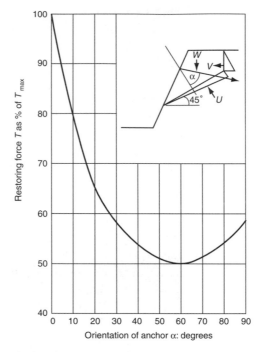

Fig. 11.9 The variation in restoring force T *with angle of inclination* α *for two example cases of rock bolting*

Fig. 11.10 Regular array of rock bolts used to stabilize a rock slope. Note the different sizes of bolt used, indicating different capacities (photograph reproduced with permission of the John Grimes Partnership)

Fig. 11.11 Beams used to spread the load imposed on the face by rock bolt heads in weak rock. Note the use of spray concrete to protect the face (photograph reproduced with permission of the John Grimes Partnership)

described in BS 8081: 1989 and BA80 (1999). It is important when designing protection systems to remember that the bolt/anchor head must also be protected against corrosion. If this becomes badly corroded it may be impossible to re-tension the bolt or anchor which may result in unwanted or dangerous deformations taking place within the rock mass.

To minimize the loosening effects associated with recently excavated rock slopes, rock bolts should be installed and tensioned as soon as possible after each lift of an excavation and preferably before the next lift is blasted.

The method used for transferring load from the head of the rock bolt to the rock depends largely on the condition of the rock. Broken rock may weather away from the bolt head and cause the bolt to lose tension. A choice must be made between protecting the rock face around the head with wire mesh and sprayed concrete, retaining the rock in place by steel strapping, or distributing the point load by use of concrete pads.

When grouting rock bolts, extreme care should be taken to ensure that the grout does not spread. Excessive grout spread will reduce the permeability of the rock mass and retard drainage by plugging joints through which ground water would normally have free access. The grout must be pumped into the borehole under low pressure and measures must be taken to retard grout spread.

Anchored cable nets

Anchored cable nets can be used to restrain masses of small loose rocks or individual blocks that protrude from a rock face. In principle, an

Fig. 11.12 Anchored cable net used to support a 140 tonne block of rock (photograph reproduced with permission of the John Grimes Partnership)

anchored cable net performs like a sling or reinforcement net, which extends around the surface of the unstable broken rock to be supported. Figure 11.12 shows an anchored cable net supporting a large block of rock (approx. 140 tonnes). The cable net strands are gathered on each side by main cables leading to rock anchors, as shown in Fig 11.12.

Beam and cable walls

Beam and cable walls can also be used to prevent smaller blocks of rock from falling out of the slope and may be used on slopes as high as 20 m. Beam and cable walls consist of cable nets fastened to vertical ribs formed of steel beams laid against the slope at 3 to 6 m intervals. The beams are held by anchored cables at the top of the slope and by concrete footings at ditch level.

Cable lashing

Cable lashing is a simple, economical type of installation for restraining large rocks. This method involves tying or wrapping unstable blocks with individual cable strands anchored to the slope. Eckert (1966) describes a case in France in which cables with a capacity of 227 tonnes and a length of more than 90 m were extended around a mass of broken rock and anchored into sound rock at each side.

Protection of slope

Slopes prone to excessive weathering and erosion will require some form of protection in order to maintain stability particularly with respect to rockfalls. Slopes may be protected from:

- deterioration by weathering and erosion by covering with sprayed concrete, masonry, retaining walls, etc.
- erosion by benching to provide drainage.

These measures have been described earlier.

Protection of property and people

When a highway, railway or buildings are located at the base of a rock slope the risk of rockfall must be minimized by the incorporation of appropriate protection measures. The primary objective here is to prevent loose blocks from moving. This may be achieved using netting or anchored wire mesh. Where movement of blocks is unavoidable then it is necessary to control the movement as far as possible such that damage is minimized. Netting can be used to prevent blocks from falling or bouncing. Ditches, fences and walls (often gabion walls) are used to catch blocks. The use of benched slopes can be used effectively to minimize the risk of rockfall. Slope ditches are incorporated into the berms to intercept rockfalls partway up the slope, and shaped berms are used at the top of the slope. These methods ensure that rockfalls get caught either before they start to roll or while they are on their way down the slope. In some locations, ditches can be made to intercept rocks and guide them laterally into disposal areas. In high slopes were relatively large blocks are likely to become loose and the above methods are impractical the highways and railways may be protected by highly reinforced concrete structures known as rock sheds.

Anchored wire mesh

Wire mesh is a versatile and economical material in providing protection against rockfall where small blocks (blocks smaller than 0.6 to 1 m) are involved. Layers of mesh are often pinned onto the rock surface to prevent small loose blocks of rock becoming dislodged. Mesh can also be used essentially as a blanket draped over the rock surface (Fig. 11.13) to guide falling rock into the ditch at the base of the slope. The same

Fig. 11.13 Anchored wire mesh used to prevent small blocks from becoming dislodged, Cinque Terre coastal path, Italy (photograph: Bruce Menzies)

Fig. 11.14 Mesh combined with rock bolts to provide reinforcement
(photograph reproduced with permission of the John Grimes Partnership)

arrangement can be used on stony overburden slopes to prevent dislodged stones from rolling down the slope. This practice is commonly
used on talus slopes in steep mountainous terrain. Mesh can be combined
with rock bolts to provide a generally deeper reinforcement (Fig. 11.14).
Mesh in combination with both sprayed concrete and rock bolts provides
general reinforcement and support and retards the deleterious effects of
weathering.

In general mesh should only be used if the slope is uniform enough for it
to be in almost continuous contact with the face. If large falling blocks in
certain areas are likely to dislodge or tear the mesh and present a hazard,
rock bolts should be used to reinforce these particular areas. For wire
mesh blankets, the bottom end of the mesh is usually left a metre or so
above ditch level and only a narrow ditch is required. The mesh normally
used is 9 or 11 gauge galvanized, standard chain-link or gabion wire
mesh. Gabion mesh appears to have an advantage over the standard
chain-link materials in that the gabion mesh has a double-twist hexagonal
weave that does not unravel when broken.

Ditches

Depth, width, and steepness of the inside slope and storage volume of the
ditch are important factors in the design of ditches to contain rockfalls. The
choice of ditch geometry should take into consideration the angle of

Table 11.2 Design criteria for catch ditches, after Ritchie (1963)

Slope	Height: m	Fallout area width: m	Ditch depth: m
Near vertical	5 to 10	3.7	1.0
	10 to 20	4.6	1.2
	>20	6.1	1.2
0.25 or 0.3 : 1	5 to 10	3.7	1.0
	10 to 20	4.6	1.2
	20 to 30	6.1	1.8*
	>30	7.6	1.8*
0.5 : 1	5 to 10	3.7	1.2
	10 to 20	4.6	1.8*
	20 to 30	6.1	1.8*
	>30	7.6	2.7*
0.75 : 1	0 to 10	3.7	1.0
	10 to 20	4.6	1.2
	>20	4.6	1.8*
1 : 1	0 to 10	3.7	1.0
	10 to 20	3.7	1.5*
	>20	4.6	1.8*

* May be 1.2 m if a catch fence is used

the slope that influences the behaviour of falling rocks. Ritchie (1963) evaluated the mechanics of rockfalls from cliffs and talus slopes and developed design criteria for ditches. The criteria, which involve the height and angle of slope, depth of ditch, and width of fallout area, are given in Table 11.2 and in Fig. 11.15, which also shows the nature of rock trajectories for different slope angles.

If lack of space prevents the ditch from being as wide as indicated, limited protection against small rolling rocks can still be provided inexpensively by excavating a catchment area, as shown in Fig. 11.16, or by installing a low barrier formed of gabions at the shoulder of the road-way, as shown in Fig. 11.17. Bedrock should not remain exposed in the bottom of ditches, but should be covered with small broken rock, gravel or sand to keep falling rocks from bouncing or shattering.

Wire mesh catch nets and fences

Wire mesh can be effective in intercepting or effectively slowing bouncing rocks as large as 0.6 to 1 m when the mesh is mounted as a flexible catch net rather than as a standard, fixed wire fence (Fig 11.18). If suspended on a cable, the mesh will absorb the energy of flying rocks with a minimum of

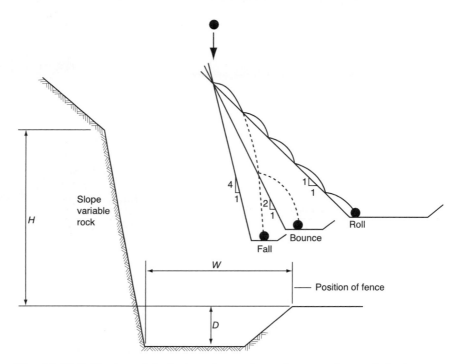

Fig. 11.15 Path of rock trajectory for various slope angles and design criteria for shaped ditches, after Ritchie (1963)

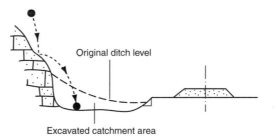

Fig. 11.16 Shaped ditch with excavated catchment area

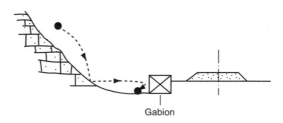

Fig. 11.17 Shaped ditch with gabions

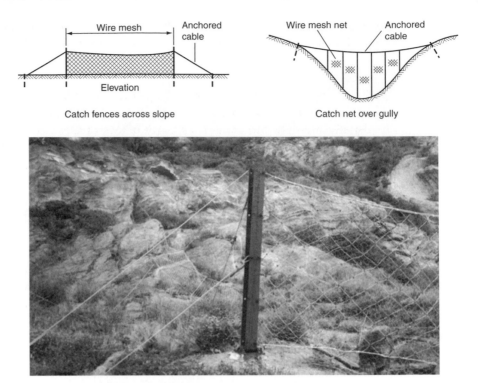

Catch fences across slope   Catch net over gully

Fig. 11.18 Examples of catch fences, Cinque Terre coastal path, Italy (photograph: Bruce Menzies)

damage to the wire catch net. Catch nets can consist of standard chain-link wire mesh or gabion mesh.

The principle of the catch fence is similar to that of a catch net. Its purpose is to form a flexible barrier to dissipate the energy of rapidly moving rocks. Various arrangements are shown in Fig. 11.19 for catch

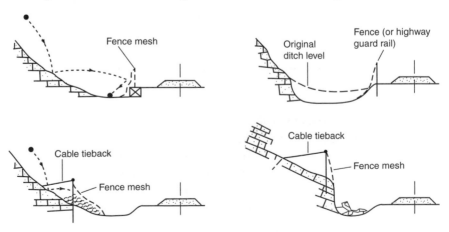

Fig. 11.19 Examples of shaped ditches with catch fences

fences located at or near ditch level. The fence should be suitably situated so that accumulated rocks can be removed easily.

Catch walls

Catch walls can be used to form a barrier to stop rolling or bouncing rocks as large as 1.5 to 2 m from reaching the right-of-way. If effective, they usually increase the storage capacity of the ditch so that maintenance intervals can be extended. In many locations, large ditches themselves are not effective for intercepting large rolling rocks and the use of catch walls is advised.

To achieve maximum protection and storage capacity, the catch walls should be located on the side of the ditch closest to the road.

Catch walls can be constructed using reinforced concrete or gabions. The gabion catch wall is a flexible structure that, upon settling or being hit by impact, tends to deflect and deform instead of break. Because gabion walls are highly deformable, differential settlement is not important. Like the gabion mesh, the gabion does not unravel if broken. Gabions therefore make good inexpensive catch wall structures.

Rock sheds and tunnels

Rock sheds and tunnels can be used for protection against rockfalls and slides when warranted and when other forms of stabilization and protection are not effective. Although expensive, they can give complete protection and should be considered in areas with serious problems. Maintenance costs are normally negligible. The methods of design and construction of tunnels are dealt with thoroughly in the literature, but the design of rock sheds is not so adequately covered, and experience is required to decide on the most suitable type of structure and the loads to be carried. A rock shed should be able to resist the energy transmitted by the largest rock mass likely to pass over it during its life; therefore, probability analysis should be involved. The energy transmitted will depend on whether rocks are falling free, bouncing, or rolling. High stress concentrations in the structure can be reduced by the provision of a thick cover of loose sand.

Methods of warning

Although warning systems do not prevent slope failure or rockfalls, they are necessary where other measures are too expensive or impractical or where a new hazard has developed. In North America, warning methods have been used on railways in mountains to detect rockfalls on tracks so that trains can stop before hitting the material. In the UK a warning system employing tiltmeters is in use on the south coast of Isle of Wight where a highway passes close to the top of a 100 m high chalk cliff in which deep tension cracks have developed (McInnes, 1983). The types of warning system which are in common use include the following.

- *Human patrols*. This simple warning method has the advantage of being reliable and flexible, and their frequency can be adjusted to the demands of traffic and weather conditions. The disadvantages are that they incur continuing costs and require personnel who are willing to work in uncomfortable and often hazardous conditions.
- *Electric fences*. Electric fences are based on the principle that a falling rock large enough to endanger traffic will break or pull out one of the wires and thus actuate a signal to warn approaching traffic. This principle is particularly adaptable to railways on which a signal system to control traffic is already in use in North America. This consists of a row of poles spaced along the uphill ditch line and wires strung between them at a vertical spacing of about 250 mm.
- *Wires*. A particularly effective type of electric warning system consists of a single wire, anchored at both ends and linked to a warning signal. Such a wire may be fastened around a large unstable rock or across a rock slope above the highway or railway across a gully where large rocks roll down, or on top of a protective catch wall.

Both fences and wires can suffer from numerous false alarms which severely reduces confidence in the system. In addition these systems may impede snow-clearing operations.

Maintenance

The management of bridges involves regular inspection, programmed monitoring and a planned maintenance programme. In many ways rock slopes should be treated in the same manner. Whatever stabilization and protection measures have been incorporated into the design of a rock slope it will not prevent the rock mass from deteriorating with time and may not remain fully effective in the long term. Provision for slope inspection, monitoring and maintenance should be included at the design stage of any rock slope.

Regular inspection of rock slopes is an essential precaution in protecting a substantial capital investment. Inspection should include observations of loose and unstable blocks, excessive weathering and erosion and the development of tension cracks. The frequency of inspection will depend on the likely rate of deterioration of the rock mass and the consequences of rockfall or slope failure. In rock masses prone to rapid weathering the frequency of inspection may be less than one year.

Rock slope design should include a program of monitoring slope movements and identifying characteristics that indicate changing stability where necessary. Detection of general creep or slow translation of highly fractured or soft slope-forming material is important in decisions regarding remedial work. Horizontal and vertical movements of points plotted against time or depth or both provide important information concerning the behaviour of a rock slope. Graphs provide a clear indication of the onset of slope failure when plots representing the change in position do not remain linear. When such accelerated movements are evident, the slope must be approaching failure, and measures should be taken to analyse and remedy the situation.

Maintainance of rock slopes should include:

- scaling and trimming
- clearing catch ditches and berms of fallen material
- clearing weep holes of any blockages (e.g. vegetation)
- re-tensioning rock bolts and anchors.

Recommended list of units, unit abbreviations, quantity symbols and conversion factors for use in soil and rock mechanics

Part 1. SI base units, derived units and multiples.

Quantity and symbol	Units and multiples	Unit abbreviations	Conversion factors for existing units	Remarks
Length (various)	kilometre metre millimetre micrometre	km m mm μm	1 mile = 1.609 km 1 yard = 0.9144 m 1 ft = 0.3048 m 1 in = 25.40 mm	 1 micrometre = 1 micron
Area (A)	square kilometre square metre square millimetre	km^2 m^2 mm^2	$1\,mile^2 = 2.590\,km^2$ $1\,yd^2 = 0.8361\,m^2$ $1\,ft^2 = 0.09290\,m^2$ $1\,in^2 = 645.2\,mm^2$	
Volume (V)	cubic metre cubic centimetre cubic millimetre	m^3 cm^3 mm^3	$1\,yd^3 = 0.7646\,m^3$ $1\,ft^3 = 0.02832\,m^3$ $1\,in^3 = 16.39\,cm^3$ 1 UK gallon = 4546 cm^3	To be used for solids and liquids
Mass (m)	megagram (or tonne) kilogram gram	Mg (t) kg g	1 ton = 1.016 Mg 1 lb = 0.4536 kg	Megagram is the SI term
Unit weight (γ)	kilonewton per cubic metre	kN/m^3	$100\,lb/ft^3 = 15.708\,kN/m^3$ $(62.43\,lb/ft^3$ pure water $= 9.807\,kN/m^3$ = specific gravity 1.0 approx.)	Unit weight is weight per unit volume
Force (various)	Meganewton kilonewton Newton	MN kN N	1 tonf = 9.964 kN 1 lbf = 4.448 N 1 kgf = 9.807 N	
Pressure (p, u)	Meganewton per square metre Megapascal	MN/m^2 MPa	$1\,tonf/in^2 = 15.44\,MN/m^2$ $(1\,MN/m^2 = 1\,N/mm^2)$	To be used for shear strength, compressive strength, bearing capacity, elastic moduli and laboratory pressures of rock
Stress (σ, τ) and Elastic moduli (E, G, K)	kilonewton per square metre kilopascal	kN/m^2 kPa	$1\,lbf/in^2 = 6.895\,kN/m^2$ $1\,lbf/ft^2 = 0.04788\,kPa$ $1\,tonf/ft^2 = 107.3\,kPa$ 1 bar = 100 kPa $1\,kgf/cm^2 = 98.07\,kPa$	Ditto for soils

Quantity and symbol	Units and multiples	Unit abbreviations	Conversion factors for existing units	Remarks
Coefficient of volume compressibility (m_v) or swelling (m_s)	square metre per meganewton square metre per kilonewton	m^2/MN m^2/kN	$1\,ft^2/tonf =$ $9.324\,m^2/MN$ $= 0.009324\,m^2/kN$	
Coefficient of water permeability (kw)	metre per second	m/s	$1\,cm/s = 0.01\,m/s$	This is a velocity, depending on temperature and defined by Darcy's law $$V = k_w \frac{\delta h}{\delta s}$$ V = velocity of flow $\frac{\delta h}{\delta s}$ = hydraulic gradient
Absolute permeability (k)	square micrometre	μm^2	$1\,Darcy = 0.9869\,\mu m^2$	This is an area which quantifies the seepage properties of the ground independently of the fluid concerned or its temperature $$V = \frac{kpg}{\eta}\frac{\delta h}{\delta s}$$ p = fluid density g = gravitational acceleration η = dynamic viscosity
Dynamic viscosity (η)	millipascal second (centipoise)	mPas (cP)	$1\,cP = 1\,mPas$ $(1\,Pa = 1\,N/m^2)$	Dynamic viscosity is defined by Stokes' Law. A pascal is a kilonewton per square metre
Kinematic viscosity (ν)	square millimetre per second (centistoke)	mm^2/s (cSt)	$1\,cSt = 1\,mm^2/s$	$\nu = \eta/\rho$
Celsius temperature (t)	degree Celsius	°C	$t°F = 5(t-32)/9°C$	The Celsius temperature t is equal to the difference $t = T - T_0$ between two thermodynamic temperatures T and T_0 where $T_0 = 273.15\,K$ (K = Kelvin)

Part 2. Other units

Quantity and symbol	Units and multiples	Unit abbreviations	Conversion factors for existing units	Remarks
Plane angle (various)	Degree Minute second (angle)	° ′ ″		To be used for angle of shearing resistance (ϕ) and for slopes
Time (t)	year	year	$1\,\text{year} = 31.557 \times 10^6\,\text{s}$	'a' is the abbreviation for year
	day hour second (time)	d h s	$1\,\text{d} = 86.40 \times 10^3\,\text{s}$ $1\,\text{h} = 3600\,\text{s}$	 The second (time) is the SI unit
Coefficient of consolidation (c_v) or swelling (c_s)	square metre per year	m²/year	$1\,\text{ft}^2/\text{year} = 0.09290$ m^2/year	

APPENDIX I
Bishop and Morgenstern's stability coefficients

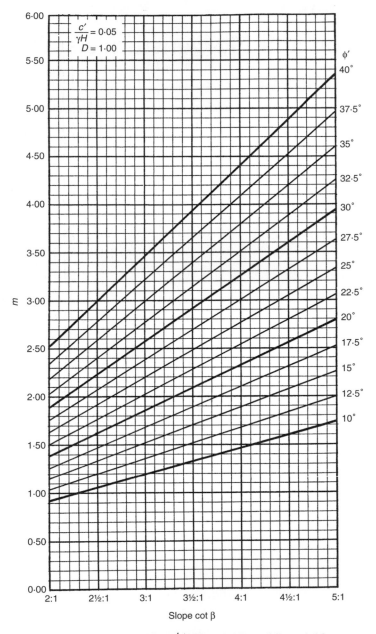

Fig. A1.1 Stability coefficient m *for* c'/γH = 0.05 *and* D = 1.00

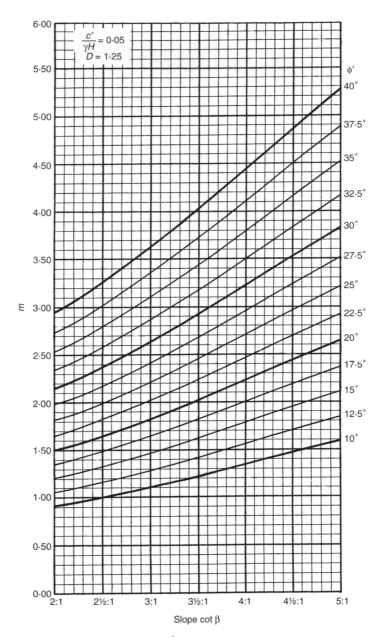

Fig. A1.2 Stability coefficient m *for* $c'/\gamma H = 0.05$ *and* $D = 1.25$

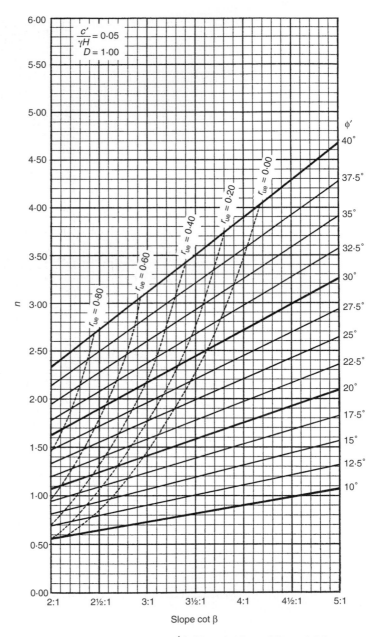

Fig. A1.3 Stability coefficient n for c′/γH = 0.05 and D = 1.00

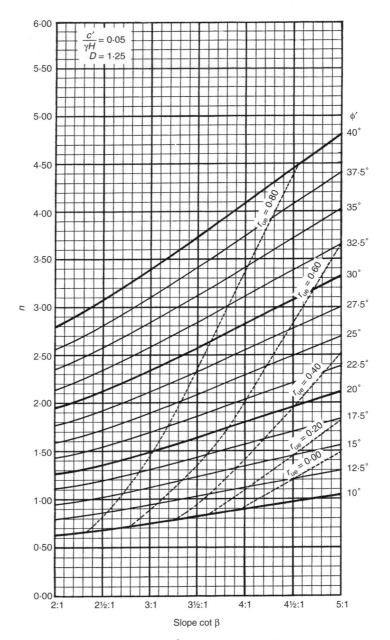

Fig. A1.4 Stability coefficient n *for* c′/γH = 0.05 *and* D = 1.25

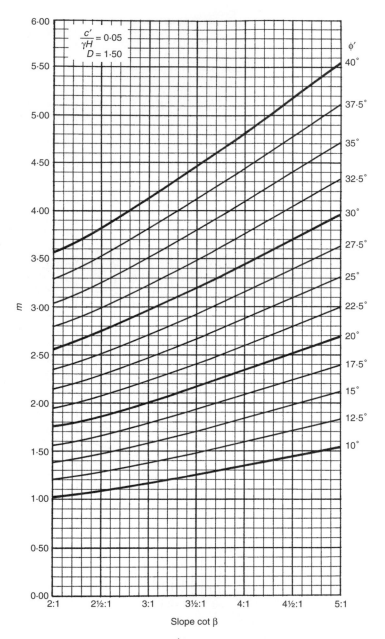

Fig. A1.5 Stability coefficient m *for* c'/γH = 0.05 *and* D = 1.50

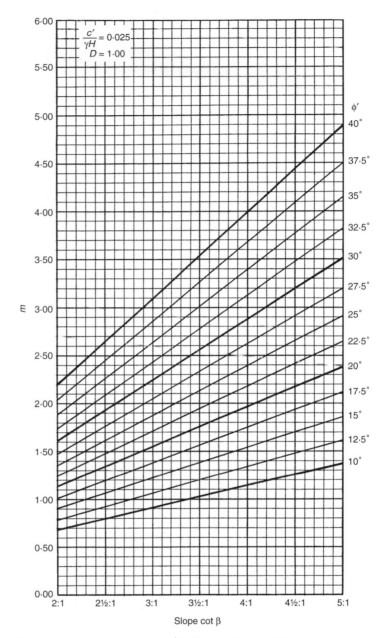

Fig. A1.6 Stability coefficient m *for* $c'/\gamma H = 0.025$ *and* $D = 1.00$

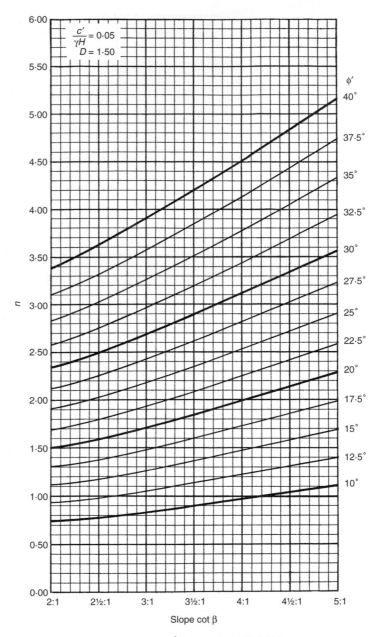

Fig. A1.7 Stability coefficient n *for* $c'/\gamma H = 0.05$ *and* $D = 1.50$

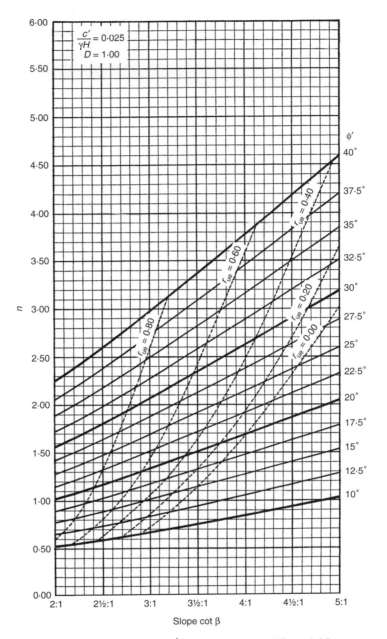

Fig. A1.8 Stability coefficient n *for* c'$/\gamma$H $= 0.025$ *and* D $= 1.00$

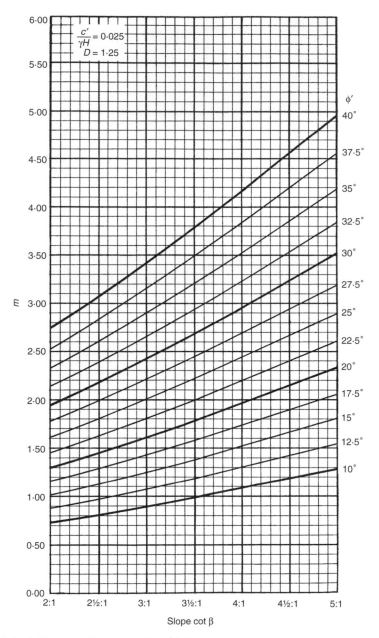

Fig. A1.9 Stability coefficient m *for* c'$/\gamma$H $= 0.025$ *and* D $= 1.25$

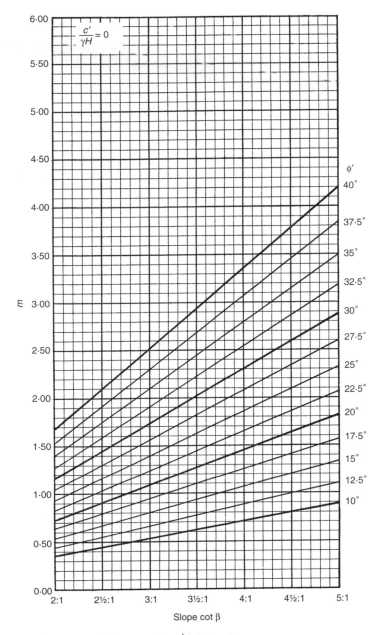

Fig. A1.10 Stability coefficient m *for* $c'/\gamma H = 0$

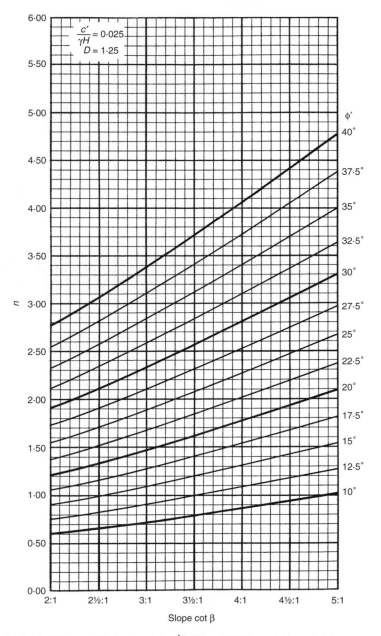

Fig. A1.11 Stability coefficient n for c'/γH = 0.025 and D = 1.25

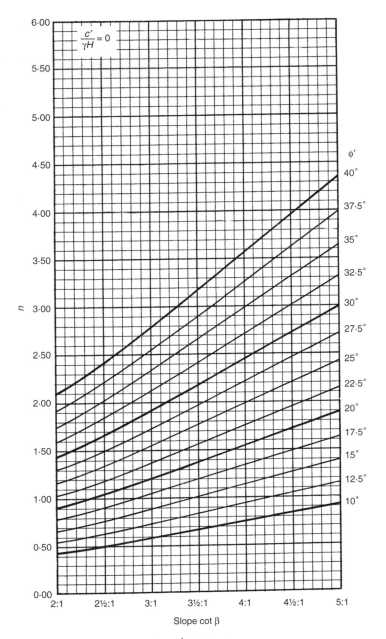

Fig. A1.12 Stability coefficient n *for* $c'/\gamma H = 0$

APPENDIX 2

Hoek and Bray's stability coefficients

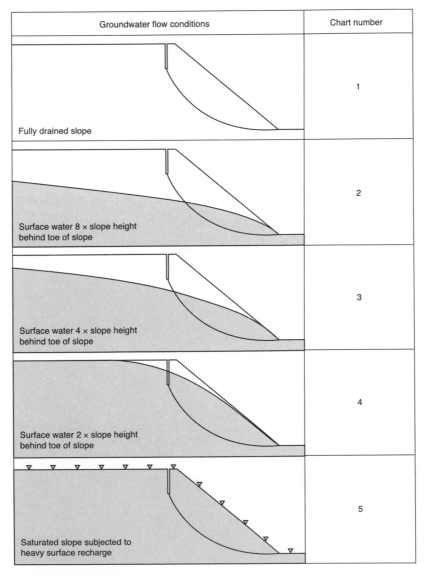

Groundwater flow conditions	Chart number
Fully drained slope	1
Surface water 8 × slope height behind toe of slope	2
Surface water 4 × slope height behind toe of slope	3
Surface water 2 × slope height behind toe of slope	4
Saturated slope subjected to heavy surface recharge	5

Fig. A2.1 Groundwater flow conditions

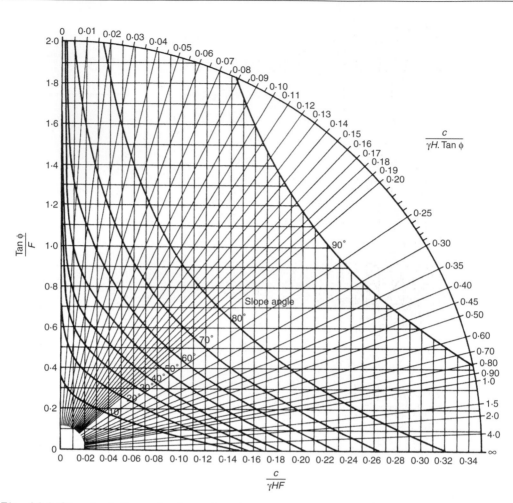

Fig. A2.2 Circular failure chart number 1

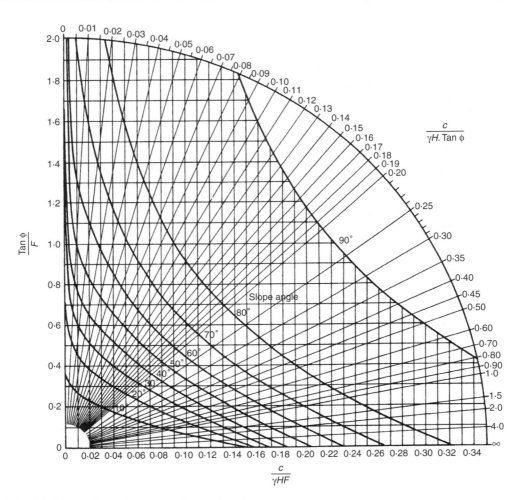

Fig. A2.3 Circular failure chart number 2

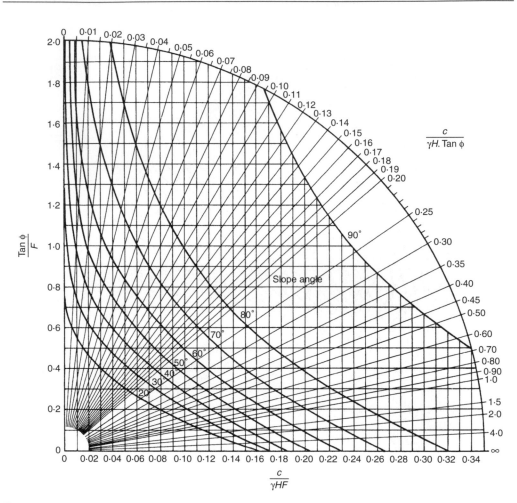

Fig. A2.4 Circular failure chart number 3

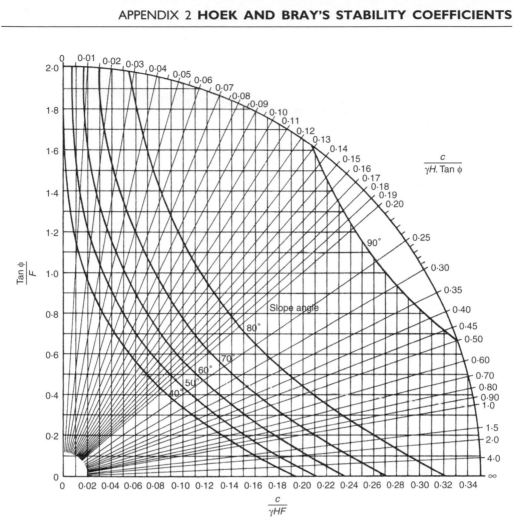

Fig. A2.5 Circular failure chart number 4

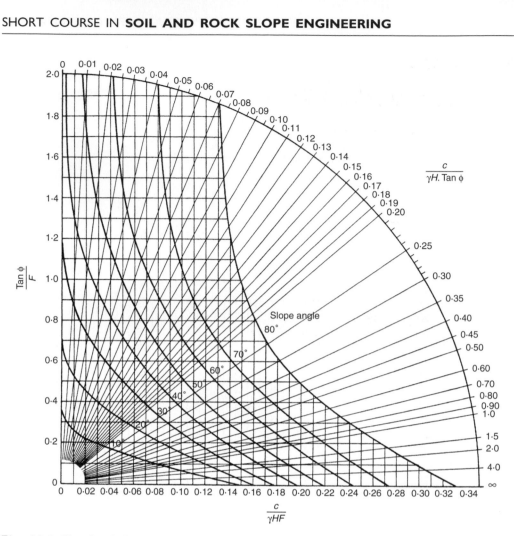

Fig. A2.6 Circular failure chart number 5

APPENDIX 3

Data for worked example

Dip Dir.	Dip	Dip Dir.	Dip	Dip Dir.	Dip	Dip Dir.	Dip
248	82	043	57	350	58	346	62
354	56	044	56	032	90	039	52
275	12	104	86	258	12	057	59
020	48	266	86	278	09	056	66
247	82	276	78	032	52	304	44
275	13	288	84	090	90	116	88
043	68	338	58	040	64	343	74
270	80	360	66	040	55	270	10
054	58	256	08	056	66	106	80
293	82	104	70	360	59	038	52
016	63	292	83	026	52	048	53
043	55	340	84	360	58	296	43
102	80	258	11	251	15	067	74
317	80	256	12	040	74	300	86
096	86	354	63	042	46	047	78
092	79	357	69	066	70	106	86
359	67	276	07	345	62	296	64
271	08	279	08	342	42	344	80
094	14	004	65	274	78	106	86
359	55	040	50	039	57	119	70
043	78	046	72	043	70	268	09
103	80	345	66	057	68	345	64
271	10	048	64	307	70	293	80
043	65	064	68	305	41	063	56
025	68	112	80	304	66	258	68
016	62	002	67	115	87	279	11
110	75	265	12	307	80	262	83
264	06	271	06	025	62	100	60
352	52	265	11	278	78	330	81
340	72	006	67	342	55	358	60
267	05	246	08	211	64	008	51
110	82	040	64	011	69	144	57
017	69	048	52	145	46	074	84
025	53	000	72	355	48	284	08
030	52	110	86	074	66	100	80
353	66	250	67	265	10	325	66

Dip Dir.	Dip	Dip Dir.	Dip	Dip Dir.	Dip	Dip Dir.	Dip
100	81	020	52	100	78	328	69
010	72	087	71	010	62	282	80
285	86	326	29	082	66	059	49
048	74	332	71	326	44	047	48
078	53	087	80	132	86	097	78
030	68	271	85	086	90	050	60
030	52	337	86	358	46	098	46
008	58	088	52	006	69	026	66
010	67	275	84	277	12	025	57
080	73	090	51	261	74	051	70
018	68	090	90	098	26	318	90
080	28	356	65	061	80	146	84
280	86	337	79	060	84	012	59
111	72	308	90	279	06	075	85
078	84	007	70	100	87	278	14
267	84	144	42	133	80	178	17
279	10	018	53	264	73	078	90
277	13						

References and Bibliography

Allen, W. and Crawhall, J. S. (1937). Shaft sinking in dolomite at Venterspost. *J. Chem., Met. and Mining Soc. of S.A.*

Allison, J. A., Mawditt, J. M. and Williams, G. T. (1991). The use of bored piles and counterfort drains to stabilize a major landslip B a comparison of theoretical and field performance. *Slope stability engineering developments and applications.* Thomas Telford, London, p. 443.

Anon, (1999). *Use of Rock Bolts.* Design Manual for Roads and Bridges Advice Note BA80/99. HMSO, London.

Arthur, J. R. F. and Menzies, B. K. (1972). Inherent anisotropy in a sand. *Géotechnique*, **22**, 115–128.

Arthur, J. R. F. and Phillips, A. B. (1972). Discussion on: inherent anisotropy in a sand. *Géotechnique*, **22**, 537–538.

Ayres, D. J. (1961). The treatment of unstable slopes and railway track formations. *Surveyor and Municipal Engr*, **125**, 25.

Ayres, D. J. (1985). Stabilisation of slips in cohesive soil by grouting. Tech. Note 9. *Proc. Int. Symp. On Failures in Earthworks, ICE, London.*

Azzoni, A., LaBarbera, G. and Zaninetti, A. (1995). Analysis and prediction of rockfalls using a mathematical model. *Int. J. Rock Mech. Min. Sci. & Geomech Abstr.*, **32**, 7, 709–724.

Bainbridge, J. R. and Roscoe, K. H. (1963). Instantaneous soil stabilisation by electro-osmosis – fact or fantasy. *R. E. Journal*, **LXXVII**(2), 109–121.

Baker, R. F. and Marshall, H. C. (1958). Control and correction. In *Landslides in engineering practice*, E. B. Eckel (ed.). Highway Research Board Special Report 29, pp. 150–187.

Bandis, S., Lumsden, A. C. and Barton, N. R. (1981). Experimental studies of scale effects on the shear behaviour of rock joints. *Int. J. Rock Mech. Min. Sci. & Geomech. Abstr.*, **18**, 1–21.

Barr, M. V. (1977). Downhole instrumentation — a review for tunnelling ground investigation. *CIRIA Technical Note 090.* Construction Industry Research and Information Association, London.

Barr, M. V. and Hocking, G. (1976). Borehole structural logging employing a pneumatically inflatable impression packer. *Proc. Symp. on Exploration for Rock Engineering*, **1**, 29–34, Balkema, Rotterdam.

Barton, N. R. (1971). A relationship between joint roughness and joint shear strength. *Proc. Int. Symp. on Rock Fracture, Nancy*, Paper 1–8.

Barton, N. R. (1973). Review of a new shear strength criterion for rock joints. *Engineering Geology*, **7**, 287–332.

Barton, N. R. (1974). *A review of the shear strength of filled discontinuities in rock.* N.G.I. Publication No. 105, Oslo.

Barton, N. R. (1976). The shear strength of rock and rock joints. *Int. J. Rock Mech. Min. Sci. & Geomech. Abstr.*, **13**, 255–279.

Barton, N. R. (1980). Discussion on Krahn & Morgenstern: The ultimate frictional resistance of rock discontinuities. *Int. J. Rock Mech. Min. Sci. & Geomech. Abstr*, **17**, 75–78.

Barton, N. R. (1986). Deformation phenomena in jointed rock. *Géotechnique*, **36**, 2, 147–167.

Barton, N. R. (1990). Scale effects or sampling bias? *Proc. Int. Workshop on Scale Effects in Rock Masses*. Balkema, Rotterdam.

Barton, N. and Bakhtar, K. (1983). Rock joint description and modelling for the hydrothermo-mechanical design of nuclear waste repositories. Contr. Rept. CANMET, Ottawa, Part IV, pp. 1–270; part V, pp. 1–108.

Barton, N. and Bandis, S. (1982). Effect of block size on the shear strength behaviour of jointed rock (keynote lecture). Issues in Rock Mechanics. *23rd US Symp. on Rock Mechanics, Berkeley, California*. Society of Mining Engineers of AIME.

Barton, N. and Bandis, S. (1990). Review of predictive capabilities of JRC-JCS model in engineering practice. In *Rock joints, int. symp. on rock joints, Oen, Norway*, N. Barton & O. Stephansson (eds), 603–610, Balkema, Rotterdam.

Barton, N. R. and Choubey, V. (1977). The shear strength of rock joints in theory and practice. *Rock Mechanics*, **10**, 1, 1–54.

Barton, N., Bandis, S. and Bakhtar, K. (1985). Strength, deformation and conductivity coupling of rock joints. *Int. J. Rock Mech. Min. Sci. & Geomech Abstr.*, **3**, 121–140.

Barton, N. R., Lien, R, and Lunde, J. (1974). Engineering classification of rock masses for the design of tunnel support. *Rock Mechanics*, **6**, 189–236.

Bayley, M. J. (1972). Cliff stability at Herne Bay. Review. Aug. 1972, 788–793. Civ. Eng. and Pub. Wks.

Bazett, D. J., Adams, J. I. and Matyas, E. L. (1961). An investigation of a slide in a test trench excavated in fissured sensitive marine clay. *Proc. 5th Int. Conf. Soil Mech. Fdn Engng*, Dunod, Paris, **1**, 431–435.

Bell, A. L. (1915). Lateral pressure and resistance of clay. *Min. Proc. Instn Civ. Engrs.*, **1**, 99, 233–272.

Bieniawski, Z. T. (1974). Geomechanics classification of rock masses and its application to tunnelling. *Proc. 3rd Int. Congr. on Rock Mech., ISRM, Denver*, **2a**, 27–32.

Bieniawski, Z. T. (1976). Rock mass classifications in rock engineering. *Proc. Symp. on Exploration for Rock Engineering*, **1**, 97–107. Balkema, Rotterdam.

Bishop, A. W. (1955). The use of the slip circle in the stability analysis of slopes. *Géotechnique*, **5**, 7–17.

Bishop, A. W. (1959). The principle of effective stress. Text of lecture to Norwegian Geotechnical Institute, 1955, *Teknisk Ukeblad*, **106**, 39, 859–863.

Bishop, A. W. (1966). Sixth Rankine Lecture. The strength of soils as engineering materials. *Géotechnique*, **16**(2), 89–128.

Bishop, A. W. (1967). Progressive failure with special reference to the mechanism causing it. *Proc. Geotech. Conf., Oslo, 1967*, **2**, 142–150.

Bishop, A. W. (1971). The influence of progressive failure on the choice of method of stability analysis. *Géotechnique*, **21**(2), 168–172.

Bishop, A. W. (1972). Shear strength parameters for undisturbed and remoulded soil specimens. *Stress–Strain Behaviour of Soils*, Proc. Roscoe Mem. Symp., Cambridge University, 1971; G. T. Foulis & Co Ltd, Henley-on-Thames, 3–58.

Bishop, A. W. (1973a). Discussion. *Proc. Symp. on plasticity and soil mechanics, Cambridge.* pp. 295–296.

Bishop, A. W. (1973b). The stability of tips and spoil heaps. *Q. J. Eng. Geol.*, **6**, 335–376.

Bishop, A. W. and Bjerrum, L. (1960). The relevance of the triaxial test to the solution of stability problems. *Research Conf. on Shear Strength of Cohesive Soils, Boulder, Colorado, 1960*, ASCE, New York, 437–501.

Bishop, A. W. and Henkel, D. J. (1962). *The Measurement of Soil Properties in the Triaxial Test*, Edward Arnold, London, p. 228.

Bishop, A. W. and Little, A. L. (1967). The influence of the size and orientation of the sample on the apparent strength of London clay at Maldon, Essex. *Proc. Geotech. Conf., Oslo, 1967*, **1**, 89–96.

Bishop, A. W. and Lovenbury, H. T. (1969). Creep characteristics of two undisturbed clays. *Proc. 7th Int. Conf. Soil Mech., Mexico*, Sociedad Mexicana de Suelos, **1**, 29–37.

Bishop, A. W. and Morgenstern, N. R. (1960). Stability coefficients for earth slopes. *Géotechnique*, **10**, 129–150.

Bishop, A. W., Kennard, M. F. and Penman, A. D. M. (1960). Pore pressure observations at Selset Dam. *Proc. Conf. Pore Pressure*, ICE, London, pp. 91–102.

Bishop, A. W., Webb, D. L. and Lewin, P. I. (1965). Undisturbed samples of London clay from the Ashford Common shaft: strength–effective stress relationships. *Géotechnique*, **15**, 1–31.

Bishop, A. W., Hutchinson, J. N., Penman, A. D. M. and Evans, H. E. (1969). *Aberfan Inquiry: Geotechnical investigation into the causes and circumstances of the disaster of 21st October, 1966.* A selection of Technical Reports submitted to the Aberfan Tribunal, HMSO, Welsh Office, London, pp. 1–80.

Bishop, A. W., Green, G. E., Garga, V. K., Andresen, A. and Brown, S. D. (1971). A new ring shear apparatus and its application to the measurement of residual strength. *Géotechnique*, **21**, 273–328.

Bjerrum, L. (1954a). Geotechnical properties of Norwegian marine clays. *Géotechnique*, **4**, 49–68.

Bjerrum, L. (1954b). Stability of natural slopes in quick clay. *Proc. European Conf. Stability of Earth Slopes, Stockholm, Balkema, Rotterdam*, **2**, 16–40 (and *Géotechnique*, 1955, **5**, 101–119).

Bjerrum, L. (1955). Stability of natural slopes in quick clay. *Géotechnique*, **5**, 101–119.

Bjerrum, L. (1967). Progressive failure in slopes of over-consolidated plastic clay and clay shales. *J. Soil Mech. and Fdn Engng Div., ASCE*, **93**(5), 3–49.

Bjerrum, L. (1969). Discussion to Main Session 5, *Proc. 7th Int. Conf. Soil Mech. Fnd. Eng., Mexico*, Sociedad Mexicana de Suelos, **3**, 410–412.

Bjerrum, L. (1971). *Kvikkleireskredg et studium av aorsaksforhold og forbygnings muligheter.* Norwegian Geotechnical Institute Pub. Nr. 89, 14p.

Bjerrum, L, (1972). Embankments on soft ground. *ASCE Proc. Conf. on Performance of Earth and Earth Supported Structures, Purdue*, **2**, 1–54.

Bjerrum, L. (1973). Problems of soil mechanics and construction on soft clays. *Proc. 8th Int. Conf. Soil Mech. & Fdn Engrg, Moscow*, **3**, 111–159.

Bjerrum, L. and Eggestad, A. (1963). Interpretation of loading tests on sand. *Proc. Eur. Conf. Soil Mech. & Fdn Engrg, Wiesbaden*, **1**, 199–204.

Bjerrum, L. and Eide, O. (1966). Anvendelse av Kompensert Fundamentering i Norge. *NGI Publication No.70.*

Bjerrum, L. and Jørstad, F., (1968). *Stability of rock slopes in Norway.* NGI Pub. Nr. 79, 1–11, Norwegian Geotechnical Institute, Oslo.

Bjerrum, L. and Øverland, A. (1957). Foundation failure of an oil tank in Fredrikstad, Norway. *Proc. Int. Conf. on Soil Mech. & Fdn Engrg*, **4**, 1, 287–290.

Bjerrum, L. and Rosenqvist, I. Th. (1957). Some experiments with artificially sedimented clays. *Géotechnique*, **7**, 124–136.

Bjerrum, L. and Simons, N. E. (1960). Comparison of shear strength characteristics of normally consolidated clays. *Proceedings of research conference on shear strength of cohesive soils*. Boulder, Colorado, 711–724.

Bjerrum, L., Johannessen, I. J, and Eide, O. (1969). Reduction of negative skin friction on steel piles to rock. *Proc. 7th Int. Conf. Soil Mech. & Fdn Engrg, Mexico*, **2**, 27–34.

Bjerrum, L., Løken, T., Helberg, S. and Foster, R. (1969). A field study of factors responsible for quickclay slides. *Proc. 7th Int. Conf. Soil Mech. Fdn Engng, Mexico*, Mexicana de Suelos, **2**, 531–540.

Bjerrum, L., Moum, J. and Eide, O. (1967). Application of electro-osmosis to a foundation problem in a Norwegian quickclay. *Géotechnique*, **17**, 214–235.

Bjerrum, L., Simons, N. and Torblaa, I. (1958). The effect of time on the shear strength of a soft clay. *Proc. Brussels Conf. Earth Pressure Prob.*, **1**, 148–158.

Blight, G. E. (1967). Observations on the shear testing of indurated fissured clays. *Proc. Geotech. Conf., Oslo*, 1967, **1**, 97–102.

Boussinesq, J. (1885) *Application des potentials à l'étude de l'équilibre et du mouvement des solides élastiques.* Gauthier-Villars, Paris.

Bozzolo, D., Pamini, R. and Hutter, K. (1988). Rockfall analysis — a mathematical model and its test with field data. *Proc. 5th Int. Symp. on Landslides, Lusanne*, **1**, July 1988, 555–560.

Brawner, C. O. (1959). The landslide problem in British Columbia highway construction. *Proc. 13th Can. Soil Mech. Conf.*, 45–60.

British Standards Institution. *BS8081:1989. British standard code of practice for ground anchorages.* BSI, London.

British Standards Institution. *BS5930:1999. Code of practice for site investigations.* BSI, London.

Broadbent, C. D. and Rippere, K. H. (1971).Fracture studies at the Kimberley Pit. *Pros. Symp. on Planning Open Pit Mines*. Balkema, Rotterdam, 171–179.

Broch, E. and Franklin, J. A. (1972). The Point-Load Test. *Int. J. Rock Mech. Min. Sci. & Geomech. Abstr.*, **9**, 669–697.

Bromhead, E. N. (1978). Large landslides in London clay at Herne Bay, Kent. *Q. J. Eng. Geol.*, **11**, 291–304.

Bromhead, E. N. (1979). *The Stability of Slopes*. E & F.N. Spon, London.

Bromhead, E. N. (1986). An analytical solution to the problem of seepage into counterfort drains. *Can. Geotech. Jnl*, **21**, 657–662.

Bromhead, E. N., Chandler, M. P. and Hutchinson, J. N. (1991). The recent history and geotechnics of landslides at Gore Cliff, Isle of Wight. *Int. Conf. Slope Stability – Developments and Applications*, Thomas Telford, London, pp. 189–196.

Broms, B. B. (1969). Stability of natural slopes and embankment foundations. Discussion, *Proc. 7th Int. Conf. Soil Mech. & Fdn Engrg, Mexico*, **3**, 385–394.

Broms, B. B. (1975). Landslides. In *Foundation engineering handbook*, H. F. Winterkorn and H-Y Fang, (eds). Van Nostrand Reinhold Co., New York.

Brown, E. T. (1981). Suggested methods for determining shear strength. In *Rock characterisation testing and measurement — ISRM suggested methods*. Pergamon Press.

Brown, E. T, Richards, L. R. and Barr, M. V. (1977). Shear strength characteristics of Delabole slates. *Proc. Conf. Rock Engng, Newcastle-upon-Tyne*, 33–51.

Brown, I., Hittinger, M. and Goodman, R. (1980). Finite element study of the Nevis Bluff (New Zealand) rock slope failure. *Rock Mechanics*, **12**, 235–245.

Brunsden, D. (1979). Mass Movements. In *Process in Geomorphology*, C. Embleton and J. Thornes (eds). Edward Arnold Ltd, London, pp. 130–186.

Brunsden, D. (1984). Mudslides. Chap. 9 in *Slope Instability*, D. Brunsden and D. B. Prior (eds). John Wiley and Sons, Chichester, New York, Brisbane, Toronto, Singapore, pp. 363–418.

Buckthorp, P., Butler, F., Dunican, P. and Morrison, J. (1974) Central Hill Lambeth. *The Structural Engineer*, **11**(52), 395–407.

Burnett, A. D. and Fookes, P. G. (1974). A regional engineering geology study of the London clay in the London and Hampshire basins. *Q. J. Eng. Geol.*, **7**, 257–295.

Butler, F. G. (1972). *Geology, Soils and Foundations.* Thames-side Symposium, Ove Arup Partnership, pp. 15–23.

Cabrera, J. G. and Smalley, I. J. (1973) Quickclays as products of glacial action: a new approach to their nature, geology, distribution and geotechnical properties. *Eng. Geology*, Elsevier, **7**, 115–133.

Cadling, L. and Odenstad, S. (1950). The vane borer. *Proc. Royal Swedish Geotech. Inst.*, **2**, 87p.

Casagrande, L. (1947). *The applications of electro-osmosis to practical problems in foundations and earthworks.* London, DSIR, Building Research, Tech. Paper 30.

Casagrande, L. (1952). Electro-osmotic stabilisation of soils. *J. Boston Soc. Civ. Engrs*, **39**, 1, 51–83

Casagrande, L. (1953). Review of past and current work on electro-osmotic stabilisation of soils. *Harvard Soil Mech. Series*, **45**.

Casagrande, A. (1971). On liquefaction phenomena: report of a lecture. *Géotechnique*, **21**, 197–202.

Casagrande, A. and Carrillo, N. (1944). Shear failure of anisotropic materials. *Proc. Boston Soc. of Civ. Engrs.*, **31**, 74–87.

Cassel, F. L. (1948). Slips in fissured clay. *Proc. Int. Conf. on Soil Mech. & Fdn Engrg*, **2**. Rotterdam, 46–50.

Cedergren, H. R. (1967). *Seepage, drainage and flow nets.* Wiley, New York.

Chan, K. C. (1975). *Stresses and strains induced in soft clay by a strip footing.* Ph.D. Thesis, Cambridge University.

Chandler, M. P. and Hutchinson, J. N. (1984). Assessment of relative slide hazard within a large, pre-existing coastal landslide at Ventnor, Isle of Wight. *Proc. 4th Int. Symp. Landslides, Toronto*, **2**, 517–522.

Chandler, R. J. (1966). The measurement of residual strength in triaxial compression. *Géotechnique*, **16**, 181–186.

Chandler, R. J. (1970a). A shallow slab slide in the Lias clay near Uppingham, Rutland. *Géotechnique*, **20**, 253–260.

Chandler, R. J. (1970b). The degradation of Lias clay slopes in an area of the East Midlands. *Q. J. Eng. Geol.*, **2**, 161–181.

Chandler, R. J. (1974). Lias clay: the long-term stability of cutting slopes. *Géotechnique*, **24**, 21–38.

Chandler, R. J. (1976). The history and stability of two Lias Clay slopes in the upper Swash valley, Rutland. *Phil. Trans. Roy. Soc. London*, **A283**, 463–490.

Chandler, R. J. (1984a). Recent European experience of landslides in over-consolidated clays and soft rocks. *Proc. 4th Int. Symp. Landslides, Toronto*, **1**, 61–81.

Chandler, R. J. (1984b). Delayed failure and observed strengths of first-time slides in stiff clays – a review. *Proc. 4th Int. Symp. Landslides, Toronto*, **2**, 19–25.

Chandler, R. J. and Skempton, A. W. (1974). The design of permanent cutting slopes in stiff fissured clays. *Géotechnique*, **24**, 457–466.

Chang, C. U. and Duncan, J. M. (1970). Analysis of soil movements around a deep excavation. *J. Soil Mech. and Fdn Engng Div.*, ASCE, **96**(5), 1655–1681.

Chen, C. L. (1987). Comprehensive review of debris flow modeling concepts in Japan. In Debris flows/avalanches: process, recognition and mitigation, J. E. Costa and G. F. Wieczorek (eds). *Geol. Soc. Amer., Reviews in Engineering Geology*, **7**, 13–29.

Chen, R. H. and Chameau, J.-L. (1983). Three-dimensional limit equilibrium analysis of slopes. *Géotechnique*, **33**, 31–40.

Christian, J. T. and Whitman, R. V. (1969). A one dimensional model for progressive failure. *Proc. 7th Int. Conf. Soil Mech. Fdn Engng, Mexico*, Sociedad Mexicana de Suelos, **2**, 541–545.

Christian, J. T. and Wong, I. H. (1973). Errors in simulating excavation in elastic media by finite elements. *Soil and Foundations*, **13**(1), 1–10.

Clark, A. R., Fort, D. S. and Davis, G. M. (2000). The strategy, management and investigation of coastal landslides at Lyme Regis, Dorset. *Proc. 8th International Symposium on Landslides, Cardiff*. Thomas Telford Publishing, London, pp. 279–286.

Clayton, C. R. I., Matthews, M. C. and Simons, N. E. (1995). *Site investigation*. 2nd edn. Blackwell Science, Oxford, p. 584.

Clayton, C. R. I., Milititsky, J. and Woods, R. (1991). *Earth pressure and earth retaining structures*. 2nd edn., E. & F.N. Spon, London.

Colback, P. S. B. and Wiid, B. L. (1965). The influence of moisture content on the compressive strength of rock. *Proc. 3rd Candian Rock Mech. Symp.*, 65–83.

Collin, A. (1846). *Glissements spontanés des terrains argileux*. Carilian-Goeury and Dalmont, Paris. (English translation published 1956, University of Toronto Press.)

Coppin N. J. and Richards, I. G. (1990). *Use of vegetation in civil engineering*. CIRIA. Butterworths, London, 282 p.

Coulomb, C. S. (1773). Essai sur une application des regies de maximis et minimis à quelques problèmes de statique, relatifs à l'architecture. *Mémoire, Academie Royale, Say. etrang.*, **7**, 343–382, Paris (1776)

Crawford, C. B. (1963). Cohesion in an undisturbed sensitive clay. *Géotechnique*, **13**, 132–146.

Crisp, D. T., Rawes, M. and Welch, D. (1964). A Pennine peat slide. *Geogr. J.*, **130**, 519–524.

Cullen, R. M. and Donald, I. B. (1971). Residual strength determination in direct shear. *Proc. 1st Australia–NZ Conf. on Geomechanics, Melbourne, 1971*, **1**, 1–9.

Cundall, P. A., Voegele, M. D. and Fairhurst, C. (1975). Computerised design of rock slopes using interactive graphics for the input and output of geometrical data. *Design Methods in Rock Mechanics, 16th US Symp. on Rock Mechanics.* American Society of Civil Engineers, 5–14.

D'Appolonia, E., Alperstein, R. and D'Apolonia, D. J. (1967). Behaviour of a colluvial slope. *Jnl Soil Mech. Fdn Div. ASCE*, **93**, No. SN4, 447–473.

Da Silveira, A. F., Rodrigues, F. P., Crossman, N. F. and Mendes, F. (1966). Quantitative characterisation of the geometrical parameters of jointing in rock masses. *Proc. 1st Congress, International Society of Rock Mechanics, Lisbon*, **1**, 225–233.

Davis, E. H. and Poulos, H. G. (1967). Laboratory investigations of the effects of sampling. *Trans. IE Aust.*, **CE9**(1), 86–94.

De Beer, E. (1967). Clay strength characteristics of the Boom Clay. *Proc. Geotech. Conf., Oslo, 1967*, **1**, 83–88.

De Beer, E. (1969). Experimental data concerning clay slopes. *Proc. 7th Int. Conf. Soil Mech. Fdn Engng*, **2**, 517–525.

De Freitas, M. H. and Watters, R. J. (1973). Some field examples of toppling failure. *Géotechnique*, **23**, 495–514.

De Jong, K. A. and Scholten, R. (eds) (1973). *Gravity and Tectonics.* John Wiley and Sons, New York, London, Sydney, Toronto.

Deangeli, C. and Ferrero, A. M. (2000). Rock mechanics studies to analyse toppling failure. *Proc. 8th International Symposium on Landslides, Cardiff.* Thomas Telford Publishing, London, pp. 409–414.

Deere, D. U. (1964). Technical description of rock cores for engineering purposes. *Rock Mechanics and Engineering Geology*, **1**(1), 17–22.

Denness, B. (1972). A revised method of contouring stereograms using variable curvilinear cells. *Geol. Mag.*, **109**(2), 157–163.

Dimitrijevic, M. D. and Petrovic, R. S. (1965). The use of sphere projection in geology. *Geoloski Zavod Ljubjana.*

Duncan, J. M. (1971). Prevention and correction of landslides. *Sixth Ann. Nevada Street and Highway Conf.*, Section II, 1–42.

Duncan, J. M. (1972). Finite element analysis of stresses and movements in dams, excavations and slopes. *Proc. Symp. Applications of the Finite Element Method*, US Army Waterways Expt. Station, Vicksburg, 1972, 267–326.

Duncan, J. M. (1976). *Notes on slope stabilisation.* University of California, Berkeley (unpublished).

Duncan, J. M. and P. Dunlop, (1969). Slopes in stiff fissured clays and shales. *J. Soil Mech. and Fdn Engng Div.*, *ASCE*, **26**, 2, 467–92.

Dunlop, P. and Duncan, J. M. (1970). Development of failure around excavated slopes. *J. Soil Mech. and Fdn Engng Div.*, *ASCE*, **96**(2), 471–493.

Eckert, O. (1966). Consolidation de massifs rocheaux par ancrage de cables/ Consolidation of Rock Masses by Cable Anchors. Sols-Soils (Paris), **5**(18), 33–40 (in English).

Eide, O. (1983). Summary of slide in stiff fissured clay at Sandnes near Stavanger in Norway. *Swedish Geotechnical Institute, Report* No. 17, 155–160.

Eide, O. and Bjerrum, L. (1954). The slide at Bekkelaget. *Proc. European Conf. on Stability of Earth Slopes, Stockholm, Balkema, Rotterdam*, **2**, 1–15 (and *Géotechnique*, 1955, **5**, 88–100).

Eide, O. and Eggestad, Å. (1963). Foundation conditions for the new head-quarters building of the Norwegian Telecommunications Administration, Oslo. *NGI Pub. 55.* Norwegian Geotechnical Institute, Oslo.

Eide, O. and Holmberg, S. (1972). Test fills to failure on the soft Bangkok clay, *Proc. ASCE Spec. Conf. Performance of Earth and Earth-Supported Structures*, Purdue University, 1972, **1**(1), 159–180.

Eide, O., Hutchinson, J. N. and Landva, A. (1961). Short and long-term loading of a friction pile in clay. *Proc. 5th Int. Conf. Soil Mech. and Foundn. Engng, Paris.* Dunod, Paris, **2**, 45–53.

Ergun, M. U. (2000). Stabilization of landslides using piles. *Proc. 8th International Symposium on Landslides, Cardiff.* Thomas Telford Publishing, London, pp. 513–518.

Esser, A. J. (2000). Case of a slope failure in Lacustrine deposits. *Proc. 8th International Symposium on Landslides, Cardiff.* Thomas Telford Publishing, London, pp. 531–536.

Evans, R. S. (1981). An analysis of secondary rock toppling failures – the stress redistribution method. *Q. J. Eng. Geol.*, **14**, 77–86.

Evans, R., Valliappan, S., McGuckin, D. and Raja Sekar, H. L. (1981). Stability analysis of a rock slope against toppling failure. In Akai, K., Hayashi, M. and Nishimatsu, Y. (eds), *Weak Rock Proc. Int. Symp., Tokyo.* Balkema, Rotterdam.

Ewan, V. J. and West, G. (1981). Reproducibility of joint orientation measurements in rock. *Department of the Environment, Department of Transport TRL Report SR702.* TRL, Crowthorne, Berks.

Ewan, V. J., West, G. and Temporal, J. (1981). Reproducibility of joint spacing measurements in rock. *Department of the Environment, Department of Transport TRL Report SR702.* TRL, Crowthorne, Berks.

Faure, R. M. and Maïolino, S. (2000). Evaluation of rock slope stability using fuzzy logic. *Proc. 8th International Symposium on Landslides, Cardiff.* Thomas Telford Publishing, London, pp. 543–548.

Fecker, E. and Rengers, N. (1971). Measurement of large scale roughness of rock planes by means of profilograph and geological compass. *Proc. Symp. on Rock Fracture, Nancy*, France, Paper 1–18.

Fellenius, W. (1936). Calculation of the stability of earth dams. *Trans. 2nd Congress on Large Dams, Washington*, **4**, 445–459.

Fillunger, P. (1915). Versuche ober die zugfestigkeit beiailsetigen wassersruck. *Osterr. Wochenschr Offentlich Bandienst.* Vienne. 443–448.

Flaate, K. and Preber, T. (1974). Stability of road embankments on soft clay. *Canadian Geotechnical Journal*, **11**, 72–89.

Fort, D. S., Clark, A. R., Savage, D. T. and Davis, G. M. (2000). Instrumentation and monitoring of the coastal landslides at Lyme Regis, Dorset, UK. *Proc.*

8th International Symposium on Landslides, Cardiff. Thomas Telford Publishing, London, pp. 573–578.

Garga, V. A. (1970). *Residual shear strength under large strains and the effect of sample size on the consolidation of fissured clay.* Ph.D. Thesis, University of London.

Garland, R. J. (2000). A method of rapid risk assessment in slopes using stereo oblique aerial photography. *Proc. 8th International Symposium on Landslides, Cardiff.* Thomas Telford Publishing, London, pp. 609–614.

Geological Society of London (1977). The description of rock masses for engineering purposes. Report by the Geological Society Engineering Group Working Party, *Q. J. Eng. Geol.*, **10**(4), 355–388.

Gens, A., Hutchinson, J. N. and Cavounidis, S. (1988). Three-dimensional analysis of slides in cohesive soils. *Géotechnique*, **38**, 1–23.

Giani, G.P. (1992). *Rock slope stability analysis.* Balkema, Rotterdam.

Gibson, R. E. (1963). An analysis of system flexibility and its effects on time-lag in pore water pressure measurements. *Géotechnique*, **13**, 1, 1–11.

Gibson R. E. (1966) A note on the constant head test to measure permeability in situ. *Géotechnique*, **16**, 256–259.

Gibson, R. E. and Shefford, G. C. (1968). The efficiency of horizontal drainage layers for accelerating consolidation of clay embankments. *Géotechnique*, **18**(3), 327–335.

Gibson, A. D., Murphy, W. M. and Petley, D. N. (2000). Quantitative landslide analysis using archive airborne thematic mapper imagery. *Proc. 8th International Symposium on Landslides, Cardiff.* Thomas Telford Publishing, London, pp. 621–626.

Golder, H. Q. and Palmer, D. J. (1955). Investigation of a bank failure at Scrapsgate, Isle of Sheppey, Kent. *Géotechnique*, **5**, 55–73.

Goodman, R.E. (1976). *Methods of geological engineering.* West, New York.

Goodman, R. E. and Bray, J. W. (1976). Toppling of rock slopes. *Proc. Speciality Conf. on Rock Engineering for Foundations and Slopes. Boulder, Colorado,* ASCE, **2**.

Goodman, R. E. and Shi, G. (1985). *Block theory and its application to rock engineering.* Prentice Hall, Englewoods Cliffs, New Jersey.

Goodman, R. E., Bray, J. W. and Boyd, J. M. (1973). Toppling of rock slopes. *Proc. Speciality Conf. on Rock Engineering for Foundations and Slopes, Boulder, Colorado,* ASCE, **2**, 201–234.

Gordon, G. (1937). Freezing arch across toe of East Forebay slide, Grand Coulee Dam. *Reclamation Era*, **27**, 12–16.

Gould, J. P. (1960). A study on shear failure in certain tertiary marine sediments. *Proc. ASCE, Res. Conf., Boulder, Colorado,* 1960, 615–641.

Grant, D. I., Cooper, M. R. and Patel, T. (2000). Stabilisation of an oversteepened slope by means of balanced multiple remediation. *Proc. 8th International Symposium on Landslides, Cardiff.* Thomas Telford Publishing, London, pp. 658–664.

Gray, D. H. (1970). Effects of forest clear-cutting on the stability of natural slopes. *Bull. Assoc. Eng. Geol.*, **7**, 45–66.

Gray, D. H. (1974). Reinforcement and stabilisation of soil by vegetation. *J. Geotech. Eng. Div., ASCE*, 100/GT6, 695–699.

Gray, D. H. (1978). The role of woody vegetation in reinforcing soils and stabilising slopes. *Proc. Symp. Soil Reinforcing and Stabilising Techniques*, NSWIT/NSW Univ., 273–306.

Gregory, C. H. (1884). On railway cuttings and embankments with an account of some slips in London clay, on the line of the London and Croydon Railway. *Minut. Proc. I.C.E.*, **3**, 135–145 (also in *A Century of Soil Mechanics*, I.C.E., 1969).

Haefeli, R. (1951). Investigations and measurements of the shear strengths of saturated cohesive soils. *Géotechnique*, **2**, 186–208.

Hanna, T. H. (1973). *Foundation Instrumentation*. Trans Tech Publishers. 372p.

Hansen, J. B. and Gibson, R. E. (1949). Undrained shear strengths of anisotropically consolidated clays. *Géotechnique* **1**(3), 189–204.

Haupt, R. S. and Olson, J. P. (1972). Case history––embankment failure on soft varied silt, *Proc. ASCE Spec. Conf. Performance of Earth and Earth-Supported Structures, Purdue University*, 1972, **1**, 29–64.

Hawkins, A. B. (1986). Rock descriptions. *Geol. Soc. Eng. Geol. Special Publication No. 2, Site Investigation Practice: Assessing BS5930*, A. B. Hawkins Ed., Geol. Soc. London.

Helland, A. (1894). Opdyrkning af lerfaldet i Vaerdalen. Norges Geologiske Undersøkelse, 14.

Hencher, S. R. and Richards, L. R. (1982). The basic frictional resistance of sheeting joints in Hong Kong Granite. *Hong Kong Engineer*, **11**(2), 21–25.

Hencher, S. R. and Richards, L. R. (1989). Laboratory direct shear testing of rock discontinuities. *Ground Engineering*, **22**, 24–31.

Henkel, D. J. (1957). Investigation into two long term failures in London clay slopes at Wood Green and Northolt. *Proc. 4th Int. Conf. Soil Mech. Fdn Engng*, Butterworth, London, **2**, 315–320.

Henkel, D. J. and Skempton, A. W. (1955). A landslide at Jackfield, Shropshire, in a heavily over-consolidated clay. *Géotechnique*, **5**, 131–137.

Henkel, D. J. and Yudbhir (1966). The stability of slopes in the Siwalik rocks in India. *Proc. 1st Conf. Int. Cong. Rock Mech.*, Lisbon, **2**, 161–165.

Hennrich, K. (2000). Modelling critical water contents for slope stability and associated rainfall thresholds using computer simulations. *Proc. 8th International Symposium on Landslides, Cardiff*. Thomas Telford Publishing, London, pp. 713–719.

Herrmann, H. G. and Wolfskill, L. A. (1966). Residual shear strength of weak shales. *MIT Soil Mechanics Publication*, No. 200.

Hinds, D. (1974). A method of taking an impression of a borehole wall. *Imperial College Rock Mechanics Research Report No. 28*, Imperial College, London.

Hobbs, N. B. (1970). Discussion, Session A: Properties of Rocks: Foundations of Surface Structures, *Proc. Conf. on In-situ Investigations in Soils and Rocks*, B.G.S., London, 47–50.

Hocking, G. (1976). A method for distinguishing between single and double plane sliding of tetrahedral wedges. *Int. J. Rock Mech. Min. Sci. & Geomech. Abstr.*, **13**, 225–226.

Hoek. E. (1973). Methods for the rapid assessment of the stability of three-dimensional rock slopes. *Q. J. Eng. Geol.* **6**(3).

Hoek, E. (1983). Strength of jointed rock masses. 23rd Rankine Lecture, *Géotechnique*, **33**, 3, 187–223.

Hoek, E. (1986). *Rockfall: a computer program for predicting rockfall trajectories.* Unpublished internal notes, Golder Associates, Vancouver.

Hoek, E. (1990). Estimating Mohr–Coulomb friction and cohesion values from the Hoek–Brown failure criterion. *Int. J. Rock Mech. Min. Sci. & Geomech. Abstr.*, **27**(3), 227–229.

Hoek, E. (1998). *Rock Engineering — The Application of Modern Techniques to Underground Design.* Notes from a short course by Dr Evert Hoek. Kochen & Cella, Brazil (also available from http:www.rocscience.com).

Hoek, E. and Bray, J. (1977). *Rock Slope Engineering* (2nd ed.), The Institution of Mining and Metallurgy, London.

Hoek, E. and Bray, J. W. (1981). *Rock slope engineering.* The Institution of Mining and Metalurgy, London, 358pp.

Hoek, E., Bray, J. W. and Boyd, J. M. (1973). The stability of a rock slope containing a wedge resting on two intersecting discontinuities. *Q. J. Eng. Geol.*, **6**(1).

Hoek, E., Bray, J. W. and Boyd, J. M. (1980). The stability of a rock slope containing a wedge resting on two intersecting discontinuities. *Q. J. Eng. Geol.*. **6**(1).

Hoek, E. and Brown, E. T. (1980). Empirical strength criterion for rock masses. *J. Geotech. Engng Div. Am. Soc. Civ. Engrs*, **106**, GT9, 1013–1035.

Hoek, E. and Brown, E. T. (1980). An empirical strength criterion and its use in designing slopes and tunnels in heavily jointed weathered rock. *Proc. 6th Southeast Asian Conf. On Soil Engineering*, Taipei, Taiwan, **2**.

Hoek, E. and Londe, P. (1974). Surface Workings in Rock. *Proc. 3rd Congress, Denver, Int. Soc. Rock Mech.*, **1**(A), 613–654.

Holmsen, P. C. (1953). Landslips in Norwegian quickclays. *Géotechnique*, **3**, 187–194.

Hong, W. P. and Park, N. S. (2000). A design of slope stabilisation using piles; a case study on the slopes of the Hwangryung Mountain in Pusan, Korea. *Proc. 8th International Symposium on Landslides, Cardiff.* Thomas Telford Publishing, London, pp. 725–730.

Horne, M. R. (1964). The consolidation of stratified soil with vertical and horizontal drainage. *Int. J. Mech. Sci.*, **6**, 187–197.

Hungr, O. and Evans, S. G. (1989). Engineering aspects of rockfall hazard in Canada, Geological Survey of Canada, *Open File 2061*, 102pp.

Hutchinson, J. N. (1968). Mass movement. *Encyclopaedia of Geomorphology*, 688–695 (*Encyclopaedia of Earth Sciences*, Series III, Ed. R. W. Fairbridge). Reinhold Publishers, New York, Amsterdam, London.

Hutchinson, J. N. (1978). Assessment of the effectiveness of corrective measures in relation to geological conditions and types of slope movement. *NGI Pub. 124.* Norwegian Geotechnical Institute, Oslo.

Hutchinson, J. N. (1982). The geotechnics of coastal cliff stabilisation. In *Shoreline Protection.* Thomas Telford, London, pp. 215–222.

Hutchinson, J. N. (1983). Methods of locating slip surfaces in landslides. *Bulletin of the Association of Engineering Geologists*, **XX**(3), 235–252.

Hutchinson, J. N. (1984). An influence line approach to the stabilisation of slopes by cuts and fills. *Canad. Geotech. J.*, **21**, 363–370.

Hutchinson, J. N. (1984). Landslides in Britain and their countermeasures. *Journal of Japan Landslide Society*, **21**(1), 1–23.

Hutchinson, J. N. (1988). General Report, Morphological and geotechnical parameters of landslides in relation to geology and geohydrology. *Proc. 5th. Int. Symp. Landslides, Lausanne*, **1**, 3–35. Balkema, Rotterdam.

Hutchinson, J. N. (2001). Fourth Glossop Lecture 'Reading the ground: morphology and geology in site appraisal'. *Quarterly Journal of Engineering Geology and Hydrogeology*. In press.

Hutchinson, J. N. and Bhandari, R. K. (1971). Undrained loading, a fundamental mechanism of mudflows and other mass movements. *Géotechnique*, **21**, 353–358.

Hvorslev, M. J. (1949). *Subsurface exploration and sampling of soils for civil engineering purposes*. Waterways Experiment Station, Vicksburg, Mississippi, p. 521.

Hvorslev, M. J. (1956). *A review of the soils studies. Potamology investigations*, Report No. 12-5. Vicksburg: Waterways Experiment Station.

Hvorslev, M. J. (1960). Physical components of the shear strength of saturated clays. *Proc. ASCE, Res. Conf. Shear Strength of Cohesive Soils, Boulder, Colorado, 1960*, 437–501.

Ingold, T. S. (1982). *Reinforced earth*. Thomas Telford, London.

Institution of Civil Engineers, London, (1969). *A Century of Soil Mechanics*.

Ireland, H. O. (1954). Stability analysis of the Congress Street open cut in Chicago. *Géotechnique*, **4**, 163–168.

Ishida, T., Chigira, M. and Hibino, S. (1987). Application of the distinct element method for analysis of toppling observed on a fissured rock slope. *Rock Mechanics and Rock Engineering*, **20**, 273–283.

ISRM (1978). Suggested methods for the quantitative description of discontinuities in rock masses. (Co-ordinator N. Barton) *Int. J. Rock Mech., Mining Sci & Geomech. Abstracts*, **15**(6), 319–368.

Jaeger, J. C. (1971). Friction of rocks and stability of rock slopes. *Géotechnique*, **21**, 97–134.

Jamaludin, A. and Hussein, A. N. (2000). Selection of slope stabilisation remedial works for a mountainous road in Malaysia. *Proc. 8th International Symposium on Landslides, Cardiff*. Thomas Telford Publishing, London, pp. 769–774.

James, P. M. (1971). The role of progressive failure in clay slopes. *Proc. 1st Australia–NZ Conf. on Geomechanics, Melbourne, 1971*, **1**, 344–348.

Janbu, N. (1954). Stability analysis of slopes with dimensionless parameters, *Harvard Soil Mechanics Series No 46*, Cambridge, Mass., p. 81.

Janbu, N., Bjerrum, L. and Kjaernsli, B. (1956). *Veiledning ved løsning av fundamenteringsoppgaver*, NGI Pub. No. 16, 93 pp., Norwegian Geotechnical Institute, Oslo.

Janda, R. J., Scott, K. M., Nolan, K. M. and Martinson, H. A. (1981). The 1980 eruptions of Mount St. Helens, Washington. Lahar movement, effects, and deposits. *U.S. Geol. Surv. Prof. Paper*, **1250**, 461–478.

Jennings, J. E. B. (1966). Building on the dolomites in the Transvaal. Kanthack Memorial Lecture, *The Civil Engineer in South Africa*, **8**(2), 41–62.

Jennings, J. E. B., Brink, A. B. A., Louw, A. and Gowan, G. D. (1965). Sinkholes and subsidences in the Transvaal dolomites of South Africa. *Proc. 7th Int. Conf. Soil Mech. Fdn Engng*, **1**, 51–54.

John, K. W. (1969). Graphical stability analysis of slopes in jointed rock. *Proc. Amer. Soc. Civ. Engnrs. J. Soil Mech. & Found. Div.*, **95**(SM2), 497–526 and **95**(SM6), 1541–1545.

Kalkani, E. E. and Piteau, D. P. (1976). Finite element analysis of toppling failure at Hell's Gate Bluffs, British Columbia. *Bull. Assoc. Eng. Geologists.*, **13**(4), 315–327.

Kalsbeek, F. (1963). A hexagonal net for counting out and testing fabric diagrams. *Neus Jahrbuch für Mineralogie,* Monatshefte, **7**, 173–176.

Kalteziotis, N. A., Menzies, B. K. and Tarzi, A. I. (1984). A bearing capacity correction for strain-softening clays. *Ground Engineering* **17**(5), 30–31.

Kenney, T. C. (1964). Sea level movements and the geologic histories of the post-glacial marine soils at Boston, Nicolet, Ottawa and Oslo. *Géotechnique*, **14**, 203–230.

Kenney, T. C. (1967). The influence of mineral composition on the residual strength of natural soils. *Proc. Geotech. Conf., Oslo*, 1967, **1**, 123–130.

Kerisel, J. (1975) Fifteenth Rankine Lecture: Old structures in relation to soil mechanics. *Géotechnique*, **25**, 431–483.

Kjaernsli, B. and Simons, N. E. (1962). Stability investigations of the north bank of the Drammen River. *Géotechnique*, **12**, 147–167.

Knill, J. L. (1971). Collecting and *processing of geological data for purposes of rock engineering.* Univ. of Alberta, Edmonton, 35pp.

Köhler, H-J. and Schultz, R. (2000). Landslides triggered in clayey soils – geotechnical measurements and calculations. *Proc. 8th International Symposium on Landslides, Cardiff.* Thomas Telford Publishing, London, pp. 837–842.

Krahn, J. and Morgenstern, N. R. (1979). The ultimate friction resistance of rock discontinuity. *Int. J. Rock Mech. Min. Sci. & Geomech. Abstr.,* **16**, 127–133.

Krahn, J. and Morgenstern, N. R. (1980). The ultimate friction resistance of rock discontinuity: Authors' reply. *Int. J. Rock Mech. Min. Sci. & Geomech. Abstr.,* **17**, 79.

La Gatta, D. P. (1970). Residual strength of clays and clay shales by rotation shear tests, *Harvard Soil Mechanics Series*, No. 86, Cambridge, Mass.

La Rochelle, P., Trak, B., Tavenas, F. and Roy, M. (1974). Failure of a test embankment on a sensitive Champlain Clay deposit. *Canadian Geotechnical Journal*, **11**, 142–164.

Ladanyi, B. and Archambault, G. (1972). Evaluation de la restance au cisaillement dun massif rocheux fragmente. *Proc. 24th Int. Geol. Congr., Montreal*, Sec 130, 249–260.

Ladd, C. C. (1972). Test embankment on sensitive clay. *Proc. ASCE Spec. Conf. on Performance of Earth and Earth-Supported Structures, Purdue University, 1972*, **1**(1), 101–128.

Lambe, T. W. (1962). Pore pressures in a foundation clay. *J. Soil Mech. and Fdn. Engng. Div., Proc. ASCE.*, **SM2**, 19–47.

Larew, H. G. (1952). *Analysis of landslides.* Highway Research Board, Washington DC, 39p.

Laughton, A. S. (1955). *The compaction of ocean sediments.* PhD thesis, University of Cambridge.

Lee, E. M. (2000). The management of coastal landslide risks in England: the implications of conservation legislation and commitments. *Proc. 8th International Symposium on Landslides, Cardiff.* Thomas Telford Publishing, London, pp. 893–898.

Lee, E. M., Brunsden, D. and Sellwood, M. (2000). Quantitative risk assessment of coastal landslide problems, Lyme Regis, UK. *Proc. 8th International Symposium on Landslides, Cardiff.* Thomas Telford Publishing, London, pp. 899–904.

Leussink, H. and Muller-Kirchenbauer, H. (1967). Determination of the shear strength behaviour of sliding planes caused by geological features. *Proc. Geotech. Conf., Oslo, 1967,* **1**, 131–137.

Liao, H. J., Chen, C. N. and Liao, J. T. (2000). Landslides of cut-and-tieback slopes in Northern Taiwan. *Proc. 8th International Symposium on Landslides, Cardiff.* Thomas Telford Publishing, London, pp. 923–929.

Littlejohn, G. S., Norton, P. J. and Turner, M. J. (1977). A study of rock slope reinforcement at Westfield Open Pit and the effect of blasting on prestressed anchors. *Proc. Conf. On Rock Eng., Univ. Newcastle upon Tyne,* 293–310.

Lo, K. Y. (1965). Stability of slopes in anisotropic soils. *J. Soil Mech. and Fdn Engng Div., ASCE,* **91**(4), 85–106.

Lo, K. Y. (1970). The operational strength of fissured clays. *Géotechnique* **20**(1), 57.

Lo, K. Y. (1972). An approach to the problem of progressive failure. *Can. Geotech. J.,* **9**(4), 407–429.

Lo, K. Y. and Lee, C. F. (1972). Discussion. *J. Soil Mech. Fnd. Engng. Div., ASCE,* **98**(SM9), 981–983.

Lo, K. Y. and Lee, C. F. (1973a). Stress analysis and slope stability in strain softening materials. *Géotechnique,* **23**, 1–11.

Lo, K. Y. and Lee, C. F. (1973b). Analysis of progressive failure in clay slopes. *Proc. 8th Int. Conf. Soil Mech. Fdn Engng,* USSR Nat. Soc. Soil Mech. & Fdn Engng, Moscow, **1**(1), 251–258.

Lo, K. Y. and Lee, C. F. (1974). An evaluation of the stability of natural slopes in plastic Champlain clays. *Canad. Geotech. J.* **11**(1), 165–181.

Lo, K. Y. and Stermac, A. G. (1965). Failure of an embankment founded on varved clay. *Canadian Geotechnical Journal,* **2**(3), 243–253.

Locat, J., Leroueil, S. and Picarelli, L. (2000). Some considerations on the role of geological history on slope stability and the estimation of the minimum apparent cohesion of a rock mass. *Proc. 8th International Symposium on Landslides, Cardiff.* Thomas Telford Publishing, London, pp. 935–942.

Londe, P., Vigier, G. and Vormeringer, R. (1970). Stability of slopes: graphical methods. *Proc. Amer. Soc. Civ. Engnrs. J. Soil Mech. & Found. Div.,* **96**(SM4), 1411–1343.

Lu, T. D. and Scott, R. F. (1972). The distribution of stresses and development of failure at the toe of a slope and around the tip of a crack. *Proc. Symp. Applications of the Finite Element Method,* US Army Waterways Expt. Station, Vicksburg, 1972, 385–430.

Lupini, J. F., Skinner, A. E. and Vaughan, P. R. (1981). The drained residual strength of cohesive soils. *Géotechnique,* **31**, 2, 181–213.

Lyell, C. (1871). *Students elements of geology.* Murray, London, pp. 41–42.

Lyndon, A. and Schofield, A. N. (1976). Centrifugal model tests of the Lodalen landslide. *Proc. 29th Canadian Geotech. Conf., Vancouver, 1976,* **3**, 25–43.

McInnes, R. G. (1983). The threat to highways from coastal erosion. *The Highway Engineer,* 2–7.

Maddison, J. D. and Jones, D. B. (2000). Long-term performance of wells and bored drains used in landslide remediation. *Proc. 8th International Symposium on Landslides, Cardiff.* Thomas Telford Publishing, London, pp. 975–980.

Maddison, J. D., Siddle, H. J. and Fletcher, C. J. N. (2000). Investigation and remediation of a major landslide in glacial lake deposits at St Dogmaelss, Pembrokeshire. *Proc. 8th International Symposium on Landslides, Cardiff.* Thomas Telford Publishing, London, pp. 981–986.

MacGregor, F., Fell, R., Mostyn, G. R., Hocking, G. and McNally, G. (1994). The estimation of rock rippability. *Q. J. Eng. Geol.*, **27**(2), 123–144.

Madhloom, A. A. W. A. (1973). *The undrained shear strength of a soft silty clay from King's Lynn, Norfolk*, M.Phil. thesis, University of Surrey.

Markland, J. T. (1972). A useful technique for estimating the stability of rock slopes when the rigid wedge type of failure is expected. *Imperial College Rock Mechanics Research Report No. 19*, 10pp.

Marsland, A. (1971a). Large in situ tests to measure the properties of stiff fissured clays. *Proc. 1st Australia-New Zealand Conf. on Geomechanics, Melbourne*, 180–189.

Marsland, A. (1971b). The shear strength of stiff fissured clays. *Proceedings of Roscoe Memorial symposium on stress-strain behaviour of soils, Cambridge*, 59–68.

Marsland, A. (1971c). Laboratory and in situ measurements of the deformation moduli of London clay. *Proc. symp. on the interaction of structure and foundation*, Midland Soil Mechanics and Foundation Engineering Society, Department of Civil Engineering, University of Birmingham, 7–17.

Marsland, A. (1972). The shear strength of stiff fissured clays, *Stress–Strain Behaviour of Soils*, Proc. Roscoe Mem. Symp., Cambridge University, 1971; G. T. Foulis & Co Ltd, Henley-on-Thames, 59–68.

Marsland, A. P. (1967). Discussion. *Proc. Geotech. Conf. Oslo*, **2**, 160–161.

Marsland, A. P. and Butler, M. E. (1967). Strength measurements on stiff fissured Baston clay from Fawley, Hampshire. *Proc. Geotech. Conf. Oslo*, **1**, 139–146.

Martin, P. L. and Kelly, J. M. H. (2000). Design and performance of rock anchored bearing pads installed in 1980 for the stabilization of an unstable soil slope at Nantgarw, South Wales, UK. *Proc. 8th International Symposium on Landslides, Cardiff.* Thomas Telford Publishing, London, pp. 1023–1028.

Matherson, G. D. (1983). Rock stability assessment in preliminary site investigations — graphical methods. *Department of the Environment, Department of Transport TRL Report LR1039*. TRL, Crowthorne, Berks.

Menzies, B. K. (1976a). An approximate correction for the influence of strength anisotropy on conventional shear vane measurements used to predict field bearing capacity. *Géotechnique*, **26**(4), 631–634.

Menzies, B. K. (1976b). Strength, stability and similitude. *Ground Engineering*, **9**(5), 32–36.

Menzies, B. K. and Mailey, L. K. (1976). Some measurements of strength anisotropy in soft clays using diamond-shaped shear vanes. *Géotechnique*, **26**, 535–538.

Menzies, B. K. and Simons, N. E. (1978). Stability of embankments on soft ground. *Developments in soil mechanics – 1*, Chapter 11, Applied Science Publishers Ltd, London, pp. 393–436.

Mitchell, J. K. (1976). *Fundamentals of Soil Behaviour.* John Wiley and Sons, New York.

Mitchell, J. K., Vivatrat, V., and Lambe, T. W. (1977). Foundation performance of Tower of Pisa. *J. Geol. Eng. Div., Proc. ASCE,* **103,** 6T3, 227–249.

Morgan, H. D. (1944). The design of wharves on soft ground. *J. Instn Civil Engrs,* **22,** 5–25.

Morgenstern, N. R. (1977). Slopes and excavations in heavily over-consolidated clays. *Proc. 9th Int. Conf. Soil Mech. and Foundn Engng, Tokyo,* State-of-the-Art Volume, pp. 567–581.

Morgenstern, N. R. and Price, V. E. (1965). The analysis of the stability of general slip surfaces. *Géotechnique,* **15,** 79–93.

Moum, J., Løken, T. and Torrance, J. K. (1971). A geochemical investigation of the sensitivity of a normally consolidated clay from Drammen, Norway. *Géotechnique,* **21,** 329–340.

Murphy, V. A. (1951). A new technique for investigating the stability of slopes and foundations. *Proc. NZ Instn of Engrs,* **37,** 222–285.

Muller, L. (1959). The European approach to slope stability problems in open-pit mines. *Colorado School of Mines Quarterly.* **54**(3), 117–133.

Nichol, D. and Lowman, R. D. W. (2000). Stabilization and remediation of a minor landslide affecting the A5 trunk road at Llangollen, North Wales. *Proc. 8th International Symposium on Landslides, Cardiff.* Thomas Telford Publishing, London, pp. 1099–1104.

Nichols, T. C. Jr. (1980). Rebound, its nature and effect on engineering works. *Q. J. Eng. Geol.,* **13,** 135–152.

Nixon, J. K. (1949) $\phi = 0$ analysis. *Géotechnique,* **b**1, 3, 208–209; 4, 274–276.

Noble, H. L. (1973). Residual strength and landslides in clay and shale. *J. Soil Mech. Fnd. Div., ASCE,* **99**(SM9), 705–719.

Odenstad, S. (1949). Stresses and strains in the undrained compression test. *Géotechnique,* **1,** 4, 242–249.

Operstein, V. and Frydman, S. (2000). The influence of vegetation on soil strength. *Ground Improvement.* Thomas Telford, **4,** No. 2, pp. 81–89.

Pacher, F. (1959). Kennziffern des flachengefuges. *Geologies and Bauwesen,* **24,** 233–227.

Pahl, P. J. (1981). Estimating the mean length of discontinuity traces. *Int. J. Rock Mech. Min. Sci. & Geomech. Abstr.,* **18,** 221–228.

Palmer, A. C. and Rice, J. R. (1973). The growth of slip surfaces in the progressive failure of over-consolidated clay. *Proc. Roy. Soc. London,* No. 1591, 527.

Papaliangas, T., Lumsden, A. C., Hencher, S. R. and Manolopoulou, S. (1990). Shear strengthy of modelled filled rock joints. *Proc. Int. Symp. on Rock Joints, Loen, Norway,* Barton, N. and Stephansson, O. (eds), Balkema, Rotterdam, 275–282.

Parkhurst, S. and Flavell, R. (2000). Risk assessment and quantification of slope conditions based upon site inspection surveys. *Proc. 8th International Symposium on Landslides, Cardiff.* Thomas Telford Publishing, London, pp. 1171–1176.

Parry, R. H. G. (1972). Stability analysis for low embankments on soft clays. *Stress–Strain Behaviour of Soils.* G. T. Foulis and Co Ltd, Henley-on-Thames, pp. 643–668.

Parry, R. H. G. and McLeod, J. H. (1967). Investigation of slip failure in flood levee at Launceston, Tasmania. *Proc. 5th Australia–NZ Conf. Soil Mech. Fdn. Engng., Auckland, 1967*, 249–300.

Patton, F. D. (1966). Multiple modes of shear failure in rock. *Proc. 1st Int. Congr. of Rock Mech., Lisbon*, **1**, 509–513.

Peck, R. B. (1967). Stability of natural slopes. *J. Soil Mech. and Fdn Engng Div., ASCE*, **93**(4), 403–417.

Peck, R. B. (1969). Deep excavations and tunneling in soft ground. *Proc. 7th Int. Conf. Soil Mech. and Foundn Engng, Mexico*, Sociedad Mexicana de Suelos, State-of-the-Art Volume, pp. 225–290.

Peck, R. B. and Bryant, F. G. (1953). The bearing capacity failure of the Transcana elevator. *Géotechnique*, **3**, 5, 201–208.

Penning, W. H. (1884). The Transvaal Goldfields – their past, present and future. *J. Soc. Arts, London*, 608–621.

Peterson, R., Iverson, N. L. and Rivard, P. J. (1957). Studies of several dam failures on clay foundations. *Proc. 4th Int. Conf. Soil Mech. and Fdn. Engng., London, 1957*, **2**, 348–352.

Pettifer, G. S. and Fookes, P. G. (1994). A revision of the graphical method for assessing the excavatability of rock. *Q. J. Eng. Geol.*, **27**(2), 145–164.

Phillips, F. C. (1971). *The use of stereographic projection in structural geology*. Arnold, London.

Pierson, L. A., Davis, S. A. and Van Vickle R. (1990). Rockfall Hazard Rating System Implementation Manual. *Federal Highway Administration (FHWA) Report FHWA-OR-EG-90-01*. FHWA, US Department of Transportation.

Pilot, G., Meneroud, J. P., Mangan, B. and Pescond, B. (1985). Failure of a large highway cutting and remedial works. *Proc. Symp. On Failures in Earthworks*. ICE, London, 141–154.

Piteau, D. R. (1971). Geological factors significant to the stability of slopes cut in rock. *Proc. Symp. on Planning Open Pit Mines, Johannesburg*. Balkema, Rotterdam, 33–53.

Piteau, D. R. (1973). Characterizing and extrapolating rock joint properties in engineering practice. *Rock Mechanics*, **2**, 5–31.

Preece, R. C., Kemp, R. A. and Hutchinson, J. N. (1995). A Late-glacial colluvial sequence at Watcombe Bottom, Isle of Wight, England. *J. Quaternary Science*, **10**, 107–121.

Priest, S. D. (1980). The use of inclined hemisphere projection methods for the determination of kinematic feasibility, slide direction and volume of rock block. *Int. J. Rock Mech. Min. Sci. & Geomech. Abstr.*, **17**, 1–23.

Priest, S. D. (1985). *Hemispherical Projection Methods in Rock Mechanics*. George Allen & Unwin, London, 124pp.

Priest, S. D. (1993). *Discontinuity analysis for rock engineering*. Chapman & Hall, London, 473pp.

Priest, S. D. and Brown, E. T. (1983). Probabilistic stability analysis of variable rock slopes. *Trans. Instn Mining & Metallurgy*, (sect A), **92**, 1–12.

Priest, S. D. and Hudson, J. A. (1976). Discontinuity spacings in rock. *Int. J. Rock Mech. Min. Sci. & Geomech. Abstr.*, **13**, 135–148.

Priest, S. D. and Hudson, J. A. (1981). Estimation of discontinuity spacing and trace length using scanline surveys. *Int. J. Rock Mech. Min. Sci. & Geomech. Abstr.*, **18**, 183–197.

Rankine, W. J. M. (1862). *A Manual of Civil Engineering*. Griffin and Bohn, London.

Reynolds, O. (1885). On the dilatancy of media composed of rigid particles in contact, with experimental illustrations. *Phil. Mag.*, **20**, 469–481.

Reynolds, O. (1886). Experiments showing dilatancy, a property of granular material, possibly connected with gravitation. *Proc. Roy. Inst. Gt. Brit.*, 354–363

Richardson, A. M., Brand, E. W. and Memon, A. (1975). *In situ* determination of anisotropy of a soft clay. *Proc. Conf. In Situ Measurement of Soil Properties, ASCE, Raleigh, 1975*, 336–349.

Robertson, A. M. G. (1970). The interpretation of geological factors for use in slope stability. In Van Rensburg, P. W. J. (ed), *Planning open pit mines, Proc. Symp., Johannesburg*. Balkema, Rotterdam.

Rocha, M. (1971). A method of integral sampling of rock masses. *Rock Mechanics*, **3**(1), 1–12.

Romero, S. U. (1968). In situ direct shear tests on irregular surface joints filled with clayey material. *Proc. Int. Symp. on Rock Mechanics, ISRM*, Madrid, **1**, 189–194.

Root, A. W. (1953). California experience in correction of landslides and slipouts. *Proc. ASCE*, **79**, No. 235, Sept., 1–18.

Root, A. W. (1958). Prevention of landslides. In *Landslides and Engineering Practice* Ed. E. B. Eckel, Highway Research Board Special Report 29, pp. 113–148.

Rosenqvist, I. Th. (1953). Considerations on the sensitivity of Norwegian quickclays. *Géotechnique*, **3**, 195–200.

Rosenqvist, I. Th. (1966). Norwegian research into the properties of quickclay – a review. *Eng. Geology*, Elsevier, **1**, 445–450.

Ross-Brown, D. M. and Walton, G. (1975). A portable shear box for testing rock joints. *Rock Mechanics*, **7**(3), 129–153.

Rowe, P. W. (1957). $C = 0$ hypothesis for normally loaded clays at equilibrium. *Proc. 4th Int. Conf. Soil Mech. Fdn Engng*, Butterworth, London, **1**, 189–192.

Rowe, P. W. (1964). The calculation of the consolidation rates of laminated, varved or layered clays, with particular reference to sand drains. *Géotechnique*, **14**(4), 321–340.

Rowe, P. W. (1972). The relevance of soil fabric to site investigation practice. *Géotechnique*, **22**, 193–300.

Roy, M. (1975). *Predicted and observed performance of motorway embankments on soft alluvial clay in Somerset*, M.Phil. Thesis, University of Surrey.

Sactren, G. (1893). Kart over skredet i Vaerdalen. *Teknisk Ukeblad*, **26**, 199–206.

Sandroni, S. S. (1977). *The Strength of London clay in Total and Effective Stress Terms*. Ph.D. Thesis, University of London.

Sanger, F. J. (1968). Ground freezing in construction. *Jnl Soil Mech & Fdns Div. ASCE*, **94**, SM 1, 131–158.

Schmertman, J. H. (1955). The undisturbed consolidation behaviour of clay. *Trans. ASCE*, **120**, 1201–1233.

Schofield, A. N. and Wroth, C. P. (1968). *Critical State Soil Mechanics*. McGraw-Hill, London

Schweizer, R. J. and Wright, S. G. (1974). A survey and evaluation of remedial measures for earth slope stabilisation. Res. Report 161- 2F CHR, University of Texas, Austin.

Sembelli, P. and Ramirez, A. (1969). Measurement of residual strength of clays with a rotation shear machine. *Proc. 7th Int. Conf. Soil Mech. Fdn. Engng., Mexico*, Sociedad Mexicana de Suelos, **3**, 528–529.

Serota, S. (1966). Discussion. *Proc. ICE*, **35**(9), 522.

Sevaldson, R. A. (1956). The slide at Lodalen, October 6, 1954. *Géotechnique*, **6**, 167–182.

Simons, N. E. (1967). Discussion. *Proc. Geotech. Conf. Oslo*, **2**, 159–160.

Simons, N. E. (1976). Field studies of the stability of embankments on clay foundations, *Laurits Bjerrum Memorial Volume*, Norwegian Geotechnical Institute, Oslo, pp. 183–209.

Simons, N. E. (1977). *Slips, settlements and sinkholes*. Professorial Inaugural Lecture. University of Surrey, England.

Simons, N. E., and Menzies, B. K., (1974): A note on the principle of effective stress. *Géotechnique* **34**(2), 259–261.

Simons, N. E. and Menzies, B. K. (1978). The long term stability of cuttings and natural clay slopes. *Developments in soil mechanics – 1*, Chapter 10, Applied Science Publishers Ltd, London, pp. 347–392.

Simons, N. E. and Menzies, B. K. (2000). *A Short Course in Foundation Engineering*, 2nd Edition, Thomas Telford, London, p. 250.

Skempton, A. W. (1942). An investigation of the bearing capacity of a soft clay soil. *J. Inst. Civil Engrs*, **18**, 307–321.

Skempton, A. W. (1948). The rate of softening of stiff fissured clays with special reference to London clay. *Proc. 2nd Int. Conf. Soil Mech. Fdn Engng*, **2**, 50–53.

Skempton, A. W. (1948a). The geotechnical properties of a deep stratum of post-glacial clay at Gosport. *Proc. Int. Conf. on Soil Mech. & Fdn Engrg, Rotterdam*, **2**, 1, 145–150.

Skempton, A. W. (1948b). The rate of softening in stiff fissured clays with special reference to London Clay. *Proc. Int. Conf. on Soil Mech. & Fdn Engrg, Rotterdam*, **2**, 2, 50–53.

Skempton, A. W. (1950). Discussion on: Watson, G. The bearing capacity of screw piles and screwcrete cylinders. *J. Instn Civil Engrs*, **34**, 76.

Skempton, A. W. (1951). The bearing capacity of clays. *Building Research Congress, London*. Papers, Div. 1, Part 3, pp. 180–189.

Skempton, A. W. (1954). The pore pressure coefficients A and B in saturated soils. *Géotechnique*, **4**, 143–147.

Skempton, A. W. (1959). Cast in situ bored piles in London Clay. *Géotechnique*, **9**, 4, 153 173.

Skempton, A. W. (1961). Horizontal stresses in over-consolidated Eocene clay. *Proc. 5th Int. Conf. Soil Mech. Fnd. Engng., Paris*, Dunod, Paris, **1**, 351–357.

Skempton, A. W. (1964). Long-term stability of clay slopes. *Géotechnique*, **14**, 77–101.

Skempton, A. W. (1965). Discussion. *Proc. 6th Int. Conf. Soil Mech. Fdn. Engng., Montreal, 1965*, University of Toronto Press, **3**, 551–552.

Skempton, A. W. (1966). Some observations on tectonic shear zones. *Proc. 1st Int. Conf. Rock Mech., Lisbon, 1966*, **1**, 328–335.

Skempton, A. W. (1970). First-time slides in over-consolidated clays. *Géotechnique*, **20**, 320–324.